Learn AutoCAD LT 2000i

Ralph Grabowski

Wordware Publishing, Inc.

Library of Congress Cataloging-in-Publication Data

Grabowski, Ralph
Learn AutoCAD LT 2000i / by Ralph Grabowski.
p. cm.
Includes index.
ISBN 1-55622-808-2 (pbk.)
1. Computer graphics. 2. AutoCAD. I. Title.

T385.G6925197 2000
620'.0042'023855369--dc21

00-050345
CIP

2320 Los Rios Boulevard
Plano, Texas 75074

Printed in the United States of America

ISBN 1-55622-808-2

10 9 8 7 6 5 4 3 2 1
0011

All inquiries for volume purchases of this book should be addressed to Wordware Publishing, Inc., at the above address. Telephone inquiries may be made by calling:

(972) 423-0090

Contents

Introduction

AutoCAD LT 2000i is Autodesk's low-cost, introductory CAD (computer-aided design) software. You made an excellent choice in acquiring LT. It is many times cheaper than AutoCAD 2000i, yet allows you to do all the same 2D drafting tasks.

This is an introductory book, designed to get you started quickly with AutoCAD LT. Once you have the program installed on your computer, I recommend that you work through the first five chapters of the book. These chapters are written so that you can complete each one in an hour or less. In five hours, you'll learn how to set up a new drawing, add details, make changes, and print the drawing on paper.

Don't worry if you have difficulty completing a chapter. The companion CD-ROM contains a copy of the project drawing as it stands at the end of each chapter. That lets you start with an accurate copy at the start of the next chapter.

Later chapters delve into more advanced functions, such as the DesignCenter, blocks and attributes, regions, isometric drafting, and connecting with the Internet.

Updated for AutoCAD LT 2000i

This book is the seventh edition. Each time Autodesk releases a new version of AutoCAD LT, this book is updated to reflect the changes in user interface, commands, and features. Earlier books covered AutoCAD LT, LT Release 2, LT for Windows 95, LT 97, LT 98, and LT 2000.

For AutoCAD LT 2000i, this book discusses these new and updated features:

- Updated user interface features (Chapter 1).
- Updated screen images (all chapters).

- New access to Autodesk portals (Chapter 11).
- New embedded hyperlinks, and updated Internet access (Chapter 12).
- New AutoCAD LT 2000 commands, as well as commands removed or renamed (Appendix C).
- Updated compatibility with AutoCAD 2000i (Appendix D).

About This Book

Each chapter begins with a summary of the topics and commands you will be learning. Every chapter contains tips, notes, and warnings that help you become more proficient with the software.

Chapter 1—"AutoCAD LT 2000i Quick Tour" introduces you AutoCAD LT 2000i and its user interface. To get acquainted, you draw a few lines. You also learn how to access the online help system.

Chapter 2—"Setting Up the Drawing" starts you on a project drawing—the lot around your home. You learn how to set up the layers, colors, and drawing aids. You also learn how to set up AutoCAD LT so that it automatically saves your work.

Chapter 3—"Creating Your First Drawing" gets you drawing with lines as well as making changes, such as moving objects and filleting corners. You also learn how to print your drawing on paper.

Chapter 4—"Adding Details to a Drawing" makes the drawing look better by adding hatch patterns and plant symbols. You also learn how to perform non-modal editing.

Chapter 5—"Making Changes to the Drawing" shows you how to make changes to the drawing, such as changing the linetype or color of objects. You also learn about advanced coordinate input via direct distance entry and tracking.

Chapter 6—"Adding Notes and Dimensions" completes the project drawing by adding text and dimensions.

Chapter 7—"The AutoCAD DesignCenter" describes the multifaceted DesignCenter that lets you find and insert symbols from other drawings, as well as text styles, linetypes, layers, and so on. You also learn how to view entire other drawings with DesignCenter.

Chapter 8—"Creating Symbols and Attributes" covers the important and time-saving topic of blocks (symbols) and attributes (database-like data).

You learn how to construct a second project drawing that consists almost entirely of symbols—the electrical diagram of a car.

Chapter 9—"Bills of Material" is the continuation of learning to use blocks and attributes. In this chapter, you learn how to export attribute data to an external file, and import the data file into Excel. You also learn how to bring the Excel spreadsheet into the AutoCAD drawing.

Chapter 10—"Region Modeling" introduces you to regions that allow you to create complex, 2D shapes. Boolean operations allow you to combine (union), remove (subtract), or intersect two or more regions.

New! Chapter 11—"Accessing Autodesk's Portals" shows you how to access the Point A, Buzzsaw, RedSpark, and AUGI Web sites from within AutoCAD LT.

Chapter 12—"Connecting with the Internet" teaches you all about hyperlinks (or URLs, Uniform Resource Locators), the Internet's universal file naming system. You learn how to insert hyperlinks into the drawing, how to open and save drawings on the Internet, and how to launch a Web browser from within AutoCAD.

Chapter 13—"Drawings on the Web" exports drawings from AutoCAD to DWF, the special file format created by Autodesk for displaying CAD drawings on the Internet, and in JPEG. You also learn how to embed a DWF file in a Web page.

Chapter 14—"Isometric Drafting" teaches you how to set up AutoCAD LT for isometric drawing. You create isometric text and dimension styles, then work through a tutorial to create an isometric part.

Appendix A—"Setting Up AutoCAD LT 2000i" provides step-by-step instructions for installing AutoCAD LT 2000i on your computer. You also learn about the recommended hardware requirements for running AutoCAD LT 2000i.

Appendix B—"Installing and Using the Companion CD-ROM" gives instructions for installing the files from the companion CD-ROM onto your computer's hard drive.

Appendix C—"AutoCAD LT 2000i Commands" is a summary of all of AutoCAD LT's command names, along with their aliases and shortcut keystrokes. The list also includes the names of commands removed from AutoCAD LT.

Appendix D—"Compatibility with AutoCAD 2000i" describes the differences between the two CAD siblings. You learn how LT 2000i differs from AutoCAD 2000i's commands, user interface, and DWG file format.

Appendix E—"Configuring LT 2000i for a Plotter" provides a step-by-step tutorial on configuring a new plotter with AutoCAD LT 2000i's plotter interface.

Appendix F—"Answers" lists the answers to the chapter review questions.

About the Author

Ralph Grabowski has been writing about AutoCAD since the fall of 1985, when he joined *CADalyst* magazine, the first magazine dedicated to AutoCAD users. He began as the technical editor in charge of reviewing the hundreds of hardware peripherals and software packages that work with AutoCAD. Later, he became the senior editor for the magazine.

Since 1991, Ralph has been a free-lance writer. He has authored over 50 books about CAD, graphics, and the Internet. For Wordware Publishing, he has written over a dozen books on AutoCAD and Visio.

Ralph is the editor of several publications. *UpFront.eZine* is the weekly CAD e-newsletter delivered every week by e-mail. *Visions.eZine* is the bi-weekly e-newsletter for Visio software users. *CRCeZine* is the bi-weekly e-newsletter for AutoCAD users. *AutoCAD User* is the paper-based quarterly magazine for Canadian AutoCAD users.

You can visit his Web site at http://upfrontezine.com, or write to him at ralphg@xyzpress.com.

Chapter 1

AutoCAD LT 2000i Quick Tour

In this chapter, you learn how to start AutoCAD LT 2000i for Windows 95/98/ME and Windows NT/2000, and tour the AutoCAD user interface. You get your feet wet by drawing a few lines.

Key Concepts

- Starting AutoCAD LT
- Becoming familiar with the AutoCAD user interface
- Learning to draw a line
- Finding out about right-click shortcut menus
- Knowing how to reverse a mistake
- Accessing online help for AutoCAD

Key Terms

Cursor—provides feedback from AutoCAD

Flipscreen—switching between graphics and text windows

Flyout—a button that hides two or more additional buttons

Icon—pictorial representation of a command

Pickbox—shows the point you are picking

Right-click—displays a context-sensitive shortcut menu

Toolbar button—executes a command

Layout—defines how a drawing will be plotted

Abbreviations	
Alt	Alt (or alternate) key
Ctrl	Ctrl (or control) key
F	Function key
U	Undo the last command or option
UCS	User-defined coordinate system

Command	Shortcut	Menu Selection
Help	? *or* F1	Help | Help
Line	L	Draw | Line
Quit	Alt+F4	File | Exit
TextScr	F2	View | Display | Text Window
Undo	Ctrl+Z	Edit | Undo
UcsIcon	...	View | Display | UcsIcon

Starting AutoCAD LT 2000i

Before you start AutoCAD LT, your computer must be running Windows 95, 98, ME, NT, or 2000. (If AutoCAD is not yet set up on your computer, read Appendix A, "Setting Up AutoCAD LT 2000i.")

To start AutoCAD LT, click the **Start** button, select **Programs**, select the **AutoCAD LT 2000i** group, and select **AutoCAD LT 2000i**.

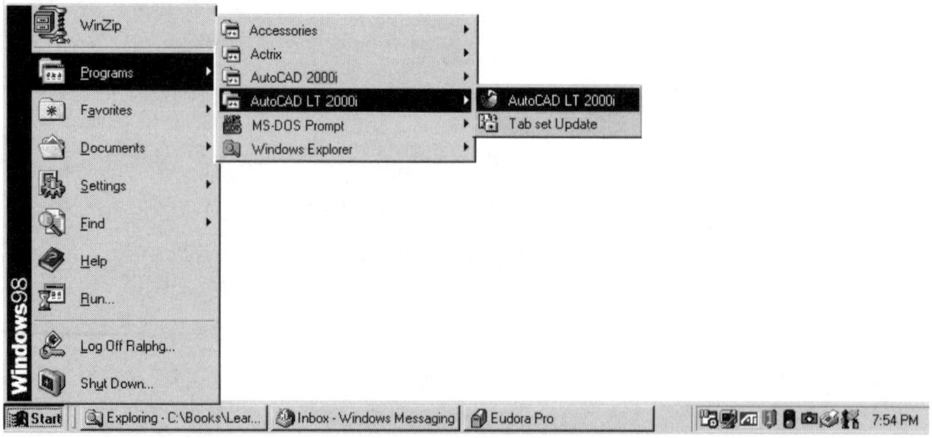

Alternatively, double-click the AutoCAD LT 2000i icon found on the Windows desktop.

Depending on the speed of your computer, it can take between 10 and 60 seconds to load AutoCAD LT. A splash screen is displayed, and the AutoCAD LT window appears.

The AutoCAD LT Window

When AutoCAD LT appears, it first displays two additional windows: the Active Assistance window and the AutoCAD LT Today window. (This replaces the Startup dialog box displayed by previous versions of LT.) For now, click the X button in both windows' upper-right corners; you use these windows in later parts of this book.

The AutoCAD LT window consists of a graphical drawing area with areas of information on three sides. At the top, you see the title bar, menu bar, and two toolbars. In the center, you see the crosshair cursor with its pickbox. A pair of toolbars is docked to the left of the drawing area. At the bottom are the command prompt area and the status bar.

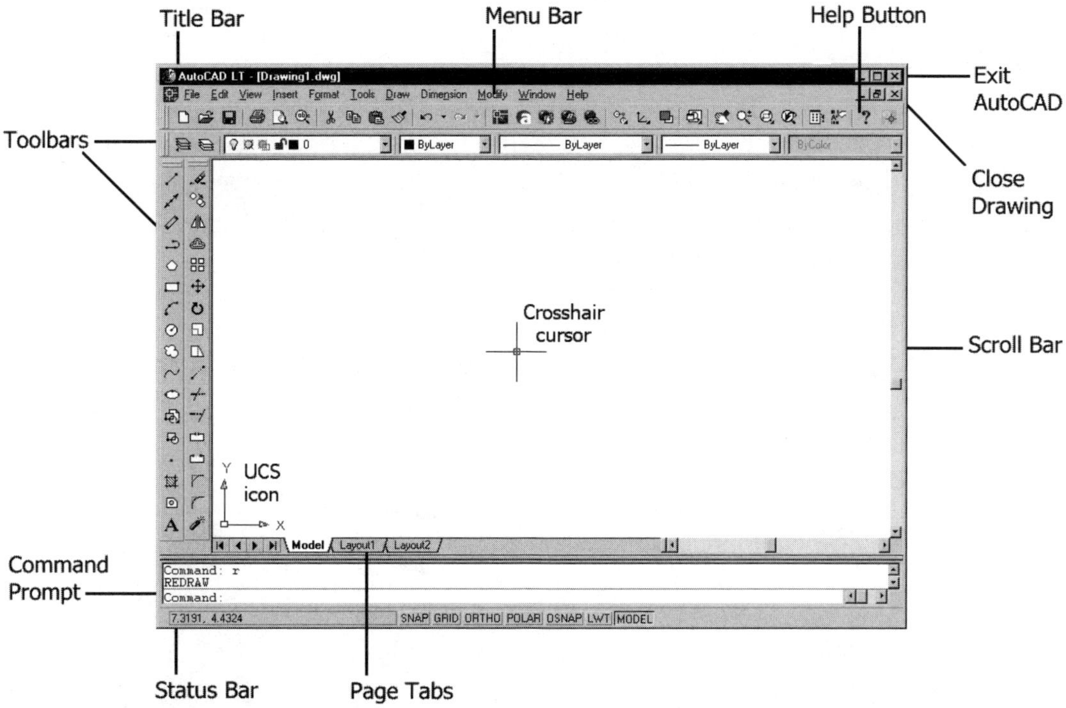

The elements of the AutoCAD user interface.

At the top of the AutoCAD LT window are four lines of information. From top to bottom, these are:

- **Title bar** reports the program name and current drawing file name, as in "AutoCAD LT - [Drawing1.dwg]."
- **Menu bar** contains the drop-down menus, such as File, Edit, and View.
- **Standard toolbar** contains a row of buttons labeled with icons.
- **Object Properties toolbar** contains a couple of buttons and several list boxes.

Let's examine each of these, one by one, to understand their functions.

The Title Bar

The *title bar* displays the name of the AutoCAD program and the drawing with which you are working. The title bar performs two other important tasks: maximizing and restoring the AutoCAD window.

To quickly open another drawing, simply drag its filename from the Exploring window (or File Explorer) to AutoCAD's title bar. AutoCAD opens the drawing in a new window. If you were to drag the drawing's filename into the current drawing, AutoCAD would insert it as a *block*.

The Menu Bar

The *menu bar* is similar to the menu bar used by other Windows programs. Notice that some of the names are identical to other Windows applications: the File, Edit, View, Insert, Format, Tools, Window, and Help menus are also found on the Word menu bar. The Draw, Dimension, and Modify items are unique to AutoCAD.

Try using the menu bar now:

1. Move your mouse so that the cursor touches the menu bar.
2. Move the mouse left or right until the cursor is over the word View.
3. Press the left mouse button to select the **View** menu. Instantly, a menu drops down that lists many of AutoCAD's view options: Redraw, Regen, Regen All, Zoom, Pan, Aerial View, and others.

4. Move the cursor down the menu, then pause over a menu item, such as Redraw. Look at the status line (at the bottom of the AutoCAD LT window). There you read a one-sentence description of the command. For example, for the Redraw command, you read, "Refreshes the display of the current viewport: REDRAW."

Each menu item displays helpful text on the status line.

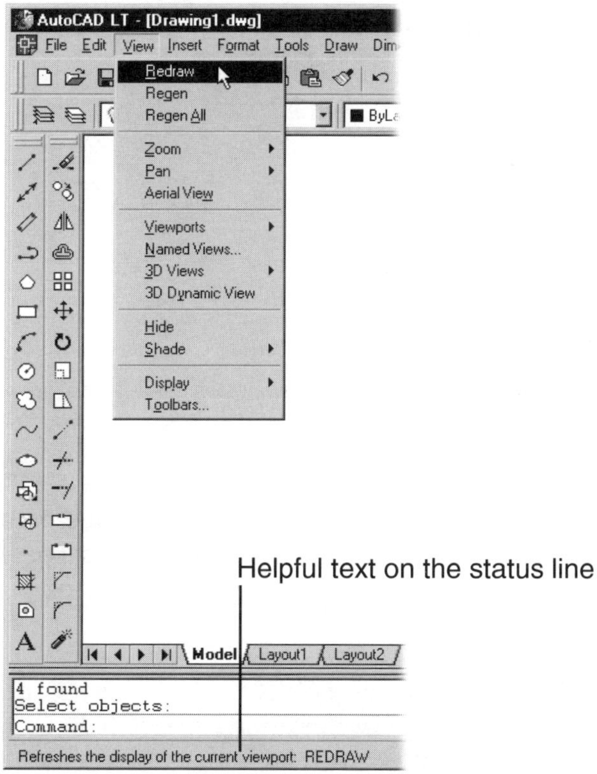

5. Now move the cursor down to the 3D Views item, located about halfway down the menu. A submenu appears, listing VPOINT, Plan View, Top, etc.

6. Move the cursor over to the submenu, and select **Plan View**. Notice that a second submenu appears.

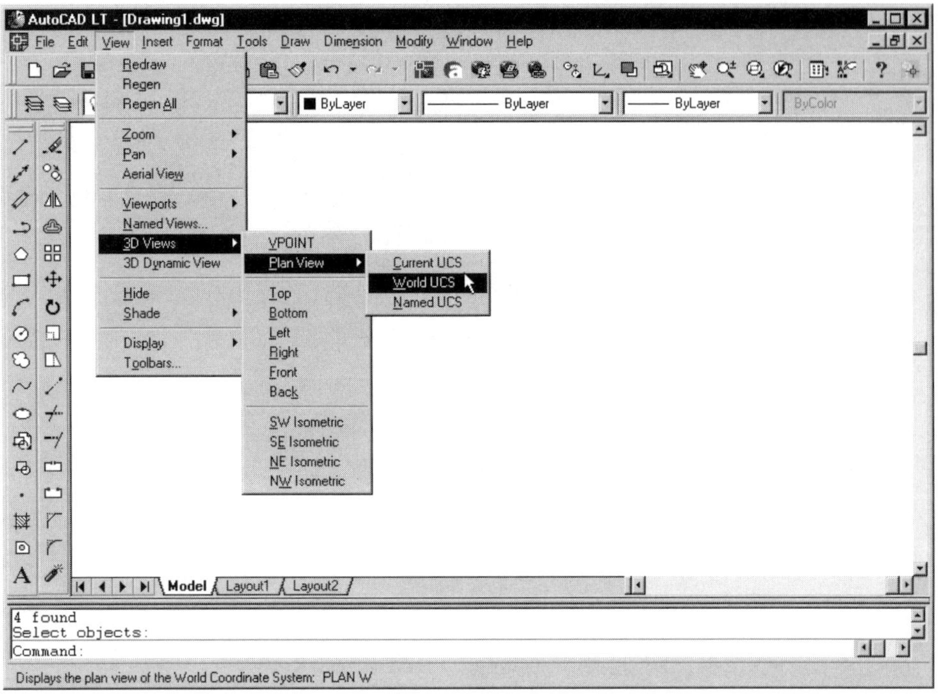

Displaying a menu and two submenus.

7. Move the cursor back to the parent menu, and move the cursor up the list. Notice that the two submenus disappear.
8. Click **View** again to dismiss the menu.

Note: The menus use several symbols that have special meaning:

Menu Symbol	Example	When Selected...
(none)	Redraw	Executes the command
... (ellipses)	Named Views...	Displays a dialog box
▸ (arrow)	3D Views ▸	Displays a submenu
✓(check)	✓Normal	Feature is turned on
(black text)	Plot Style	Feature is available
(gray text)	Plot Style	Feature is not available

Toolbars

Below the menu bar are two *toolbars*. Each toolbar consists of a row of buttons and/or list boxes. Each button has a small picture called an icon. The icon is a pictorial representation of the command. For example, the first icon of the first toolbar shows a blank sheet of paper. This represents the New command for creating a new drawing file.

Clicking a button executes an AutoCAD command or a short *macro*. A macro is a series of commands that execute automatically. The toolbars can stick to any side of the drawing area or float anywhere on the desktop. AutoCAD LT 2000i has 18 toolbars, of which you currently see only four: two along the top, and two along the left edge.

The meaning of the icon is displayed in a small yellow tag called a *tooltip*. Pass the cursor over the icon, then wait for a second. The tooltip appears. At the same time, look at the status line (at the very bottom of the AutoCAD LT window). It displays a one-sentence description of the button's meaning.

Each button on a toolbar displays a tooltip.

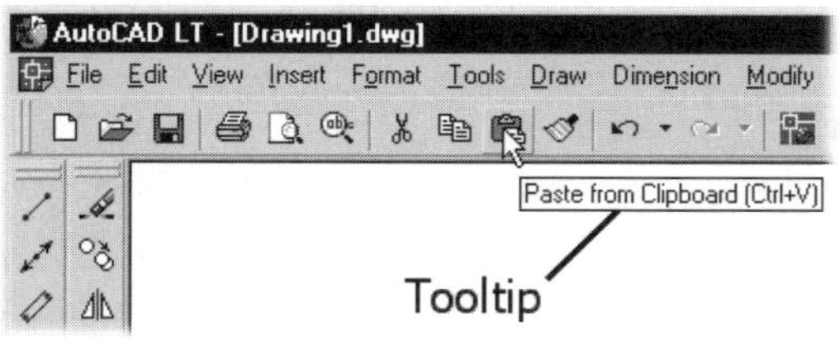

For example, when you pause the cursor over the New icon, the tooltip displays "New (Ctrl+N)" and the status line displays "Creates a new drawing file: NEW." The reference to Ctrl+N reminds you of the shortcut keystroke for starting a new drawing.

AutoCAD LT lets you change the look of the icons displayed by all toolbars, the function of the icons, and even the shape of the toolbar.

The Standard Toolbar

The top toolbar has buttons that you are probably familiar with from other Windows applications, plus several unique to AutoCAD LT. This toolbar is called the Standard toolbar because it is standard to most Windows applications. From left to right, the buttons on the Standard toolbar have these meanings (with the related AutoCAD LT command in brackets):

AutoCAD LT's Standard toolbar shares some similarities with other Windows software.

- Create a new drawing file (**New** command)
- Open an existing drawing file (**Open**)
- Save the drawing (**QSave**)
- Print or plot a drawing to a plotter, printer, or file (**Plot**)
- Plot preview shows how the drawing will look before it is printed or plotted (**Preview**)
- Find, replace, select, or zoom to specified text (**Find**)
- Cut objects to the Clipboard (**CutClip**)
- Copy objects to the Clipboard (**CopyClip**)
- Insert data from the Clipboard into the drawing (**PasteClip**)
- Apply the properties of a selected object to other objects (**MatchProp**)
- Undo (**U**)
- Redo (**Redo**)
- Open the AutoCAD Today window (**Today**)
- Display Autodesk's Point A Web site in the browser (**Browse**)
- Share the current AutoCAD session among other users across the Internet (**MeetNow**)
- Create HTML pages that include images of selected drawings (**PublishToWeb**)
- Attach a hyperlink to an object in the drawing (**Hyperlink**)
- Locate a point by temporary points (**Tracking**) and object snap flyout
- Manage user-defined coordinate systems (**UCS**) and UCS flyout
- Change the display order of overlapping objects (**DrawOrder**)
- Save and restore views by name (**View**) and viewpoint flyout
- Pan the view in real time (**Pan**)
- Zoom the view larger and smaller in real time (**Zoom**)
- Zoom to the area specified by a rectangle (**Zoom Window**) and zoom flyout
- Return to the previous view (**Zoom Previous**)
- Displays AutoCAD DesignCenter window (**AdCenter**)
- Displays Properties window (**Properties**)
- Displays online help (**Help**)
- Displays the Active Assistance window (**Assist**)

Look carefully at the Zoom Window button; it contains a tiny triangle in the lower-right corner. That indicates the button contains a *flyout*, which means two or more additional buttons are "hidden" underneath. To see how a flyout works, move the cursor over the Zoom Window button, then hold down the left button. A column of several buttons flies out.

Displaying a flyout from a toolbar button.

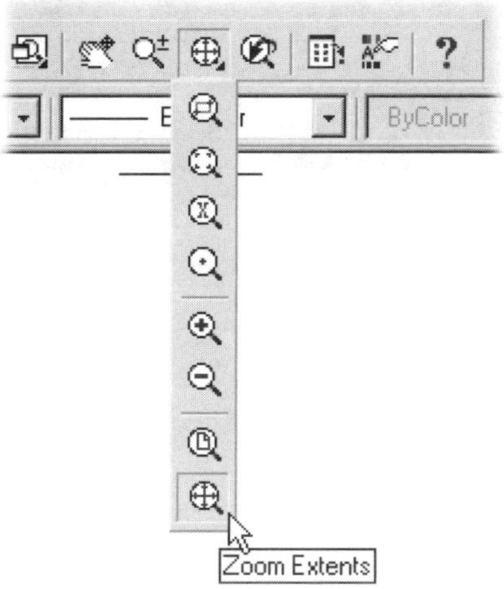

To select a flyout button, move the cursor down, and let go of the mouse button. Notice that the button you selected now appears on the toolbar.

Object Properties Toolbar

The most important toolbar in AutoCAD is the Object Properties toolbar. It has several *list boxes* that provide immediate feedback on the objects with which you are working. A list box displays a list of options, such as a variety of colors or layer names. From left to right, the toolbar buttons and list boxes have these meanings:

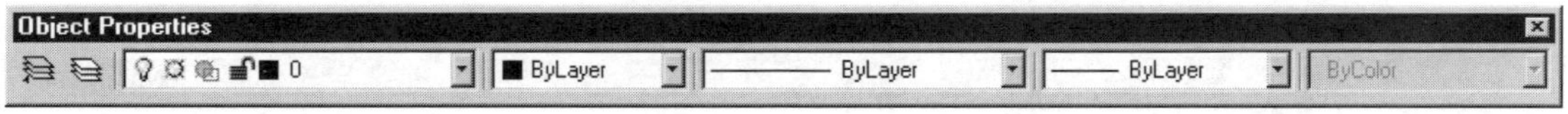

The Object Properties toolbar displays the properties of objects in the drawing.

Make Object's Layer Current button

This is a shortcut button that lets you set the *current* (working) layer by selecting an object. That object's layer becomes the current layer (**Ai_Molc** command).

Layers button

Click the **Layers** button to display a dialog box with the rather imposing name of Layer Properties Manager, which lets you create and set layer names.

The Layer Properties Manager dialog box.

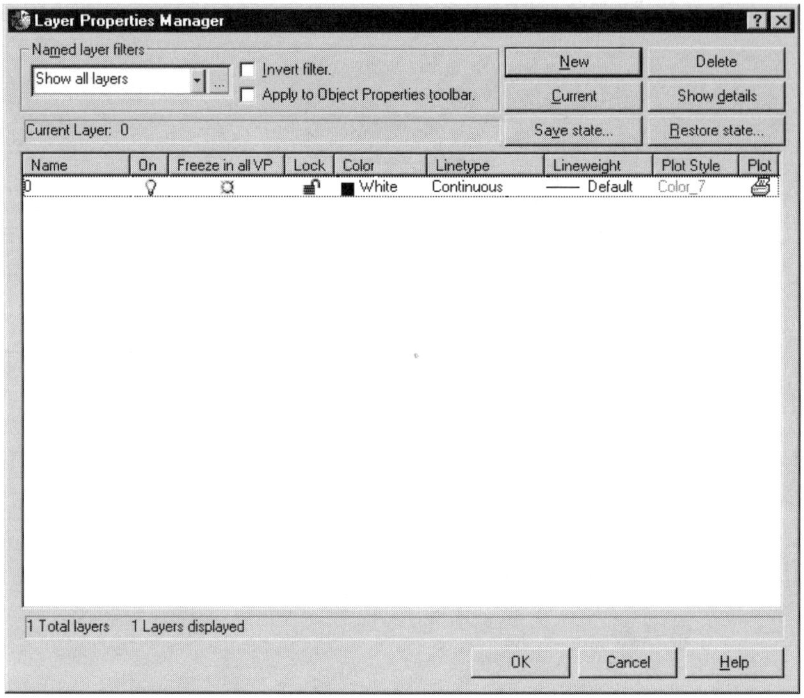

Layer Control list box

Clicking the Layer Control list box displays the name and status of every layer in the drawing. A new drawing, such as this one, has only one layer, named "0", which can never be erased or renamed. Selecting a name from the layer list causes AutoCAD to set that layer name as the current layer.

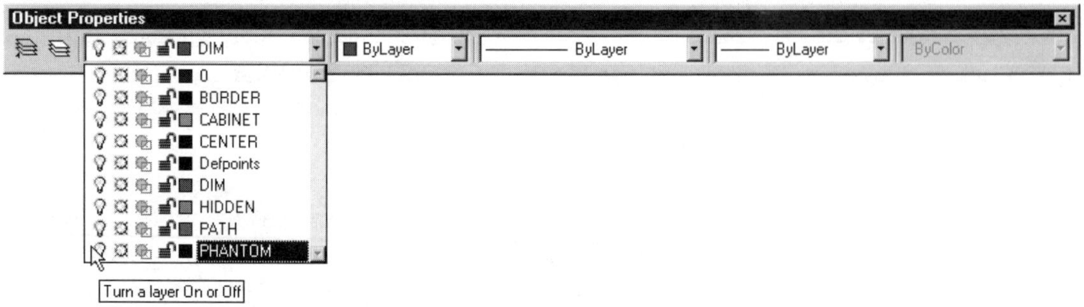

Selecting a layer from the Layer Control list box.

Each layer name is prefixed by five small symbols. In the illustration, look at the first two layer names to see the difference between the symbols, which have these meanings:

- Turns the layer on and off
- Thaws and freezes the layer in all viewports
- Thaws and freezes the layer in the current viewport
- Unlocks and locks the layer
- Specifies the color assigned to objects on the layer

Layers will be discussed in greater detail in the next chapter.

Color Control list box

Clicking the Color Control list box displays a selection of colors. Although you see just nine colors listed here, AutoCAD can work with as many as 255 colors. To select one, click **Other** (found at the end of the list). The Select Color dialog box appears. Select a color by clicking it, then click the **OK** button. The color you selected is added to the list. From now on, all objects are drawn in this new color.

Tip: The ByLayer and ByBlock colors have special meanings in AutoCAD. ByLayer means that objects take on the color defined by the layer on which they reside. ByBlock means the objects take on the color defined by the block of which they are part.

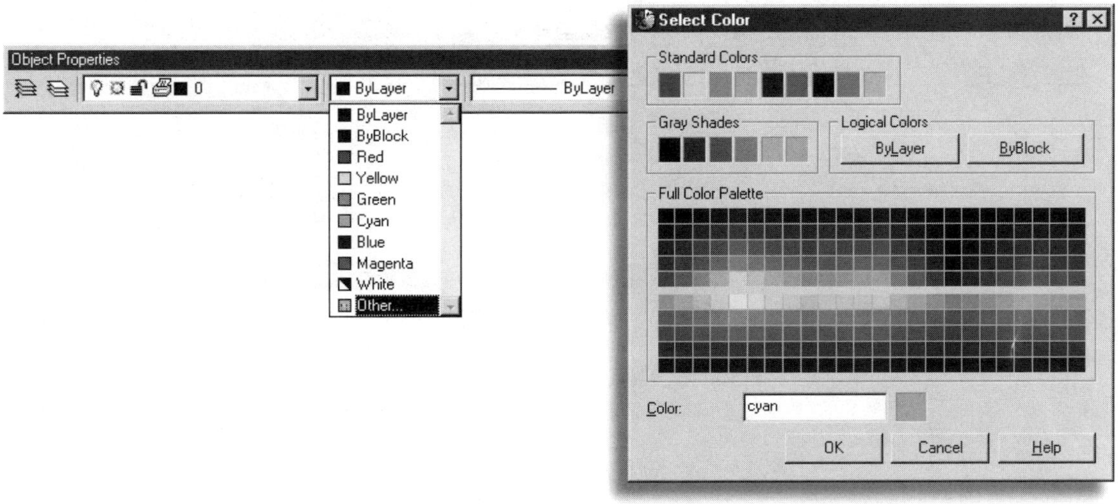

Selecting a color from the Select Color dialog box.

Linetype Control list box

Clicking the Linetype Control list box displays a list with the name and description of every linetype in the drawing. A new drawing, such as this one, has just three linetypes. Selecting a linetype causes AutoCAD to set that linetype pattern as the current linetype. To load other linetypes, click **Other** to display the Linetype Manager dialog box.

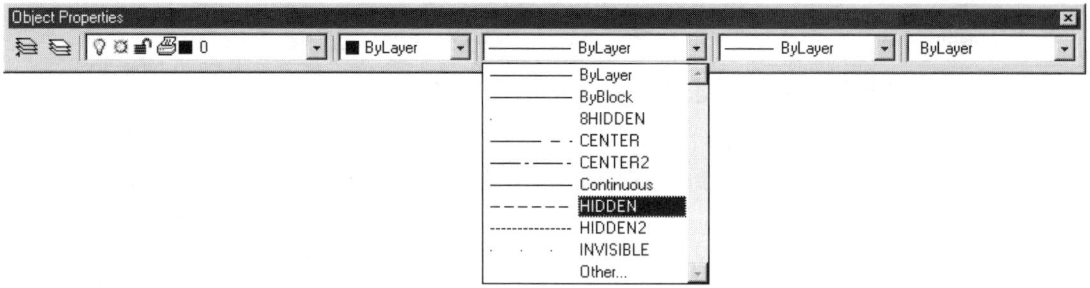

Selecting a linetype from the ones loaded into the drawing.

Lineweight Control list box

Clicking the Lineweight Control list box displays a list of line widths. Every drawing has a fixed number of widths, ranging from 0 to 2.11 mm (0.083"). Selecting a lineweight causes AutoCAD to set that width as current.

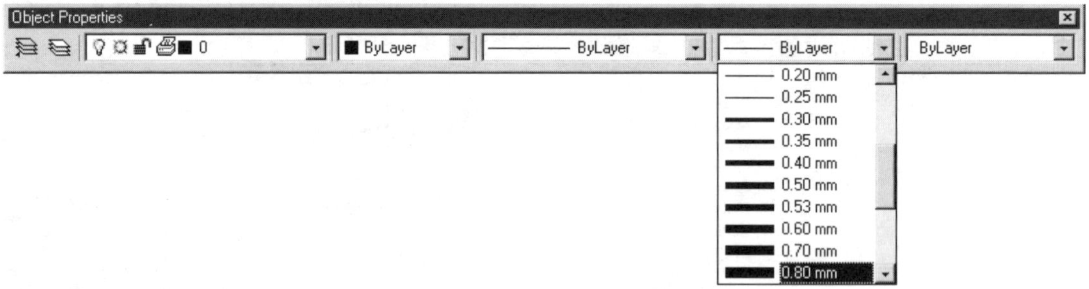

Selecting a lineweight.

Plot Style list box

Clicking the Plot Style list box displays a list of plot styles currently loaded into the drawing. A *plot style* defines how every line in the drawing should be plotted. Plot styles are too complex to be covered by this introductory book.

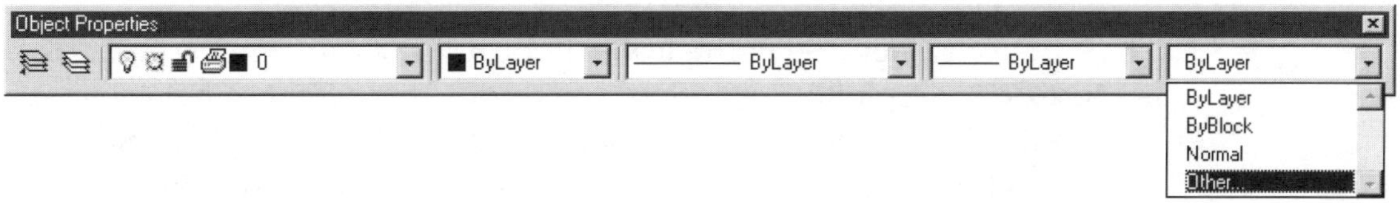

Select a plot style that has been enabled for this drawing.

> **Note:** The Plot Style list box is normally gray, meaning you cannot use it. To use plot styles, you must first turn on the command, which is only available for drawings that use named plot styles. To create a drawing using a named plot style, from the menu, select Tools | Options | Plotting, then select the Use Named Plot Styles option. Plot styles come into effect with the next new drawing you start.

Layout Tabs

Under the drawing area are three tabs, labeled Model, Layout1, and Layout2. You may have seen similar tabs in other Windows software, such as the page tabs in Excel. These tabs help you switch between layouts, which is a new feature with AutoCAD LT 2000. A *layout* lets you define how the drawing will be plotted.

The Crosshair Cursor and UCS Icon

In the drawing area, the *cursor* gives you feedback from AutoCAD. While in the AutoCAD drawing area, the cursor is a crosshair to show where you are in the drawing. Try moving the cursor around the AutoCAD window by moving your mouse.

The small box in the center of the crosshairs is called the *pickbox*, which shows you the point you are picking. You use the cursor pickbox in Chapter 4, "Adding Details to a Drawing."

When you move the crosshair out of the drawing area, the cursor changes to an arrow shape. You are probably familiar with the arrow cursor from other Windows applications. The arrow cursor lets you make menu selections and pick icon buttons.

The cursor changes to several other shapes, which you learn about in the coming chapters. For example, when the cursor turns to a double-ended arrow, you can resize the AutoCAD LT window

The UCS icon is located in the lower-left corner of the drawing area. UCS is short for "user-defined coordinate system." It is meant to help you draw in 3D. Since AutoCAD LT is not meant for 3D drawing, the best thing is to turn it off. From the menu bar, select **View | Display |UCS Icon | On**. The vertical bars (|) separate menu picks.

The Command Area

Along the bottom of the AutoCAD window is the *command prompt area*. Here you type AutoCAD commands, if you (like me) prefer typing to making menu selections. Typing command names is a fast way to draw and edit if you are a touch typist, but it is slow if you aren't. The command area is also where AutoCAD prompts you for additional information it might need to complete a command.

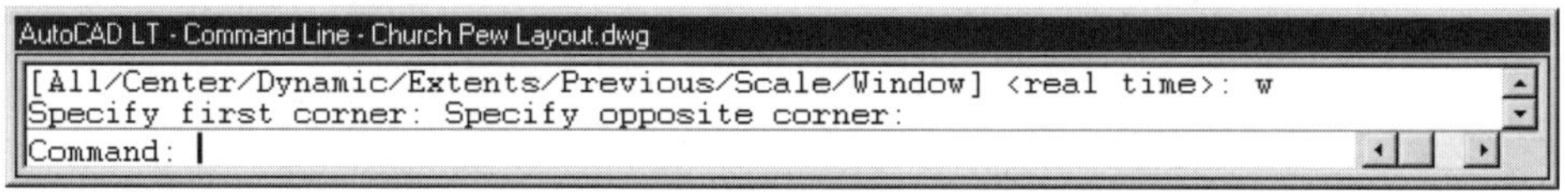

The command prompt area displays the last three lines of command-related text.

When you see the Command: prompt, like this:

Command:

it means AutoCAD is ready for you to type a command name. If you want to type a command, but text other than Command: appears, then press the Esc key once or twice to clear the command line. Try drawing a few lines now:

1. Type the **Line** command, as follows:

   Command: line [Enter]

   Type the word **line**, then press the Enter key.

2. AutoCAD changes the prompt from Command: to Specify first point:

   Specify first point:

AutoCAD is asking you where the line should start. As you move the mouse, you see the crosshair cursor move about the drawing portion of the screen.

3. Pick a point on the screen by pressing the first button (the left button) on your mouse, also known as the "pick button." AutoCAD changes the prompt to read,

   Specify next point or [Undo]:

   and a "rubberband" line stretches from the point you picked as you move the mouse around.

4. Move the mouse some more and press the pick button again. You have drawn your first line in AutoCAD!

Drawing lines with the Line command.

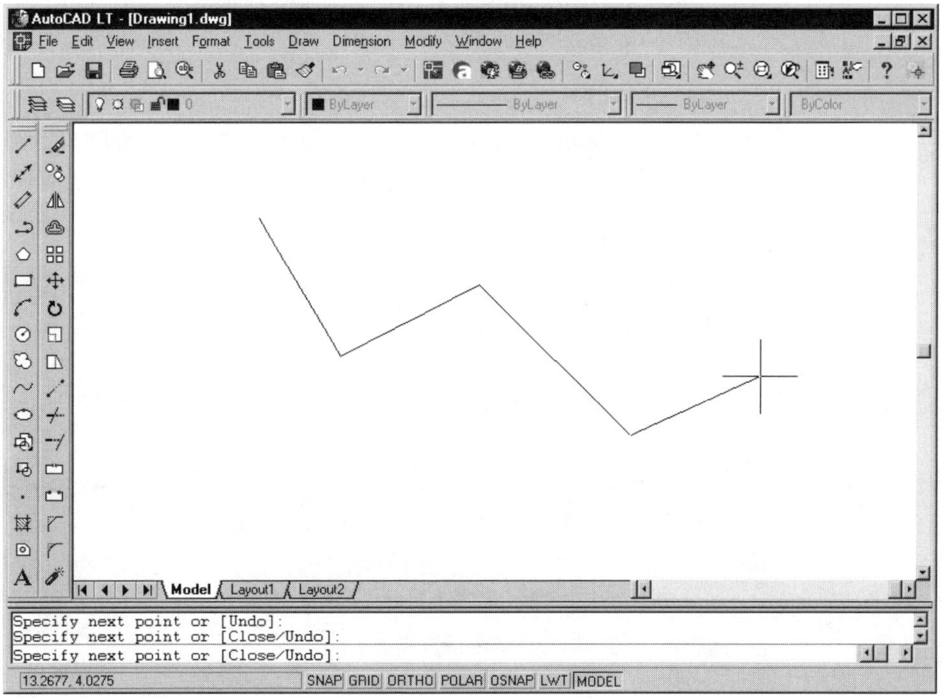

5. Draw another line by moving the mouse again and pressing the pick button.

6. You end the Line command by pressing the **Enter** key, or pressing the **Esc** key, or pressing the right mouse button, as follows:

   Specify next point or [Undo]: [Enter]

   Pressing **Esc** cancels any AutoCAD command; some commands may need a couple of presses of **Esc**.

Tip: The Line command has three options, one of which is hidden. Pressing the Enter key has three different effects, depending on the prompt:

- At the Line command's From point: prompt, pressing Enter causes AutoCAD to continue drawing from the last point, whether a line or arc. This is a great way to ensure that a line is drawn perfectly tangent to the end of an arc.
- At the Line command's To point: prompt, pressing Enter terminates the Line command.
- At the Command: prompt, pressing Enter repeats the last command, which in this case is the Line command.

7. To erase the lines you drew, type **U** at the Command: prompt to undo the lines, as follows:

   Command: u

   - Or, you can select the undo icon from the toolbar.
   - Or, you can select **Edit Undo** from the menu bar.
   - Or, you can press **Ctrl+Z**.

   As you can see from this example, AutoCAD provides you with several different ways to control the program. You will probably find yourself using a combination of keyboard typing, toolbar icons, menu picks, and keyboard shortcuts—whichever you find most convenient.

Tip: You can right-click in the drawing area at any time to display a shortcut menu. This menu is *context-sensitive*, meaning that its display changes depending on what you are doing at the time. During a command, for example, the shortcut menu displays all of the command's options. When not using a command (at the Command: prompt), the shortcut menu displays some commonly used commands.

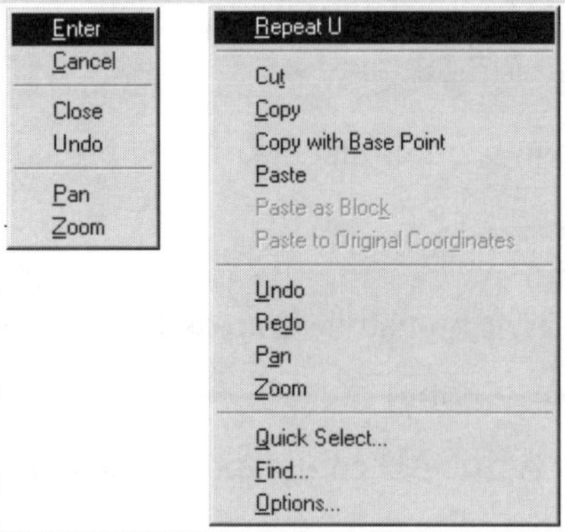

The shortcut menus displayed by the Line command (at left) and by no command (at right).

The Status Bar

Below the command prompt area is the *status bar*, or *status line*, which reports the status of the drawing:

The status bar displays 2D coordinates and several mode buttons.

From left to right, the status line displays:

- The current 2D x,y-coordinates, as in 5.4083,6.6937. Click the coordinates to change the display between static, dynamic, and polar, as explained later in this chapter.
- The setting of snap, grid, orthographic (ORTHO), polar snap, object snap (OSNAP), lineweight (LWT), and model/paper modes. Each word is on a button. When the button looks like it sticks out, the mode is turned off, as in SNAP in the figure, above. Clicking the button to turn on the mode; the button appears pressed in, as in the MODEL item, above.
- When the cursor is paused on a toolbar button or a menu item, the status bar displays a line of help text, as described earlier.

Tip: You can right-click any of the buttons (except MODEL) to display a shortcut menu. Nearly all the shortcut menus have the same three options: On, Off, and Settings. The Settings option is a great shortcut to get to the dialog box that relates to the options. For example, to change the settings for lineweight, right-click LWT and select Settings. AutoCAD displays the Lineweight Settings dialog box.

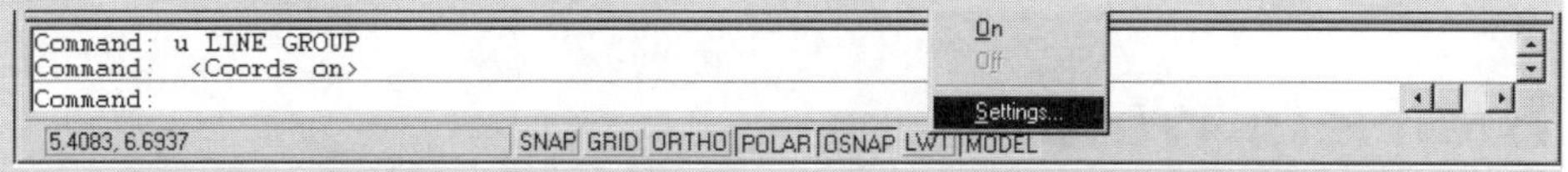

Right-click a mode button on the status bar for quick access to its settings.

The Text Window

When you need to see more than three lines of the command prompt area, switch to the text screen by pressing function key **F2**. Instantly, a second AutoCAD window appears on the screen, labeled "AutoCAD LT Text Window."

The text window displays more command text.

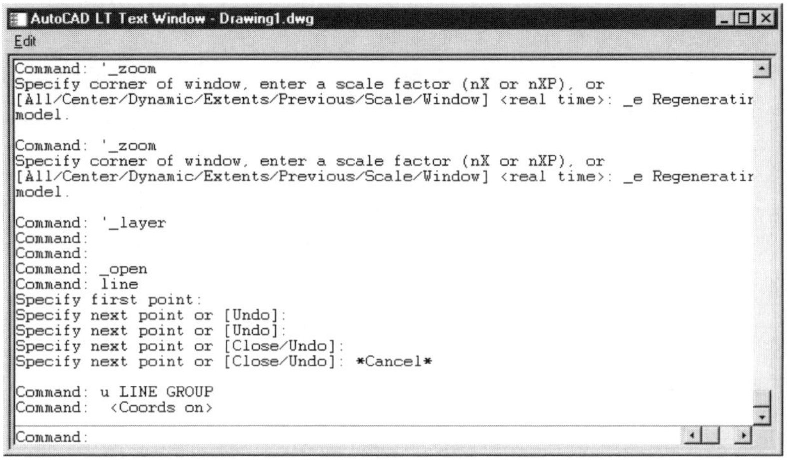

The text screen displays the most recent 1,500 lines of command text. You scroll back to earlier text by clicking on the vertical scroll bar at the right edge of the window.

By right-clicking the text window, you can copy all or part of the text to the Clipboard.

You return to the drawing screen by pressing **F2** again. Function key **F2** is called the flipscreen key.

The Help Window

AutoCAD's online help system is based on Microsoft's brand of Web browser, known as Internet Explorer. At any time, you can call up help on using AutoCAD LT as follows:

1. Select **Help | AutoCAD LT Help**. The vertical bar (|) is shorthand for the following actions: click on the menu bar's **Help** item, then select **AutoCAD LT Help** from the menu.

AutoCAD LT's online help system.

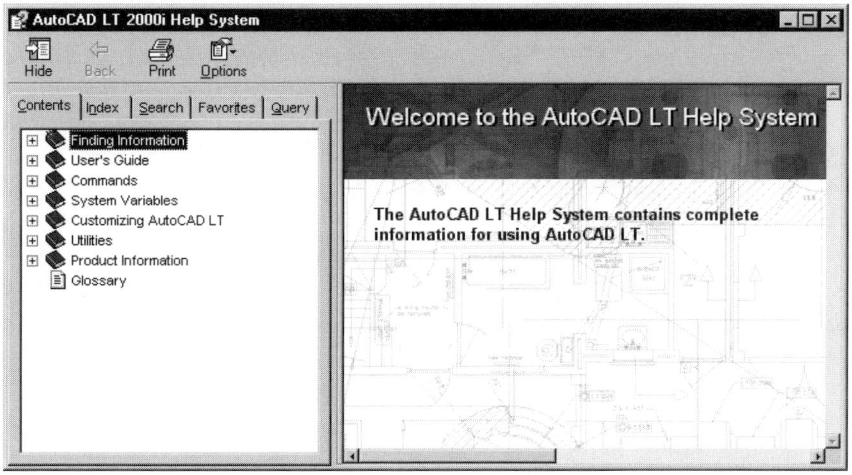

2. Select the **Contents** tab. Double-click on any book icon to find out more information about a subject.

Context-Sensitive Help

The help system is context sensitive. That means AutoCAD displays helpful information related to the current AutoCAD command.

To access help, type the Line command, then press the **F1** function key. Pressing **F1** invokes context-sensitive help, as follows:

Command: line
Specify first point: [F1] '_help
Resuming LINE command
Specify first point:

AutoCAD displays helpful information about using the Line command.

Press F1 during a command to display help for the command.

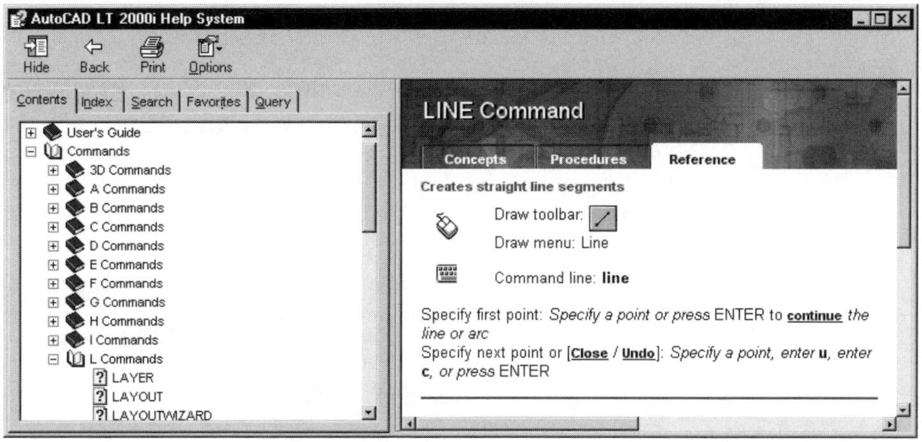

Whenever you see blue underlined text, click on the text to display the definition of the word or to find out more about an option.

Press **Esc** to end the Line command.

Resuming LINE command
From point: [Esc]
Command:

Active Assistance Window

New to LT 2000i is the Active Assistance window. By default, this window pops up when AutoCAD LT starts, and continually displays a couple of paragraphs of help for whatever command you select.

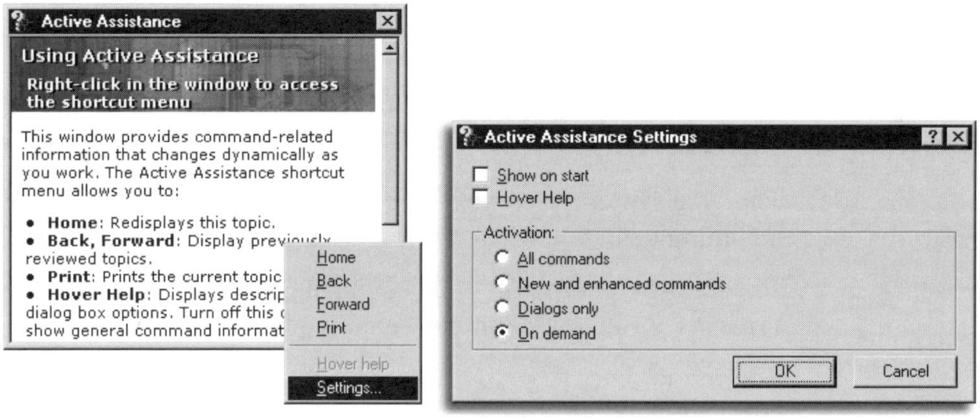

You can control the display of this window, as follows:

1. Right-click the Active Assistance window.
2. From the shortcut menu, select **Settings**.
3. In the Active Assistance Settings dialog box, select the options you prefer. To have Active Assistance always displayed, select:
   - Show on Start: **On** (check mark shows)
   - All commands: **On** (dot shows)

   To turn off Active Assistance, select:
   - Show on start: **Off** (no check mark shows)
   - On demand: **On** (dot shows)
4. Click **OK** to exit the dialog box.

If you want, you can exit AutoCAD LT with the Quit command, or press **Alt+F4** or select **File | Exit**.

Review Questions

1. To open a drawing in AutoCAD, you can drag the drawing's filename from the Exploring window to AutoCAD LT's title bar.

   True / False

2. In a menu, the ellipses (...) behind a command means:

   a. The command is not available.
   b. The command opens a dialog box.
   c. Not all of the command name fits on the menu.
   d. The command is currently active.

3. The picture on a toolbar button is called:

   a. An icon.
   b. A macro.
   c. A tooltip.
   d. A shortcut.

4. A macro is one or more commands that execute automatically.

   True / False

5. AutoCAD LT does not allow you to change the look of toolbar icons or their function.

   True / False

6. Match the AutoCAD term with the best meaning:

a.	ByLayer	i.	Defines the width of lines.
b.	Lineweight	ii.	Defines how lines are plotted.
c.	ByBlock	iii.	Objects take on the layer's color.
d.	Plot style	iv.	Objects take on the block's color.
e.	Layout	v.	Defines how a drawing is plotted.

7. The pickbox:

   a. Is at the center of the arrow cursor and shows the point you are picking.
   b. Is at the center of the crosshair cursor and shows the 3D orientation you are viewing.

c. Is at the center of the arrow cursor and shows the menu items you are picking.

d. Is at the center of the crosshair cursor and shows the point you are picking.

8. Match the abbreviation with its best meaning:

a.	U	i.	Help
b.	UCS	ii.	Control key
c.	F1	iii.	User-defined coordinate system
d.	Ctrl	iv.	Escape key
e.	Esc	v.	Undo

9. The command prompt area is where AutoCAD prompts you for additional information.

   True / False

10. The text screen is AutoCAD's interface for handicapped users.

    True / False

Chapter 2

Setting Up the Drawing

In this chapter, you learn how to prepare the CAD environment for creating a new drawing. By the end of the chapter, you know how to save your work to disk and how to get out of AutoCAD.

Key Concepts

- Preparing a drawing for first-time use
- Understanding how layers organize the drawing
- Saving the drawing
- The importance of automatic backups

Key Terms

Filename—the computer name of the drawing

Units—AutoCAD can display units in a variety of measurement styles

Snap—constrains mouse movement to discrete distances

Grid—visual guide consisting of an array of dots

Limits—specifies the nominal limits of a drawing, and constrains the limits of the grid marks

Layers—organizes the drawing, separating common elements

Zoom—enlarges the visual size of the drawing

Abbreviations

'	feet
"	inches
in	inches
ANSI	American National Standards Institute

DWG	portion of the filename that identifies AutoCAD drawings
BAK	backup of an AutoCAD drawing file

Command	Shortcut	Menu Selection
Layer	la	Format | Layer
New	Ctrl + N	File | New
Preferences	pr	Tools | Preferences
SaveAs	Alt + FA	File | Save As

Before You Begin

To learn AutoCAD, you should work with a drawing based on something you can easily find: the yard. The example used for the drafting portion of this book creates and modifies a drawing of the yard around a house.

Before you begin this tutorial, you may want to measure your yard and locate the major features, such as the house, driveway, and garden areas.

Hand sketch of the yard.

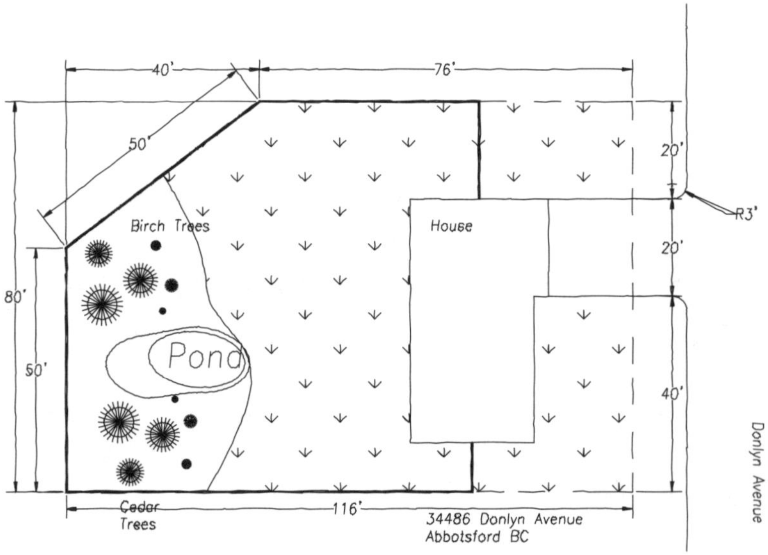

If you'd rather not measure your yard, or if your home doesn't have a yard, you can follow along with the sketch above, which is the drawing used in the next several chapters.

Preparing for Drawing the Yard

Before you draw the yard, you need to prepare AutoCAD for a new drawing. In this chapter, you learn how to name the drawing, select the units of measurement, set the snap and grid spacing, size the drawing limits, and name the layers.

If AutoCAD LT is not running, start it now by double-clicking its icon on the Windows desktop.

Start a New Drawing

When you want to start a new drawing in AutoCAD LT, you can start "from scratch" or you can enlist the services of a software wizard. The *wizard* is a series of dialog boxes that takes you through the steps needed to set up a new drawing, such as its units and the scale.

1. When AutoCAD LT appears on the screen, notice the AutoCAD LT Today window. (The AutoCAD LT Today window replaces the Startup dialog box found in versions of LT prior to 2000i.)

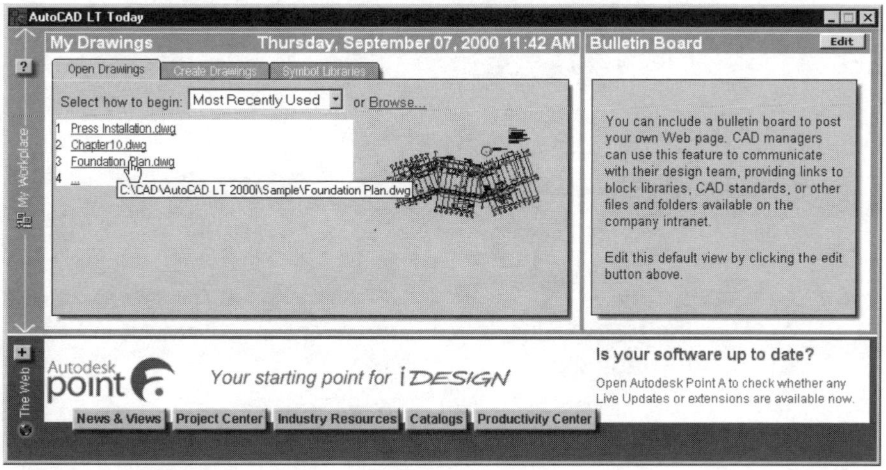

Note: The AutoCAD LT Today window consists of two halves:

My Workplace. The upper half of the AutoCAD LT Today window provides access to drawings located on your computer or on other computers via a network. In the My Drawings section, notice the three gray tabs:

- Open Drawings lists the names of drawings you previously opened in AutoCAD LT. If this is the first time you are using LT 2000i, the list is empty. When you have opened drawings, their names are stored in this list. The list can be sorted by filename, location, week opened, or in order of most recent use. Alternatively, click Browse to display the Select File dialog box.
- Create Drawings lets you start a new drawing based on a template, a wizard, or from scratch.
- Symbol Libraries lists the names of symbol library collections. When you select a library, AutoCAD opens the DesignCenter window (more about this in Chapter 7, "The AutoCAD DesignCenter"). You create your design by dragging the pre-drawn symbols into the drawing.

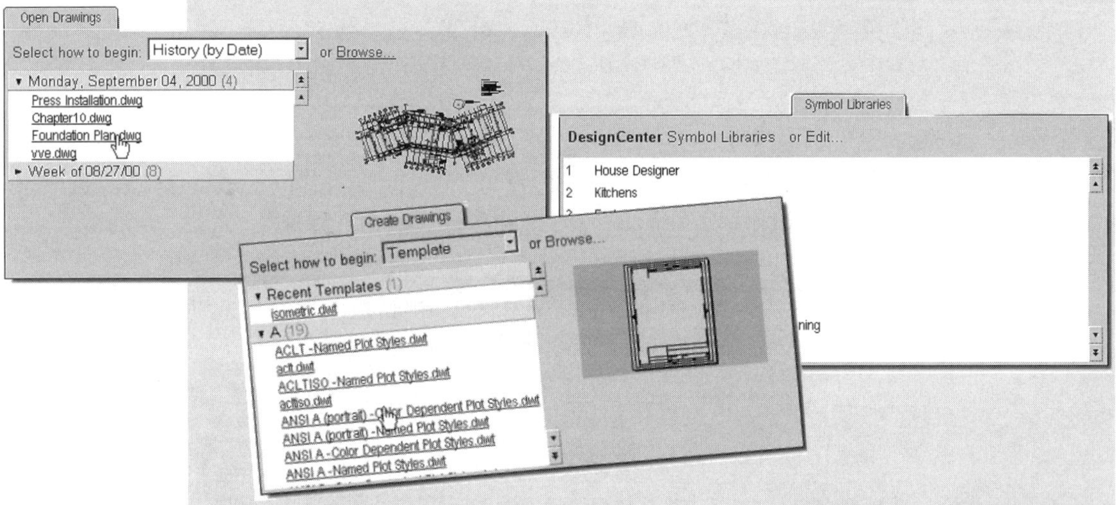

The AutoCAD LT Today window uses a different user interface from the rest of AutoCAD LT, because it is based on Web browser technology:

- To select a file: Hover the cursor over the filename. Notice that a preview of the file is displayed on the right.
- To open a file: Click once. Like a Web browser, clickable items are underlined in blue.

The Bulletin Board area can be made to display messages from a central file server, which requires a network connection.

The Web. The lower half of the window displays information from Autodesk's Point A Web site, which requires a connection to the Internet. See Chapter 11, "Accessing Autodesk's Portals."

2. In the My Drawings area, select the **Create Drawings** tab.

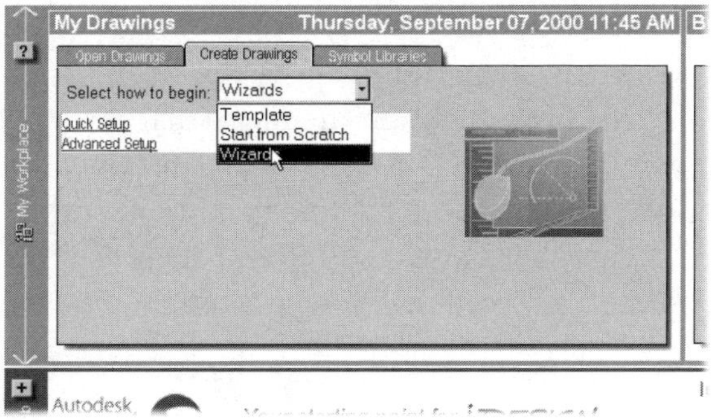

3. From the Select how to begin list box, select **Wizards**. Notice the two choices: Quick Setup and Advanced Setup.
4. Select **Advanced Setup**. Notice that AutoCAD displays the Advanced Setup dialog box. The dialog box has five stages: Units, Angle, Angle Measure, Angle Direction, and Area. Let's explore each one.

Units

AutoCAD displays units in a variety of measurement styles, such as fractional or decimal inches, decimal units (similar to metric), and exponential (used in science). Once you select a measurement style, AutoCAD displays all measurements in that style of units. You may at any time during the drawing switch the measurement style with the Units command.

1. **Select the unit of measurement.** When you measured your yard, you probably used feet and inches. For this reason, you should click the circle (called a *radio button*) next to **Architectural.**

The Units stage of the Advanced Setup dialog box.

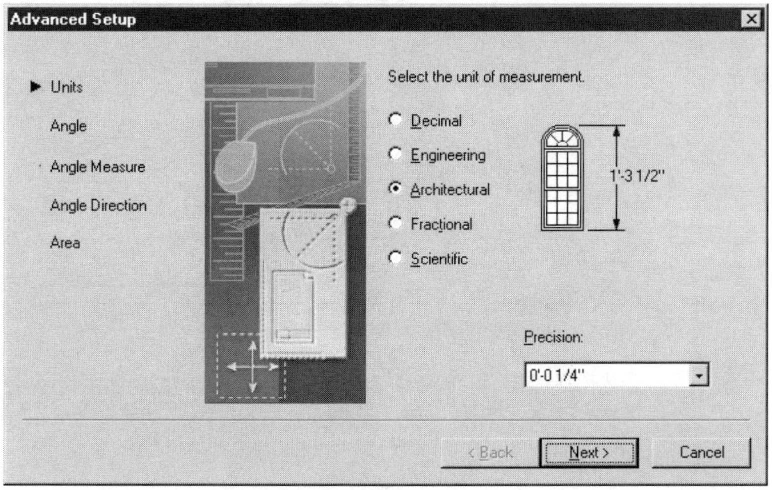

(If you measured your yard in meters and centimeters, you would click the radio button next to **Decimal**.)

2. **Precision**. Measurements made to the nearest 1/4-inch are accurate enough for this project. Click the down arrow and select the **0'-0 1/4"** item. (If necessary, click on the vertical scroll bar until 0'-0 1/4" is visible.) You are free to enter distances more accurate than 1/4-inch; AutoCAD remembers distances to full accuracy. When AutoCAD displays coordinates, however, they are rounded to the nearest 1/4".
3. Click **Next**, and AutoCAD displays the Angle settings.

Angle

The dialog box displays the five types of angle units with which AutoCAD works. By default, AutoCAD uses the Cartesian method of measuring angles, and fractional degrees are measured in decimals. If we were real land surveyors, we would use Surveyor format (which looks like N 45d23'10.9 W), but since we're not, we'll stick with the more familiar decimal degrees.

1. **Select the angle of measurement.** Ensure the default, **Decimal Degrees**, is selected.
2. **Precision.** The default precision of 0 degreees is adequate for our project.

The Angle stage of the Advanced Setup dialog box.

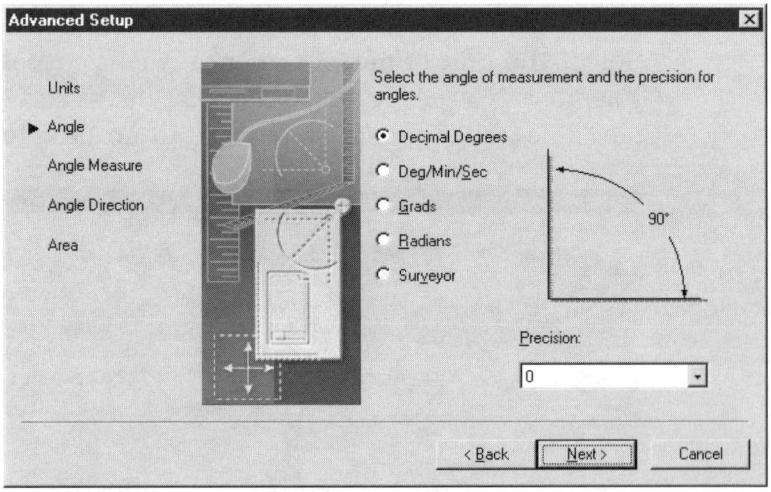

3. Click **Next**. AutoCAD displays the Angle Measure dialog box.

Angle Measure

By default, AutoCAD measures 0 degrees starting in the direction to the East (the positive x-direction).

The Angle Measure stage of the Advanced Setup dialog box.

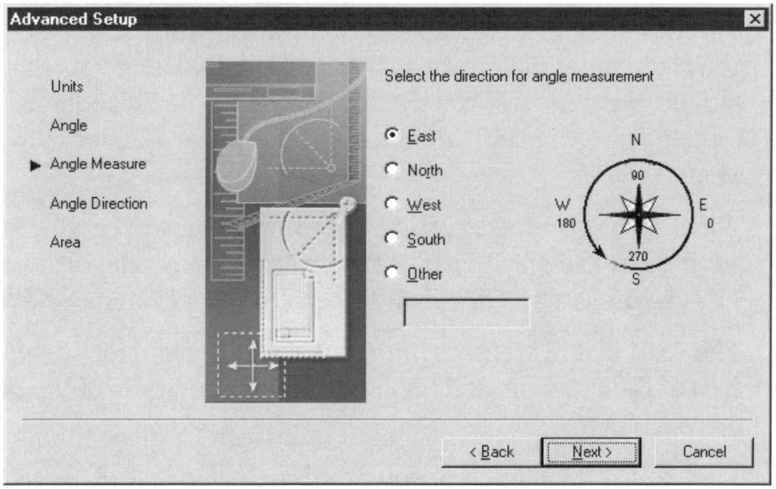

Accept the default setting by clicking **Next**. Notice that AutoCAD displays the Angle Direction dialog box.

Angle Direction

By default, positive angles are measured counterclockwise in AutoCAD.

The Angle Direction stage of the Advanced Setup dialog box.

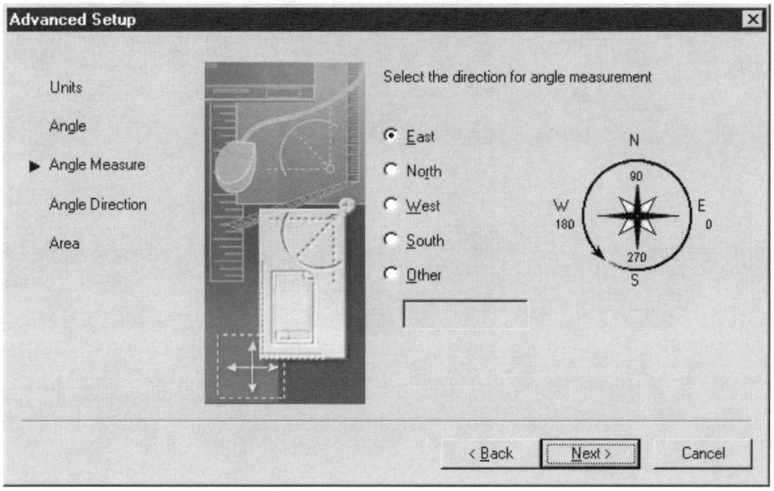

Accept the default setting by clicking **Next**. Notice that AutoCAD displays the Area dialog box.

Area

There is no limit to the size of the drawing you can create with AutoCAD. You could draw the entire solar system full-size, if you want. (An early sample drawing was exactly that; AutoCAD could zoom from the orbit of Pluto down to a plaque mounted on the Lunar Lander in a crater on the Moon.)

When it comes time to print your drawing, however, your drawing is constrained by the size of paper the printer can handle, often just 8-1/2" x 11"—a far sight smaller than the entire solar system, or even your yard.

The Area dialog box sets *limits*, which are useful for showing the nominal area of the drawing, and for constraining the limits of the grid marks; more later.

1. Look at the sketch of your yard. Leaving a bit of "breathing" room around the plan, you'll need about 130 feet of width and 100 feet of height. In the Width text box, enter **130'**.
2. In the Length text box, enter **100'**. Notice that the preview of the paper updates itself to show the drawing as 1200 x 1560 inches.

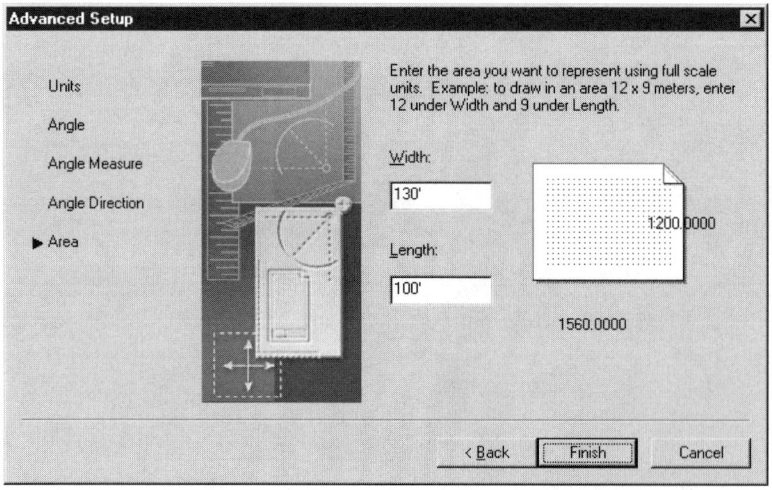

The Area stage of the Advanced Setup dialog box.

Warning It is important to type the apostrophe ('), since that tells AutoCAD you are entering the measurement in *feet*. If you leave out the apostrophe, AutoCAD assumes you typed 130 inches, which it translates into 10'-10".

Tip When entering text and numbers in a dialog box, pressing the Tab key is a fast way to get to the next *field*. A field is a button, text entry box, list box, or other dialog box element that you can change.

To return to a previous field, press Shift+Tab (hold down the Shift key and press Tab).

3. Click **Finish.** The Advanced Setup is complete, and AutoCAD displays a drawing that appears blank to you.

Drawing Aids

One great advantage to drawing with AutoCAD is that it allows you to create *very* accurate drawings. AutoCAD has several features that help you draw with perfect accuracy. One such feature is called *snap* mode; AutoCAD's snap can be thought of as setting a drawing resolution. Another is called *grid*, which helps you visualize distances. You learn about other accuracy aids in later chapters of this book.

1. Move the cursor up to the menu bar.
2. From the Tools menu, select **Drafting Settings.** AutoCAD displays the Drafting Settings dialog box with the Snap and Grid tab.
3. To draw everything to an accuracy of one inch, click the white square (called the *check box*) next to **Snap On (F9).** The (F9) acts as a reminder that you can turn the snap on and off by pressing function key **F9.**
4. Change the snap distance in the Snap X spacing and Snap Y spacing to **1**". Leave other Snap settings at their default values (Angle = 0; X and Y base = 0).
5. As a visual guide, turn on the *grid* (an array of dots) by clicking the **Grid On (F7)** check box. Again, the (F7) is a reminder that the grid can be turned on and off at any time by pressing **F7.**
6. The grid is meant to guide you; the default spacing of 1/2" would get in your way because it is too closely spaced. (In fact, AutoCAD will

not display the grid when it is too closely spaced). Change the spacing to 10 feet by changing the value of Grid X Spacing to **10'**.

> **Tip:** A grid distance of 0" has a special meaning in AutoCAD. It means that the grid spacing matches the snap spacing, 1" in our case.

7. Press **Tab**, and AutoCAD automatically set the grid's y-spacing to 10'.

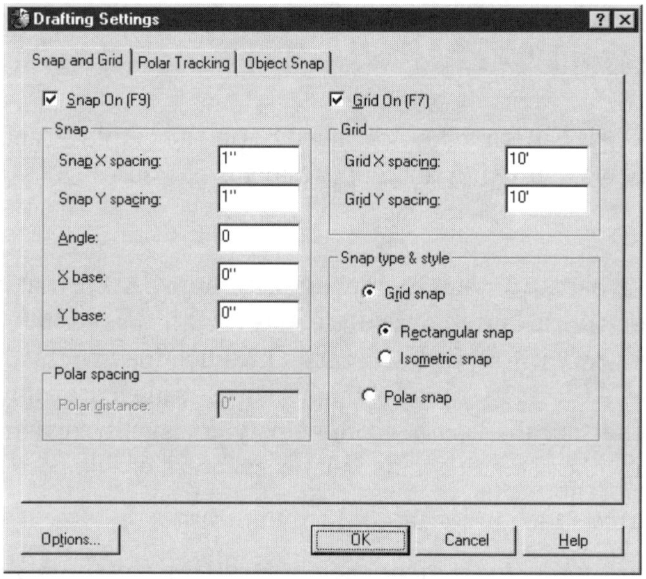

The Snap and Grid tab of the Drafting Settings dialog box.

8. Click **OK**. The dialog box disappears. Move your mouse around and note how the cursor jumps on the screen instead of moving smoothly. The cursor is jumping in one-inch increments. Look at the coordinate display and notice it is changing by the nearest inch, rather than the nearest quarter-inch, as set earlier.

9. Now use the **Zoom All** command to see the grid, which gives you a visual indication of the extents of the drawing limits, as follows:

```
Command: zoom
Specify corner of window, enter a scale factor (nX or nXP), or
[All/Center/Dynamic/Extents/Previous/Scale/Window] <real time>: a
Regenerating drawing.
```

Don't be intimidated by all those prompt options; for the most part, we use just a couple throughout this book.

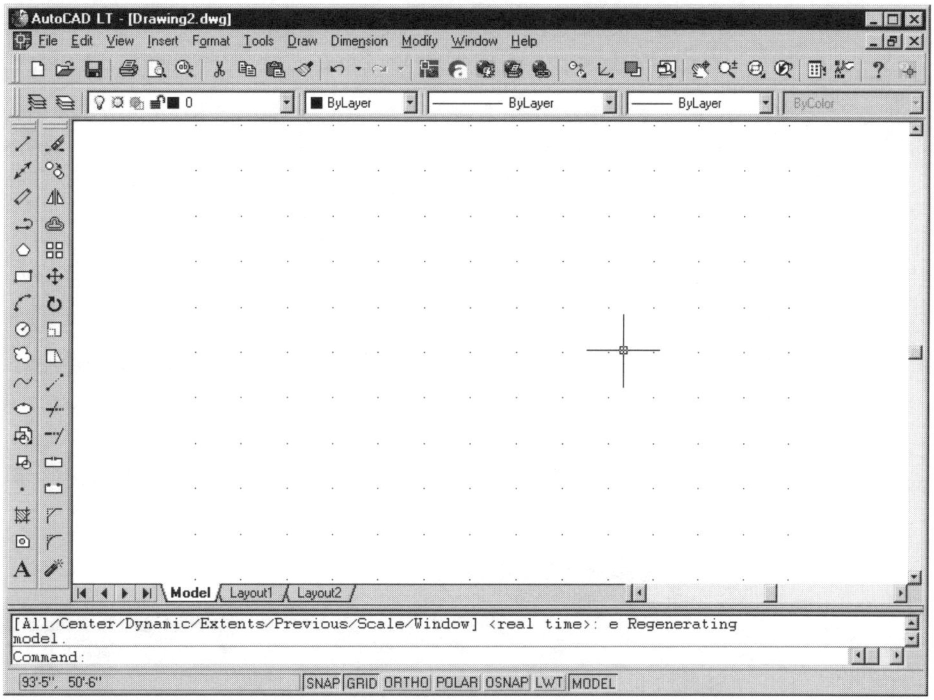

The grid is displayed as a series of dots, spaced ten feet apart.

> **Note:** For most options in AutoCAD's commands, you need only type the first letter of the option. Above, you typed "a" as the abbreviation for the "All" option. When two options begin with the same letter, you need to type the first two characters of the option.

When you began, AutoCAD displayed an area of one foot by nine inches; now AutoCAD displays an area of 130 feet by 100 feet. Thus, the Zoom command lets you see the "big picture," as well as zooming in for a detailed look.

Creating New Layers

If you have worked with overlay drafting, then you are familiar with the concept of *layers*. In overlay drafting, you draw the base plan on one clear sheet of Mylar (plastic drawing media), the electrical on another sheet, and the structural on a third. Since the Mylar is transparent, you can overlay the three drawings to create a single blueprint.

Layers in CAD operate in a similar manner. You draw parts of the drawing on different layers. Then, you turn layers off and on to display the drawing in different ways. For example, the electrical contractor might be interested in seeing only the base plan layer and the electrical layer.

While it is possible to create hundreds or thousands of layers in a drawing, it is more common to work with a few dozen layers; in this book, we work with a mere half-dozen layers.

Layers are specified by a name. AutoCAD lets you give layers names up to 255 characters long. The easiest way to set up new layers is to click the **Layer** button on the toolbar.

1. Move the mouse to the Object Properties toolbar, and click the **Layers** button (the one that looks like three stacked sheets of paper, on the left end of the toolbar). AutoCAD displays the Layer Properties Manager dialog box. The dialog box lets you control almost every aspect of AutoCAD's layers.

Click the Layers button to display the Layer Properties Manager dialog box.

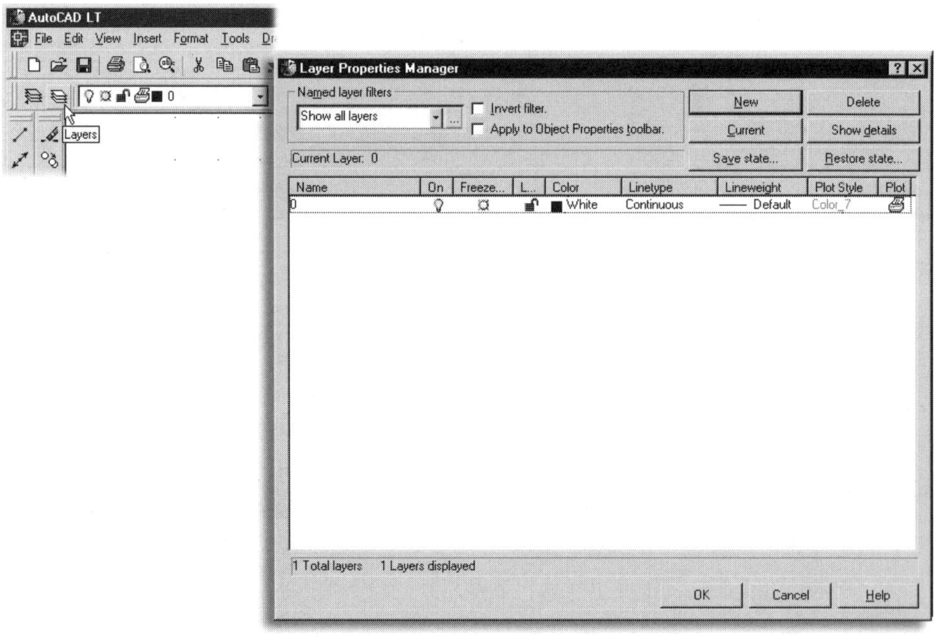

As you can see, the drawing already has one layer:

Layer	Meaning
0	Every new AutoCAD drawing has one layer, called "0", which you can never remove. Layer 0 has special properties that affect the creation of blocks, as discussed in greater detail later in this book.

2. Create a layer by clicking the **New** button. Notice that AutoCAD creates a new layer called "Layer1."
3. Change the name by clicking **Layer1**, then typing **Lot.**
4. Assign a color to each layer to make it easier to determine which lines belong to which layer. Change the color of the Lot layer to blue, as follows:

   Click the black square under the Color column, across from layer name Lot. A second dialog box appears, called Select Color.

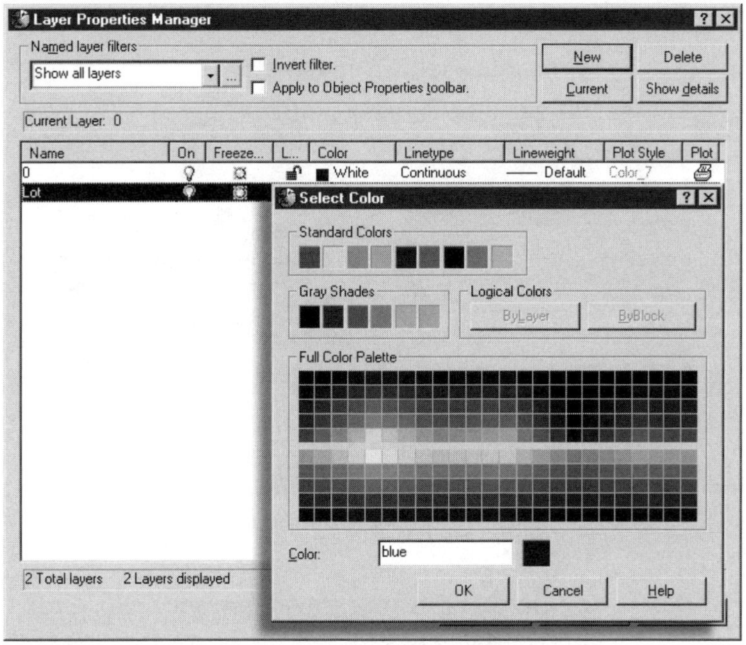

5. The dialog box displays 255 colors. At the top of the dialog box, under Standard Colors, click the blue-colored square. The word "blue" appears in the Color text box at the bottom of the dialog box.
6. Click **OK** to exit the Select Color dialog box. The color of the square across from layer Lot changes to blue.
7. Add the remaining layer names and colors, using the following table as a guide:

Layer Name	Layer Color
House	white (select the black square)
Road	red
Lawn	magenta
Plants	green
Pond	cyan (select the light blue square)

As you add the layer names, notice that AutoCAD automatically rearranges them in alphabetical order. That makes it easier to find a layer name when the list is long.

If you make a spelling mistake, just click the layer name and type the correction.

Note: There can be some confusion over the color "white"—or is it black? AutoCAD LT switches the colors, depending on the background color of the drawing area. If the background color is black, AutoCAD displays white lines; if the background is white, AutoCAD displays black lines. So, white can be black—at least in the world of AutoCAD.

8. When you are finished assigning colors to layer names, pick the Lot layer name, then click **Current**. From now on, the drawing takes place on the Lot layer until you select another name as the current layer.
9. To exit the Layer Properties Manager dialog box, click **OK** at the bottom of the dialog box. Notice how the name of the layer on the Object Properties toolbar changes from **0** to **Lot**. The color has changed from black to blue.

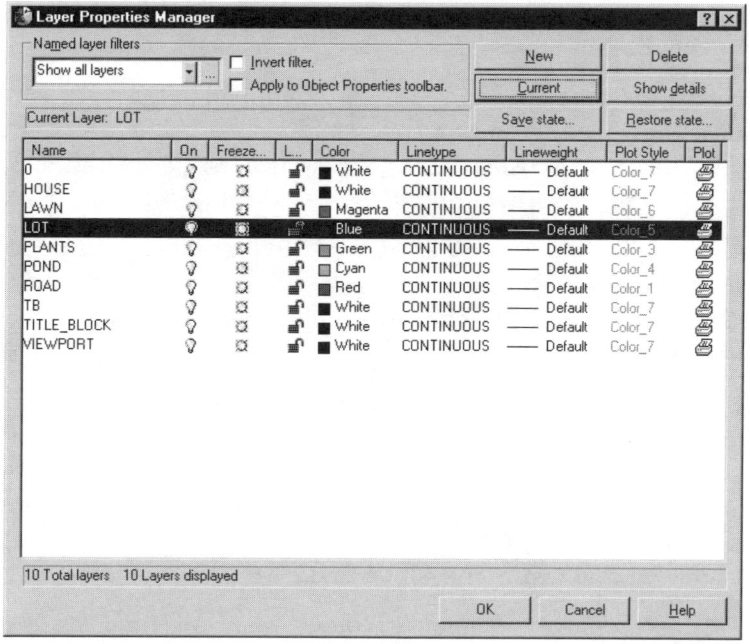

All layers have been added to the drawing.

Note: AutoCAD gives you a great deal of control over layers. Every layer can have the following properties, displayed by the header bar in the Layer Properties Manager dialog box:

Name: The layer name can be up to 255 characters long. You can use numbers, letters, and the following punctuation: dollar signs ($), hyphens (-), underscores (_), and spaces. Two layers cannot have the same name in the same drawing. There is no limit to the number of layers in a drawing.

On: When a layer is *off*, you cannot see objects on that layer. To turn *on* a layer means to see the objects again. The default is on.

Freeze in all VP: When a layer is *frozen*, you cannot see it or edit it, and it is not included when AutoCAD performs a drawing regeneration. It is usually better to freeze a layer than turn the layer off. To *thaw* a layer means to turn off its frozen status. The default is thaw; "VP" is short for viewports.

Lock: When a layer is *locked*, you can see the objects, but cannot edit them. To *unlock* a layer means to make the objects available for editing. The default is unlock.

Color: Select any of AutoCAD's 255 colors. The default color is 7, which can be either white or black, depending on the background color. Any object drawn on the layer is displayed in this color, but can be overridden.

Linetype: Choose from AutoCAD's many linetypes, or create your own linetype. The default is Continuous (solid line). Before other linetypes can be used, their definitions must be loaded into the drawing. Any object drawn on the layer is displayed by this linetype.

Lineweight: Select one of AutoCAD's predefined line widths. The default is 0.00 mm; the maximum is 2.11mm (about 0.08" or 6 points wide). Any object drawn on the layer is displayed by this lineweight.

Plot Style: Select a predefined plot style. Any object drawn on the layer is plotted with the named style, which defines properties like color, width, and percentage of black during plotting. You'll find that you cannot change the plot style, since the feature is turned off by default. You turn on the feature with the Use named plot styles option of the Plotting tab of the Options dialog box.

Plot: When on, the layer is plotted; when off, the layer is not plotted. The default is on.

When a drawing is in *layout* mode, this dialog box displays two more columns:

Active VP Freeze: You can independently freeze and thaw layers for every viewport displayed in the layout. The default is thawed, meaning you can see and edit objects in the current viewport.

New VP Freeze: When a new viewport is created in paper space, you can specify that the layer be frozen. The default is for layers to be thawed in new viewports.

To give you some control over a drawing with many layers, the Layer Properties Manager dialog box lets you sort layers, as well as display a selected group of layers. Click the header—Name, On, Freeze, etc.—to sort layers alphabetically (A to Z). Click the header a second time to sort in reverse order (Z to A).

The Named layer filters list box lets you choose from the following groups of layers:

Show all layers: Display all layer names (the default).

Show all used layers: Display layers that contain at least one object.

Show all xref dependent layers: Display layer names from attached externally-referenced drawings.

The Invert filter options reverses the above three. For example, when the Invert filter is turned on, Show all used layers shows all *unused* layers.

The Apply to Object Properties toolbar option applies this setting to the list of layers displayed by the Layer Control list on the Object Properties dialog box.

The almost unnoticeable ellipsis (...) button displays the Named Layer Filters dialog box. It lets you narrow down the list of layers, essentially creating an AND type of selection. For example, that dialog box can narrow down a long list of layer names by selecting all layers that start with Wall AND are Frozen AND are colored Red.

To erase one or more layers from the drawing, select the layer and press the Del key. As an alternative, you can use the Purge command to purge all empty layers. Note that this only works if the layer is empty; AutoCAD displays a warning dialog box otherwise.

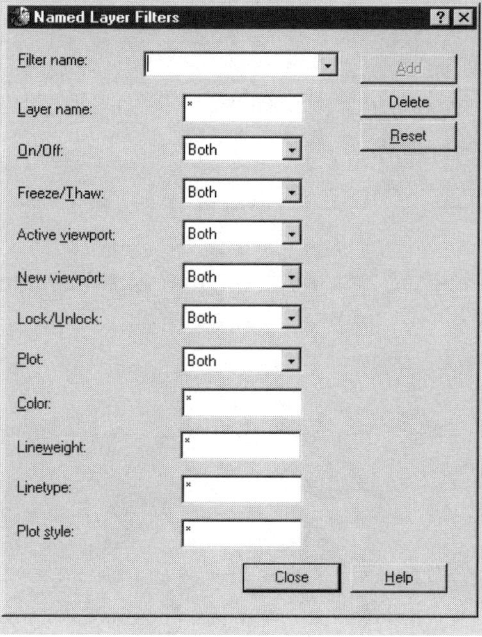

The Named Layer Filters dialog box.

Saving the Drawing

Finally, let's save this important work you have done with the Save command.

1. Click on the icon that looks like a diskette (tooltip = Save).

2. Since this is a drawing with the generic name, "Drawing1.Dwg," AutoCAD displays the Save Drawing As dialog box. By default, AutoCAD LT 2000i stores drawings in the My Documents folder. Type the name **Yard** in the text entry box.

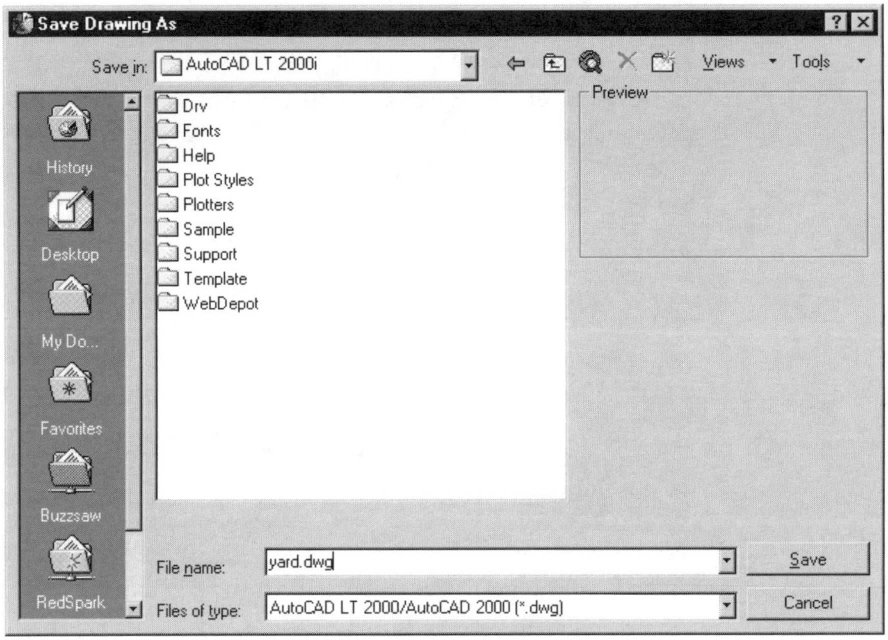

3. Click the **Save** button, and AutoCAD saves the drawing with the name Yard.Dwg. From now on, you and AutoCAD refer to this drawing as Yard.

Tip: The Save As dialog box lets you save the drawing in several dialects of AutoCAD drawing. A newer version of AutoCAD (such as LT 2000i) can always read a drawing created by an older version of AutoCAD (such as LT 98). Some versions of AutoCAD LT can read a newer version of drawing file, such as:

- AutoCAD LT 2000 and 2000i
- AutoCAD LT 97 and 98

In most cases, however, an older version of LT cannot read drawing files created by a newer version. For example, AutoCAD LT 98 cannot read the drawing file created by AutoCAD LT 2000. For this reason, you must explicitly save the drawing in an earlier format for the earlier software to read the drawing:

AutoCAD R14/LT 98/LT 97 Drawing (*.dwg): Saves the drawing in the DWG format read by AutoCAD LT 98, LT 97, and AutoCAD Release 14.

AutoCAD R13/LT 95 Drawing (*.dwg): Saves the drawing in the DWG format read by AutoCAD LT 95 and AutoCAD Release 13.

Drawing Template File (*.dwt): Saves the drawing in the DWT template format.

DXF: Drawing interchange format, read by many other CAD and graphical programs. Since very few programs understand the DXF file in newer formats, you should export in LT R2/R12 DXF format. AutoCAD imports DXF files with the undocumented DxfIn command.

AutoCAD LT has the capability to translate the drawing in one other file format. Use the File | Export command to save the drawing in WMF (Windows metafile) format, used by the Windows operating system as its default vector format. AutoCAD imports WMF files with the WmfIn command.

Warning: AutoCAD erases and alters some objects when it translates a drawing to an earlier format of DWG.

Automatic Backups

For speed, AutoCAD keeps the drawing in the computer's RAM (memory). The drawback is that if Windows crashes or if the power is cut to your computer, you lose your work. Since Windows crashes are common, it is a good idea to save your drawing every 10 or 15 minutes.

AutoCAD lets you set a time to automatically save the drawing, without needing the Save command. The default setting is turned on for every 120 minutes, which is much too much a long time between saves to do us any good. Here's how to change it:

1. Turn on the automatic save feature by selecting **Tools | Options**. AutoCAD displays the Options dialog box.

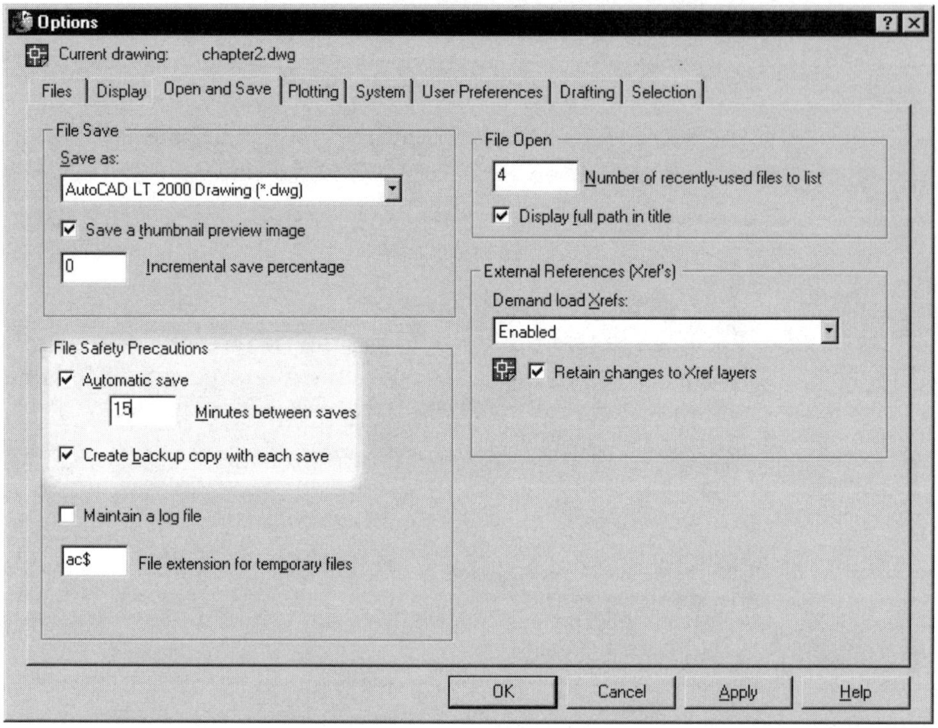

2. Click the **Open and Save** tab.
3. If necessary click the check box next to **Automatic save** to turn it on.
4. In the **Minutes between saves** box, change the value to 15.
5. Look over the Options dialog box. You might want to turn on **Create backup copy with each save.** The ✓ in the check box means the feature is turned on. This means that AutoCAD makes a copy of the drawing the way it was with the previous save.
6. Click **OK**. AutoCAD LT now saves the drawing every fifteen minutes to a file called Auto.Sv$. Although AutoCAD automatically saves the drawing, it is still a good idea for you to save your work after finishing a significant amount of editing and drawing.
7. If you need to take a break at this point, use File | Exit to leave AutoCAD. AutoCAD LT closes its window and you find yourself back at the Windows desktop.

Let's review the drawing to this point. Although you haven't drawn anything yet, the drawing file contains a fair amount of information.

The Yard drawing has been prepared.

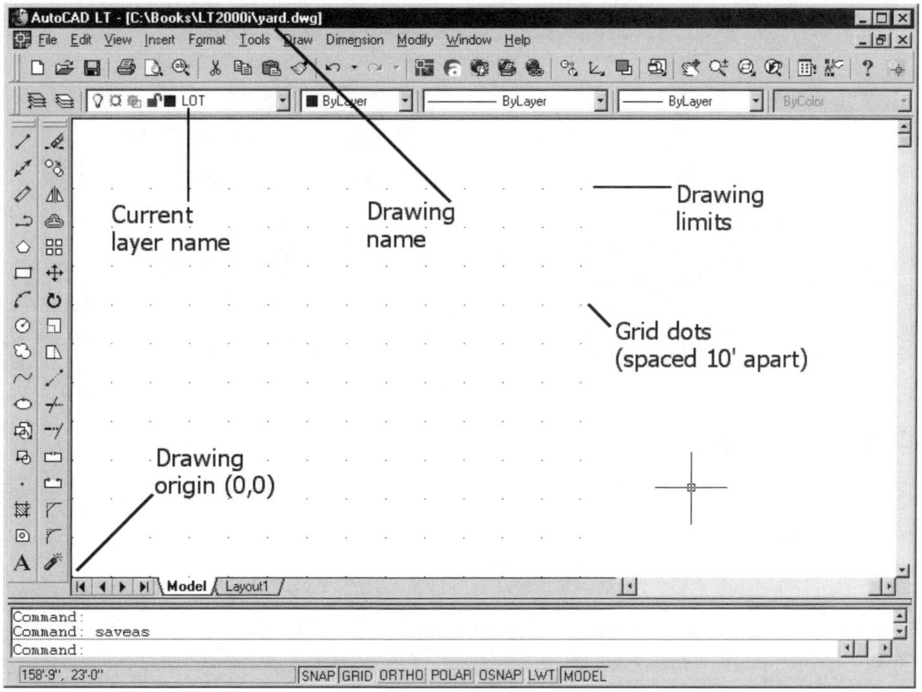

On the toolbar, you see the color of the current layer is blue and its name is Lot. On the status line, you see that the coordinates are displaying in feet and inches; and that the snap, grid, model, and tile modes are turned on. The figure above also shows you where the origin (0,0) is located. The automatic backup feature saves your drawing every quarter-hour.

Review Questions

1. Match the term with its best meaning:

a. Units	i. Specifies the nominal limits of a drawing, and constrains the grid marks.
b. Snap	ii. Visual guide consisting of an array of dots.
c. Grid	iii. Constrains mouse movement to discrete distances.
d. Limits	iv. Organizes the drawing by separating elements.
e. Layers	v. Variety of measurement styles.

2. Match the abbreviation with its best meaning:

a. '	i. inches
b. "	ii. Identifies the file as an AutoCAD drawing.
c. in	iii. Backup copy of an AutoCAD drawing file.
d. DWG	iv. feet
e. BAK	v. inches

3. To draw in metric units, you select decimal units.

   True / False

4. When units are set to the nearest 1/4", AutoCAD cannot draw more accurately than to the nearest 0.25".

   True / False

5. By default, AutoCAD's 0-degree setting points:

   a. North
   b. South
   c. East
   d. West

6. By default, AutoCAD measures positive angles in the clockwise direction.

   True / False

7. To specify 20 feet in AutoCAD, you enter:

   a. Twenty
   b. 20
   c. 20ft
   d. 20'

8. To specify 20 inches in AutoCAD, you enter:

   a. 20
   b. 20"
   c. Both of the above.
   d. None of the above.

9. When the grid spacing is set to 0, this means:
   a. AutoCAD displays the isometric grid.
   b. AutoCAD displays the grid at the same spacing as the snap distance.
   c. AutoCAD displays the snap at the same spacing as the grid distance.
   d. AutoCAD displays no grid.
10. You need only enter the capitalized portion of a command option.

    True / False

Chapter 3

Creating Your First Drawing

In the last chapter, you learned how to start AutoCAD LT, set up a new drawing, and save the drawing to the computer's hard drive. In this chapter, you learn how to draw lines accurately, make simple changes to the drawing, and produce a copy of the drawing on your printer.

Key Concepts

- Drawing with lines and polylines
- Understanding absolute and relative distances
- Using polar coordinates
- Modifying the drawing by fillet and mirror
- Plotting (printing) the drawing

Key Terms

Absolute coordinates—the measurement is made relative to the origin

Aperture—the square cursor that indicates AutoCAD's search area for object snaps

Direct Distance Entry—point in a direction, then type a distance

Extents—an invisible rectangle that encompasses the entire drawing

Fillet—rounds the intersection

Mirror—creates a mirrored copy of objects

Object Snap—snaps the cursor to a geometric feature, such as the end of a line; used for precise drafting

Origin—located at 0', 0' usually the lower-left corner of the drawing

Ortho—constrains cursor movement to the vertical and horizontal; short for "orthographic"

Pick Cursor—the small cursor that indicates AutoCAD's search area for selecting objects

Polar coordinates—specify a distance and an angle

Relative coordinates—the measurement is made relative to the last point

Abbreviations	
@	instructs AutoCAD to use relative coordinates, such as @2,3
<	specifies an angle, such as 10<45
-	hyphen prefix forces AutoCAD to use command-line version of a command, such as -layer
[*Option*]	square brackets indicate command options, such as [Undo]
<*value*>	angle brackets indicate the default (current) value, such as <LOT>
X	x-coordinate, along the horizontal axis
Y	y-coordinate, along the vertical axis

<table>
<tr><th>Command</th><th>Shortcut</th><th>Menu Selection</th></tr>
<tr><td>Fillet</td><td>f</td><td>Modify | Fillet</td></tr>
<tr><td>Mirror</td><td>mi</td><td>Modify | Mirror</td></tr>
<tr><td>Move</td><td>m</td><td>Modify | Move</td></tr>
<tr><td>Open</td><td>op or Ctrl+O</td><td>File | Open</td></tr>
<tr><td>Ortho</td><td>F8 or Ctrl+L</td><td>...</td></tr>
<tr><td>OSnap</td><td>F3 or Ctrl+F</td><td>Tools | Drafting Settings</td></tr>
<tr><td>PLine</td><td>pl</td><td>Draw | Polyline</td></tr>
<tr><td>Plot</td><td>pp or Ctrl+P</td><td>File | Plot</td></tr>
<tr><td>QSave</td><td>Ctrl+S</td><td>File | Save</td></tr>
<tr><td>Zoom</td><td>z</td><td>View |</td></tr>
<tr><td>Cancel</td><td>Esq</td><td>...</td></tr>
</table>

Bringing Back the Yard Drawing

■ File | 1

If you exited AutoCAD at the end of the last chapter, you need to restart AutoCAD LT, and load the Yard drawing. Here's how:

1. Be sure Windows is running. Load AutoCAD by selecting **AutoCAD LT 2000i** from the Windows Start menu. Alternatively, double-click the **AutoCAD LT 2000i** icon on the Windows desktop.
2. When the Start Up dialog box appears, click **Open a Drawing** to display the list of recently opened drawings.

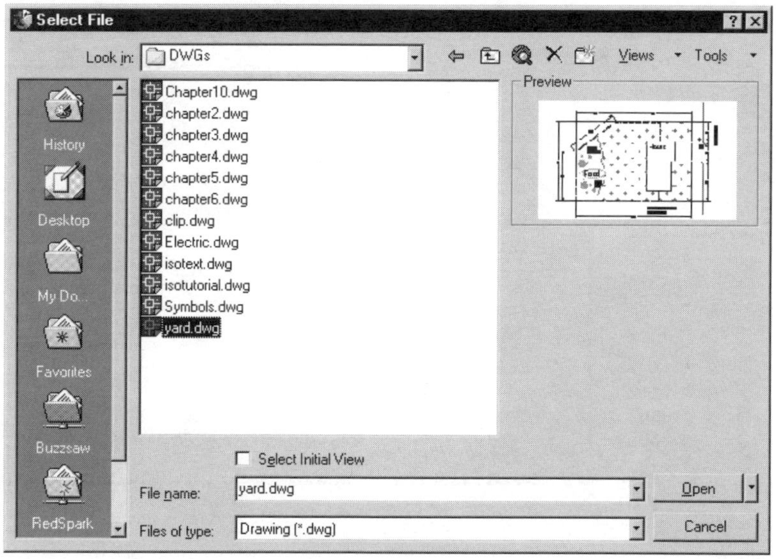

Opening a previously opened drawing.

3. From the **Select File** list, select **yard.dwg.** It should look exactly the same as when you last saw the drawing—that is to say blank, except for the grid.

> **Tip:** AutoCAD LT remembers the last four drawings you worked with and displays their names in two locations: the Start Up dialog box and the bottom of the File menu. When you select one of the drawing names, AutoCAD executes the Open command, and loads the drawing.
>
> A new behavior in LT 2000i is that it remembers the last folder in which you worked. To change this to the way LT was in previous versions, set the system variable RememberFolders to 0 at the AutoCAD command line.

Drawing the Lot Boundary

■ Draw | Line

Let's get some lines on the screen! To orient yourself, the first thing to do is to draw is the boundary of the yard. That helps you see the extent of the drawing.

The lines making up the lot boundary are drawn with the Line command. You begin drawing the lot lines at the lower left corner, the origin (0,0), then work counterclockwise around the lot boundary, as shown in the illustration.

The lot lines are drawn using a combination of drafting methods.

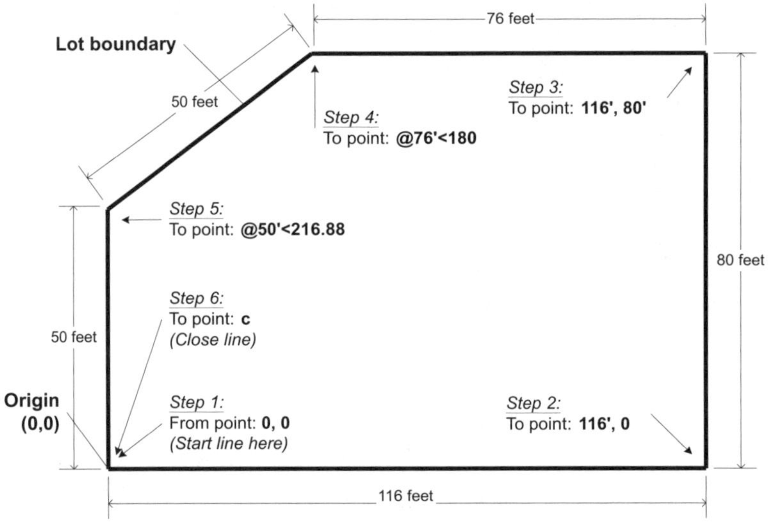

1. From the menu bar, select **Draw | Line**, or select the **Line** button from the **Draw** toolbar.

Selecting the Line command from the toolbar (at left) and menu (at right).

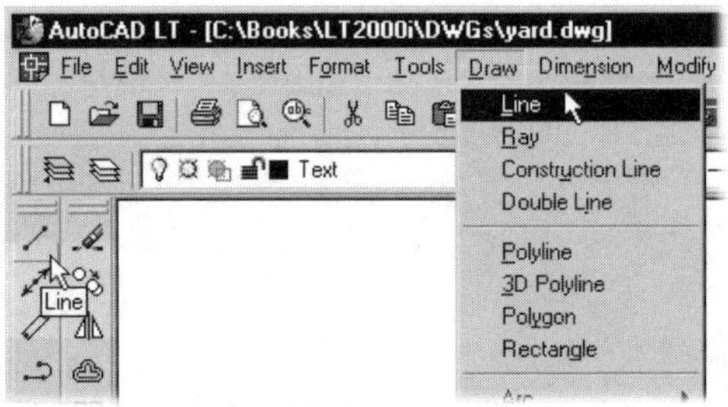

Note: When you see two words separated by a vertical line, such as Draw | Line, it means to select Draw from the menu bar, then select Line from the Draw menu. As confirmation, AutoCAD prints "_line" in the command line.

2. Respond to the "Specify first point" prompt by typing the coordinates of the origin:

   Command: _line Specify first point: 0,0

3. To draw the lower 116'-long boundary line, you know the other end of the line must be located at the x,y-coordinates of 116', 0:

   Specify next point or [Undo]: 116',0

   Notice the word Undo. Undo is an option of the Line command. If you were to type "undo" (instead of the 116,0 coordinates), AutoCAD would undraw the last line segment. The square brackets indicate that Undo is an option. Later in the Line command, you will see a second option, called Close.

4. The next line is 80 feet north, which is coordinate 116', 80':

   Specify next point or [Undo]: 116',80'

   Remember to include the apostrophe (') with each number. The apostrophe indicates feet; if you leave it out, AutoCAD interprets the numbers as inches, and you would end up with a very small yard!

5. You drew the first two lines with absolute coordinates, where you calculated the coordinates based on measurements relative to the origin (at 0', 0'). AutoCAD can, however, do these calculations for you. When you use *relative coordinates*, AutoCAD draws a line from the current point relative to the last point.

   AutoCAD can also work with *polar coordinates*, which specify a distance and an angle. Continue drawing the lot boundary using both relative and polar coordinates combined, as follows:

   Specify next point or [Close/Undo]: @76'<180

   When you tell AutoCAD to draw a line with relative polar coordinates, you enter a special notation that has the following meaning:

Notation	Meaning
@	Use relative coordinates.
76'	Distance is 76 feet from the current point.
<	Draw the line at an angle...
180	...of 180 degrees

The line is drawn relative to the current point; the angle, however, is measured in absolute degrees using the East-is-0-degrees convention. Using relative polar coordinates makes sense when you have many angled lines to draw.

Specify next point or [Close/Undo]: @50'<216.88

With all that punctuation, entering @50'<216.88 can feel like quite a typing chore. I'm getting you to type the coordinates here so that you can better appreciate other methods later. In the meantime, if you make a mistake entering the coordinate notation, simply use the Undo option (type u at the prompt) and re-enter the coordinates.

6. To finish the lot boundary, you use a shortcut. Type **c** (short for "close") instead of typing the final coordinates (0,0), as follows:

   Specify next point or [Close/Undo]: c

   AutoCAD automatically draws a line from the current end point to the beginning of the first line.

   As mentioned earlier, square brackets surround the options of a command prompt. Here [Close/Undo] means you have two options: close the polygon, or undo the last line segment.

7. To give some space around the drawing, use the Zoom command, as follows:

   Command: zoom
   All/Center/Extents/Previous/Scale(X/XP)/Window/<Realtime>: 0.9x

   This zooms the drawing 90% smaller.

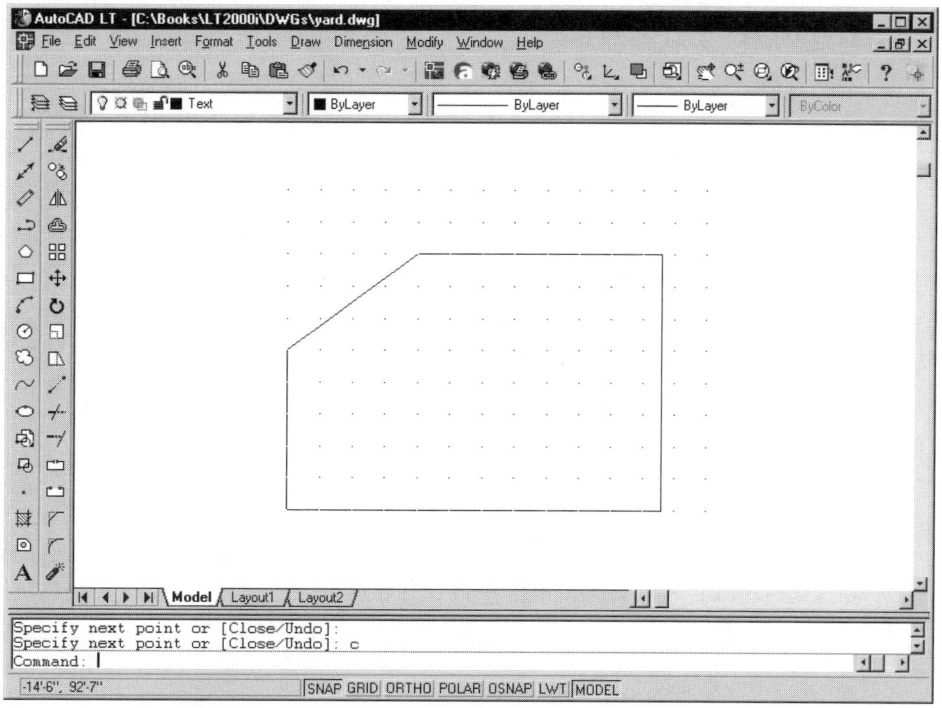

The lot lines are in place.

Instead of drafting on paper, you have created your first digital drawing! More importantly, you have drawn it full size (even if it looks kind of small on your screen). That's one of the powerful aspects of CAD. There is no need for a scale ruler, or to divide distances by a scale factor as in manual drafting. Everything is drawn full size.

The next step is to draw the outline of the house. From the sketch, you know that the lower-right corner of the house is located 10' up and 20' in from the lot corner. There are several ways to locate one object relative to another, in this case the corner of the house:

- Calculate the coordinates of the house's corner.
- Use the XLine command to draw a pair of construction lines, then start drawing from their intersection.
- Use the From option to start drawing from an offset from the lot corner.
- Use *tracking* to offset the starting point (described in Chapter 5).
- Draw the house in the corner of the lot, then move the house into position.

We'll use the last method to exercise two of AutoCAD's most powerful commands, PLine and Move.

Changing Layers

■ -Layer

Before you draw the house, you need to change the layer to House. Before I show you how to change layers, keep in mind that many of AutoCAD's commands, including the -Layer command, can be carried out in more than one way.

In the previous chapter, you used the Layer Properties Manager dialog box to create the new layers, and set the Lot layer as the current (or working) layer. Earlier in this chapter, you started the Line command via the toolbar or the menu.

This time, let's try executing a command by typing it at the keyboard. We type the -Layer command at the Command: prompt. The hyphen (-) forces the command to display its prompts in the command area, located at the bottom of the LT window.

1. To set the House layer, type the -Layer command, as follows:

   ```
   Command: -layer
   Current layer: "Lot"
   ```

   AutoCAD reminds you of the name of the current layer. The -Layer command has more than a dozen options, most of which you ignore for now, as follows:

   ```
   Enter an option
   [?/Make/Set/New/ON/OFF/Color/Ltype/LWeight/Plot/Freeze/Thaw/L
   Ock/Unlock]: s
   ```

2. Type the letter **s** to invoke the Set option.

Note: For many commands, AutoCAD presents a list of options. To select an option, you need only type its first character, such as M for the Make option and S for the Set option. When two (or more) options begin with the same first letter, such as the LType and LWeight options, you need to type the first two or three characters (LT and LW). AutoCAD shows you the minimum number of characters by capitalizing them.

3. AutoCAD prompts you to enter the name of the layer, as follows:

   Enter layer name to make current <Lot>: house

   Notice that AutoCAD lists the current layer name in angle brackets, <Lot>. This is called the *default name*, which lets you retain the current layer by simply pressing the **Enter** key. To change the working layer to House, type **house** and press **Enter**.

4. The **Layer** command repeats its 12-option prompt. Press **Esc** to cancel the command and return to the Command: prompt.

   Enter an option
   [?/Make/Set/.../LOck/Unlock]: [Esc]
   Command:

Tip: You cancel a command at any time by pressing Esc. Sometimes, you may need to press Esc two or even three times for commands that have a large number of suboptions, such as the PEdit command.

5. Look at the toolbar to confirm that AutoCAD has changed the working layer from layer Lot to layer House.

Drawing the House Outline

- Draw | Polyline
- Object Snap

Earlier in this chapter, you drew the lot boundary with the Line command. You created what looks like a polygon—one continuous line made of several segments. In fact, each line segment is independent; they only look connected. That can make it difficult to select the entire lot boundary at once.

To remedy this, AutoCAD has a special kind of line called the *polyline*. As the prefix "poly" suggests, a polyline is a line made up of many features, such as lines and arcs, or splines of varying widths, all connected together as a single object.

AutoCAD's polyline can take on many forms.

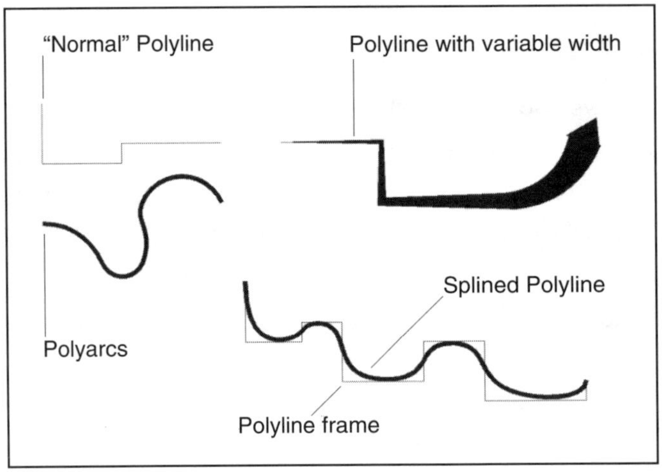

The coordinates for drawing the house outline.

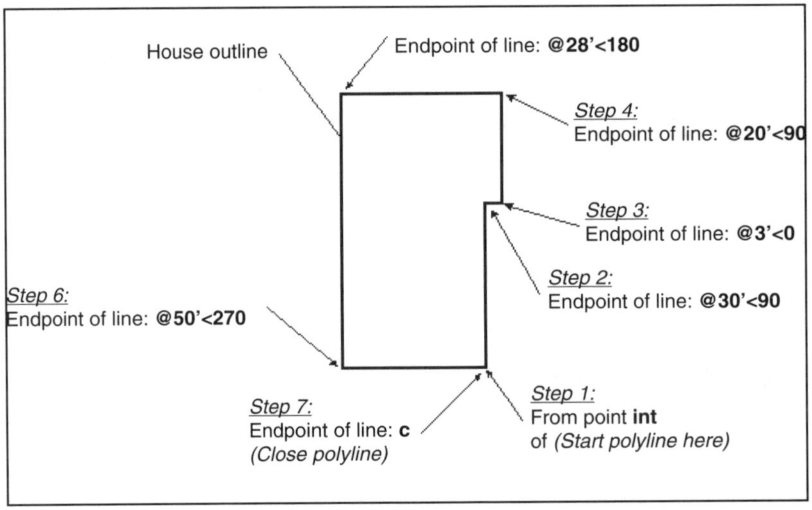

1. Draw the house outline as a polyline. Select **Draw | Polyline**. As with the Line command, AutoCAD prompts you for the point from which to begin drawing the polyline:

   Command: _pline

2. Instead of specifying a coordinate, you ask AutoCAD to find a geometric feature by using *object snap*:

   Specify start point: int
   of: [pick lower right corner of lot]

   When you type **int** (short for "intersection"), AutoCAD attempts to snap to the nearest intersection, rather than snapping to the nearest 1" you specified with the snap mode in Chapter 2. Since the intersection snap overrides the 1"-snap, this is sometimes referred to as *object snap override*.

AutoCAD has 10 geometric object snaps:

Mode	Object Snap	Meaning
cen	CENter	Snaps to the center of arc, circle, and polyarc.
end	ENDpoint	Snaps to the end of line, arc, and other open objects.
ins	INSertion	Snaps to the insertion point of block or text.
int	INTersection	Snaps to the intersection of lines, arcs, circle, and other objects.
app	APParent	Snaps to the apparent intersection of two objects.
mid	MIDpoint	Snaps to the middle of line, arc, and other open objects.
nea	NEArest	Snaps to the nearest point on an object.
nod	NODe	Snaps to a point.
per	PERpendicular	Snaps perpendicular to a line, arc, or other object.
qua	QUAdrant	Snaps to the 0-, 90-, 180-, and 270-degree points on an arc, circle, and polyarc.
tan	TANgent	Snaps tangent to arc or circle.

You get to use some of the other object snap modes later.

When you type **int**, a square appears around the crosshair cursor. The square is called the *aperture*, and shows the area in which AutoCAD hunts for the object snap point.

3. AutoCAD then curtly prompts you with "of." It is asking you to position the aperture cursor near the intersection *of* the two lines. Move the cursor until the aperture is over the lower right corner of the lot boundary and press the pick button.

Using object snap to find the precise intersection.

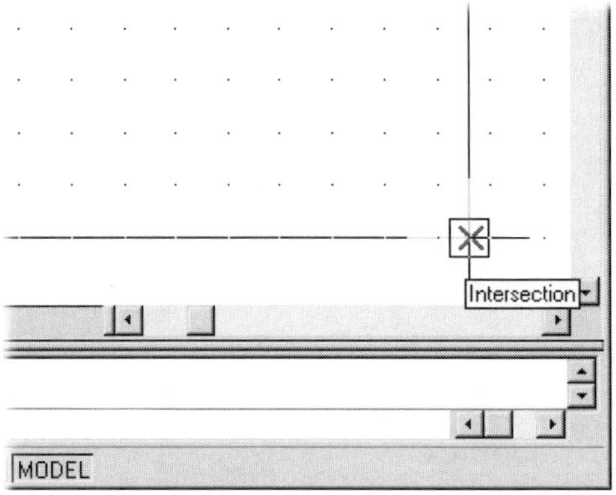

When the aperture cursor gets close to the intersection, AutoCAD displays three navigation tools:

- A tooltip that displays the name of the object snap—Intersection, in this case.
- An icon that locates the object snap position—the intersection icon.
- A *magnetic* cursor that jumps to the object snap location.

Collectively, these three are under "AutoSnap," which displays object snap locations.

4. AutoCAD then displays a long prompt:

   Current line-width is 0'-0"
   Specify next point or [Arc/Close/Halfwidth/Length/Undo/Width]: @30'<90

   The prompt reports that the polyline currently has no width (0'-0"). That does not mean the line is invisible. Rather, *zero width* in AutoCAD means that the line is drawn as narrowly as possible on the screen and by the printer (usually one pixel wide).

   The second line of the prompt displays seven of the PLine command's 22 options. Don't let it intimidate you; for now, you ignore all options except the default, Specify next point.

5. Draw the remainder of the house outline, as follows:

   Specify next point: @3'<0
   Specify next point: @20'<90
   Specify next point: @28'<180
   Specify next point: @50'<270
   Specify next point: c

You complete the polyline with the C option, as you did with the Line command.

The completed house outline.

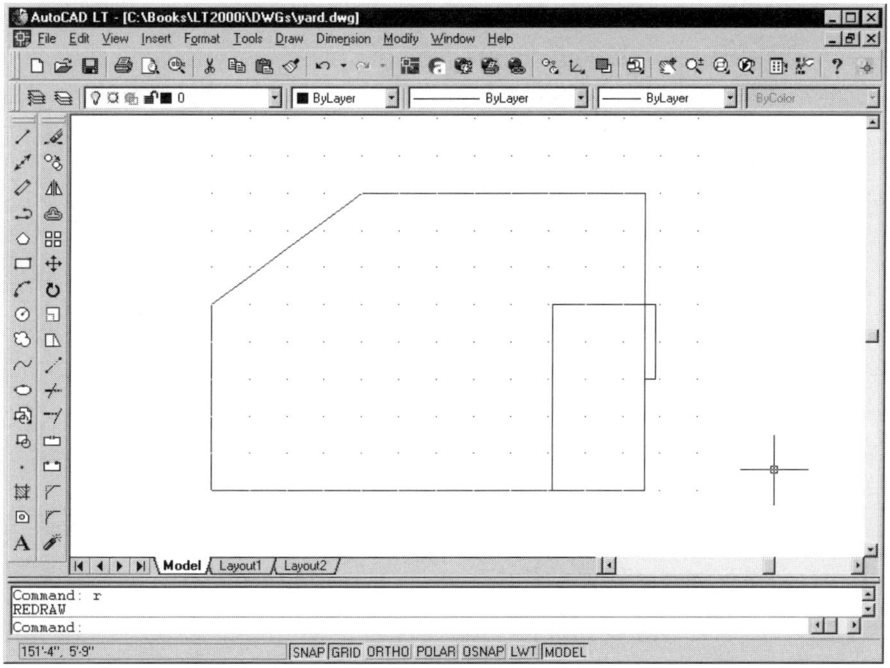

When Lot was the working layer, the lines you drew showed up in blue. You changed the layer to House, and AutoCAD automatically drew the lines in black instead of blue. Notice that the lines take on the color specified by their layer. (You can, if you need to, change colors on the fly with the Color command.)

Moving the House into Position

■ Modify | Move

Now that you've drawn the outline of the house, you move it into position. You move objects in AutoCAD with the Move command.

1. Select **Move** from the Modify toolbar. When you start the command, AutoCAD asks you what you want to move, as follows:

   Command: _move
   Select objects: [pick house]

AutoCAD changes the crosshair cursor into a small square cursor, called the *pick cursor.*

2. Move the cursor to any part of the polyline making up the outline of the house and press the pick button. The entire house outline is highlighted. The highlighting shows as a dotted line. This is how AutoCAD lets you know it found the object you picked.

Selecting objects with the pick cursor.

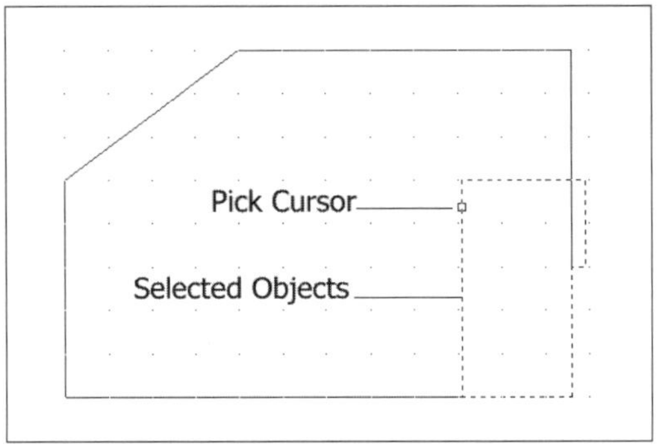

3. AutoCAD lets you select more than one object to move:

   Select objects: [Enter]

   However, you only need to move the one polyline. Press **Enter** to end the selection process.

4. Just as with drawing a line, the Move command needs to know a from-point and a to-point. The Move command calls the from-point the *base point*, as follows:

   Base point or displacement: 0,0

5. Now AutoCAD wants to know where you want to move the selected objects. The Move command calls the to-point the *second point*, as follows:

   Specify second point of displacement or <use first point as displacement>: -20',10'

   The coordinates (-20', 10') tell AutoCAD that you want the house moved left by 20 feet (X = -20') and up by ten feet (Y = 10'). AutoCAD instantly relocates the house much faster than a house mover. The Move command shows you a powerful aspect of CAD: no eraser dust!

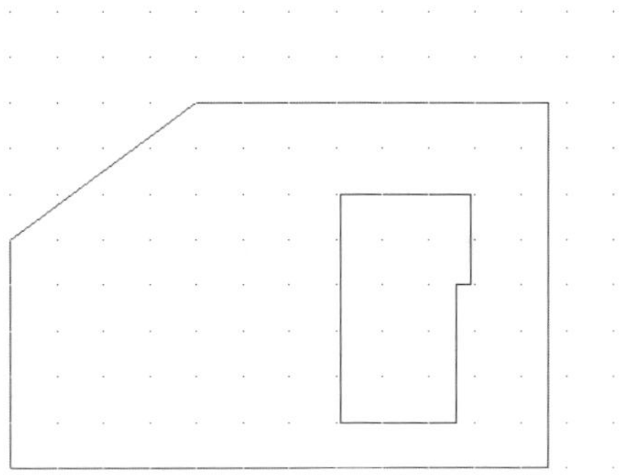

The house moved into position.

Starting On the Driveway

- Direct Distance Entry
- Ortho Mode

The final drafting you do this chapter is of the driveway and the street. Before you begin drawing them, change to the Roads layer, as follows:

1. Click on the **House** layer name on the toolbar.
2. When the list box appears, click on **Road**. AutoCAD changes the layer name from House to Road and the working color from black to red.

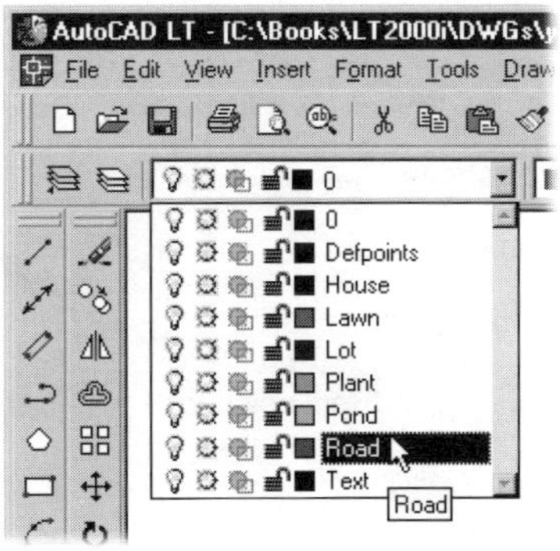

Changing layers via the toolbar.

With the Road layer set, you draw the driveway and street outlines. First you draw the upper roadwork using direct distance entry; later you duplicate the lower driveway and street outline with a single command.

Earlier, you used relative coordinates to draw part of the lot boundary (remember the < and @ notation?). This time, you draw relative distances using *direct distance entry*. This is how it works: you point the cursor in the direction you want the line to go, then type the distance. AutoCAD draws the line. As you find out, this is much easier than typing a lot of punctuation. Let's try it.

3. Since the two lines describing the driveway are at right angles, I recommend that you use *ortho mode*. Ortho mode constrains your drawing and editing to the horizontal and vertical. Turn on ortho mode by clicking the **ORTHO** button on the status bar; as confirmation, AutoCAD reports:

   Command: <Ortho on>

   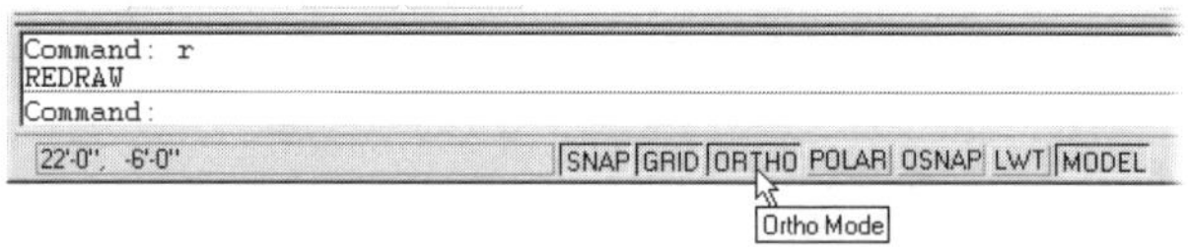

   Click ORTHO to toggle orthographic mode.

4. Select **Line** icon from the Draw toolbar to draw the upper driveway and street outline, as follows:

   Command: _line

The directions for drawing the driveway.

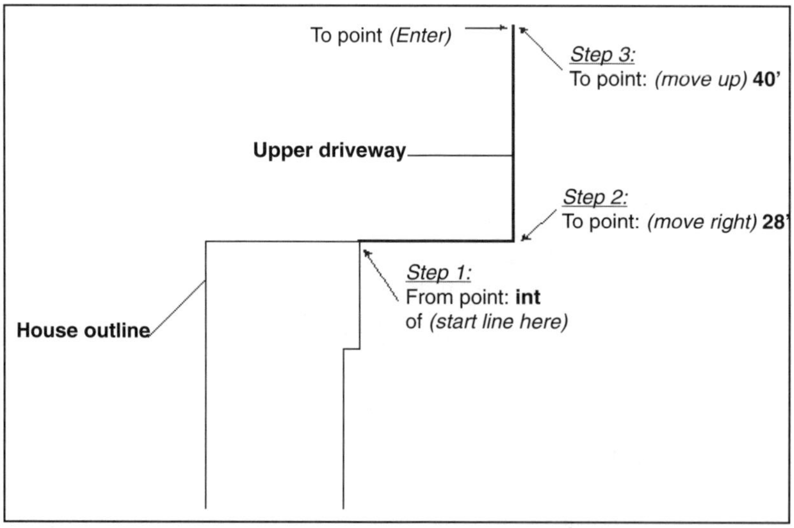

Specify first point: int
of: [pick upper right corner of house]
Specify next point or [Undo]: [move cursor left] 28'
Specify next point or [Undo]: [move cursor up] 40'
Specify next point or [Close/Undo]: [Enter]

The driveway lines are in place.

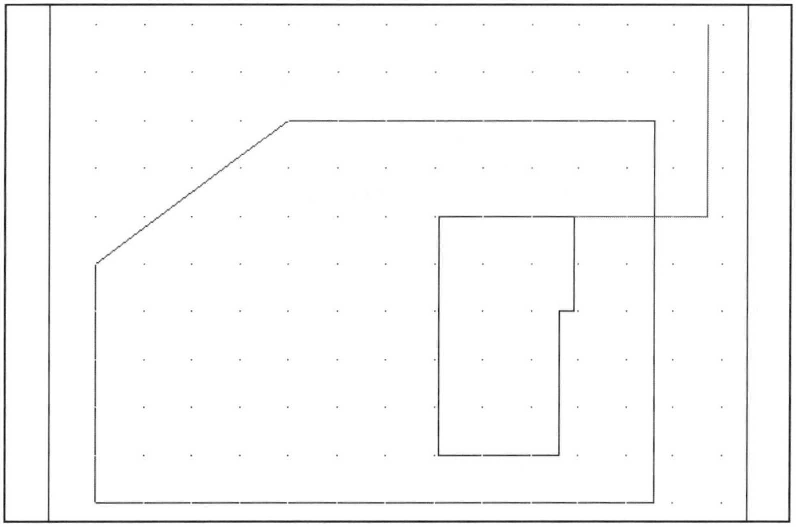

Recall that if you leave out the apostrophe ('), such as 40 instead of 40', the line is drawn only 40 inches long. You can back up and undraw the incorrect line with the "u" option (short for "undo"), as follows:

Specify next point or [Undo]: @40<0
Specify next point or [Undo]: u
Specify next point or [Undo]: @40'<0

Finishing the Driveway

- Modify | Fillet
- Modify | Mirror

To add the *curb return,* the arc joining the driveway and street, use AutoCAD's Fillet command. Fillet draws an arc between two intersecting lines. The lines don't have to physically meet; AutoCAD takes care of extending (or trimming) the lines so that the arc is drawn between them.

Unfortunately, you have to use the Fillet command twice—once to set the radius, and a second time to apply the fillet.

1. To start the Fillet command, move the cursor to the Modify toolbar. Look for the Fillet icon; it's the second to the last button. Click the **Fillet** button, and AutoCAD starts the Fillet command. (As an alternative, select **Fillet** from the Modify menu.)
2. Select the Radius option by typing R and pressing Enter:

   Command: _fillet
   Current settings: Mode = TRIM, Radius = 0'-0 1/2"
   Select first object or [Polyline/Radius/Trim]: r
3. Enter the fillet radius of three feet:

   Specify fillet radius <0'-0 1/2">: 3'
4. With the fillet radius set to three feet, perform the filleting, as follows:

   Select first object or [Polyline/Radius/Trim]: [pick driveway line]
   Select second object: [pick street line]

> **Tips:** New to LT 2000i: you no longer need to repeat the Fillet (or Chamfer) command after changing the value of the radius.
>
> By pressing the Enter key at the Command: prompt, you repeat the previous command. This AutoCAD feature lets you quickly repeat the command several times over.

The fillet is applied to the two intersecting lines.

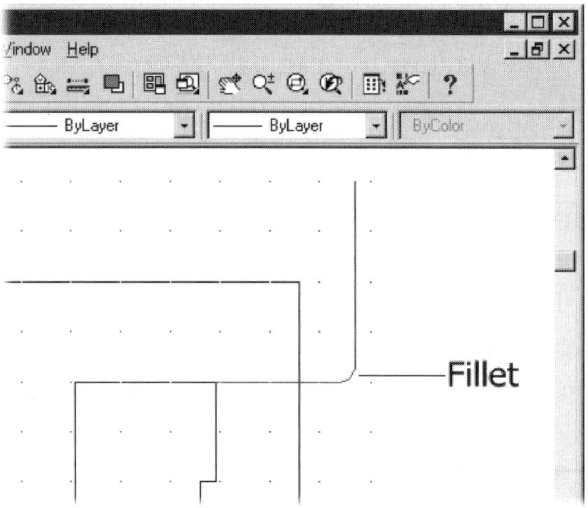

AutoCAD automatically shortens the two lines to fit the 3-foot arc between them.

You needed the Line command and two applications of the Fillet command to create part of the driveway and street. One of the most important concepts behind computer-aided anything is that you should never have to draw the same line twice. To illustrate the power of this concept, you use the Mirror command to duplicate the lower driveway and street line *without having to draw them!* The Mirror command creates a mirrored copy of a set of objects.

5. Select **Modify | Mirror**. AutoCAD asks you to select the objects you want to mirror. You use the cursor to pick the line and arc segments, as follows:

   ```
   Command: _mirror
   Select objects: [pick the driveway line]
   Select objects: [pick the curb return]
   Select objects: [pick the street line]
   Select objects: [Enter]
   ```

6. AutoCAD needs you to specify an imaginary line about which it mirrors the objects you just picked:

   ```
   First point of mirror line: mid
   of: [pick center of garage entrance]
   Second point: per
   of: [pick right-hand lot boundary]
   ```

   The length of the mirror line is not important, but the angle is crucial. For this reason, you used two new object snap modes: *mid* to find the midpoint of the garage entrance, and *per* to ensure the mirror line is perpendicular to the lot boundary, as shown.

The steps to mirroring objects.

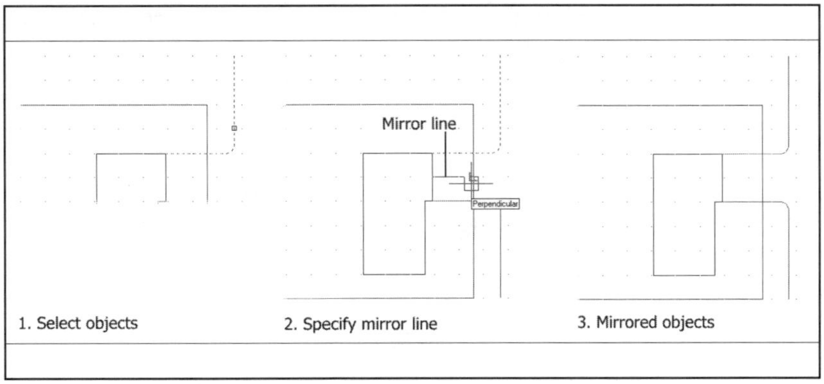

7. At this point, AutoCAD gives you the option of erasing the old objects—the two lines and arc you picked. In most cases, as in this case, you don't want them erased:

   Delete old objects? <N> n

   AutoCAD draws the lower driveway and street outline as a perfect mirror image of the upper set, as shown in step 3 of the figure.

You have now drawn the outline of the lot, house, and roadway. The work you have done is valuable and it is important that you save the drawing to disk. Use the Save button on the toolbar (the miniature picture of the diskette) to store the Yard drawing on disk.

Putting the Drawing on Paper

■ File | Plot

While it is efficient (and environmentally aware) to create and store drawings on a computer, you may want to print a copy of the drawing on paper. That lets you mark up the drawing with notes or show off your progress to your friends and family!

The Plot command sends the drawing to a printer, a plotter, or a file. The following series of dialog boxes assumes you have a printer compatible with Windows attached to your computer. The options and prompts for plotters differ slightly.

1. Press **Ctrl+P** (hold down the **Ctrl** key and press **P**) to create a check plot of the Yard drawing. (A check plot is a quick plot that allows you to see on pape the progress of the drawing.) AutoCAD displays the Plot Configuration dialog box.
2. The Plot dialog box has a confusing array of options. For a check plot, you ignore nearly all of them, except for these:

   Click on the radio button next to **Extents** (along the left side, under Plot area). The extents option ensures everything in your drawing is plotted on the paper.

Under Drawing orientation (on the right side), click **Landscape**.

Under Plot scale, ensure **Scaled to fit** has been selected.

Under Plot offset, click **Center the plot**.

Selecting options for a check plot.

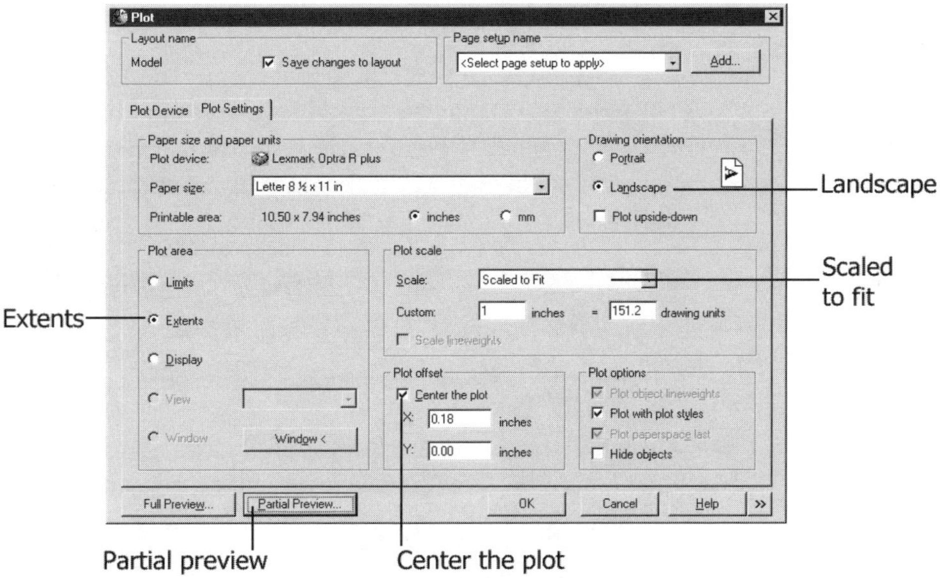

3. To check quickly that the drawing fits the paper, click the **Partial Preview** button (in the lower left). A second dialog box, called **Partial Plot Preview**, appears on the screen.

   The white rectangle represents the paper.

The partial preview gives an overview of how the drawing fits the paper.

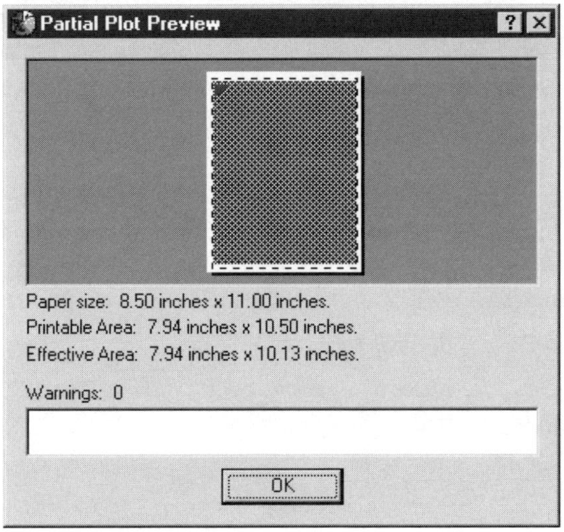

The black dashed rectangle indicates the margins, within which the drawing can be plotted.

The blue rectangle shows the extents of the drawing.

4. Click **OK** to exit plot preview. Partial preview is faster than full preview, as we see next.
5. Now let's try the full preview feature. In the Plot dialog box, click **Full Preview**. This time, you see the complete drawing. While Full Preview displays the entire drawing on its sheet of paper, it can take a long time to display when the drawing is complex. In those cases, you might want to use the Partial Preview.

The full preview shows all the drawing details.

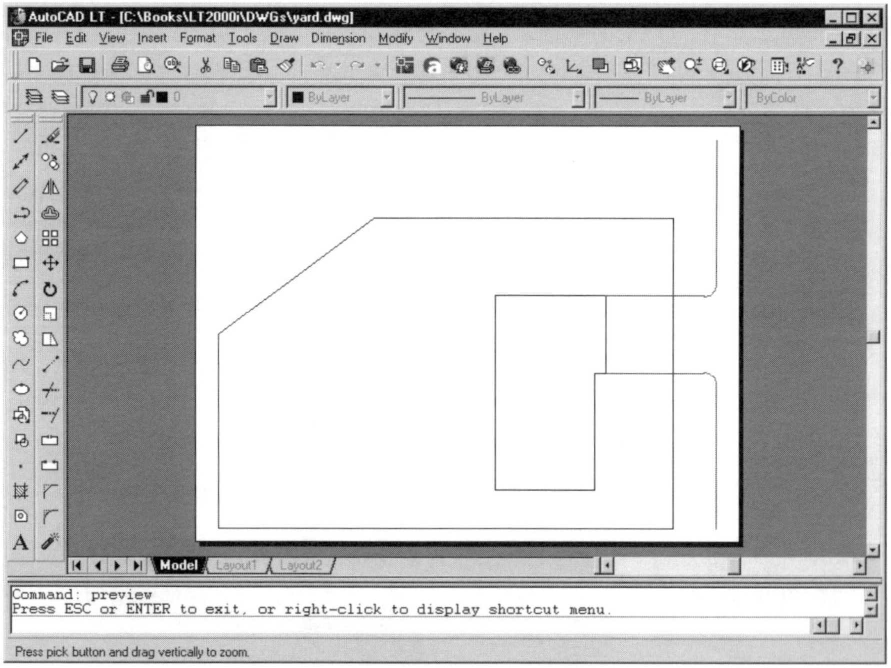

6. Press **Esc** to exit full preview mode.
7. When the **Plot** dialog box reappears, click the **OK** button. AutoCAD calculates the area the drawing takes on the paper and sends the drawing to the printer:

Effective plotting area: 7.94 wide by 10.13 high
Plotting viewport 2.

A dialog box displays the progress of the plot. AutoCAD converts its vector drawing to the format required by the printer. When done, AutoCAD returns the Command: prompt.

The bar graphs show the progress of the plot.

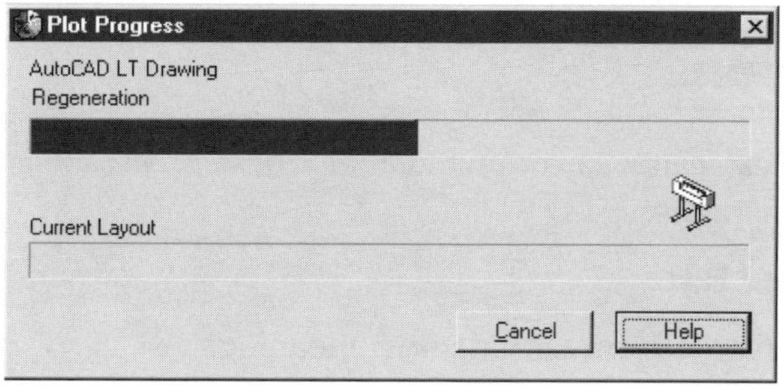

Tip: If you use a laser printer, you may want to change its resolution to lower or higher, such as 150dpi or 1200dpi, for these reasons:

- Your laser printer has insufficient memory, and only a portion of the drawing is printed. Lower the resolution or add memory.
- You find the lines too thin; a lower resolution doubles the width.
- You find the details hard to read; a higher resolution makes details finer.
- You feel the plotting time takes too long. Lower the resolution.

To change the resolution to a higher or lower value: in the Plot dialog box, click the Plot Device tab. Click Properties. In the Plotter Configuration Editor, click Custom Properties.

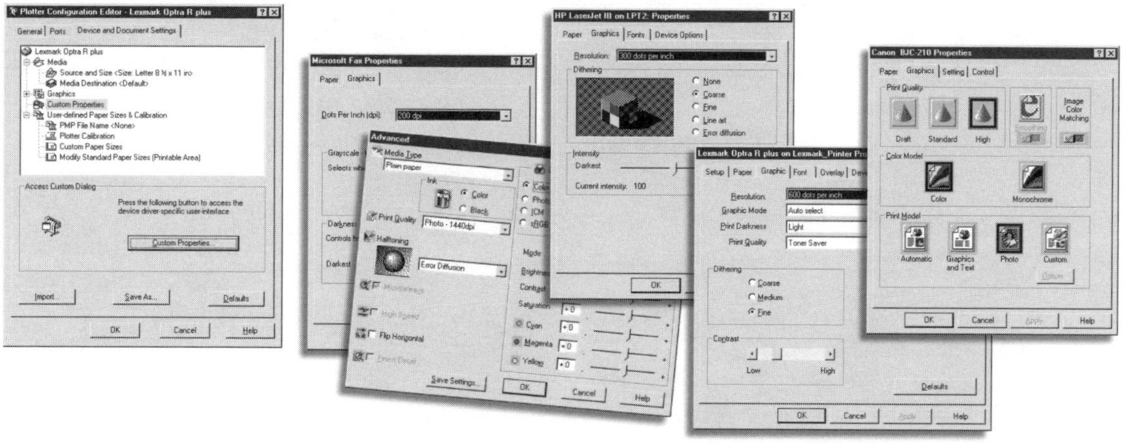

Every printer has a different dialog box for specifying resolution.

Congratulations! You've drafted your very first drawing using a computer. You can save the drawing as a memento of your introduction to computer-aided drafting.

Review Questions

1. You can cancel a command at any time by pressing the Esc key one or more times.

   True / False

2. Match the term with its best meaning:

a.	Origin	i.	Measurement is made relative to the last point.
b.	Ortho	ii.	Cursor that indicates AutoCAD's search area for selecting objects.
c.	Pick cursor	iii.	Specifies a distance and an angle.
d.	Polar coordinates	iv.	Constrains cursor movement to the vertical and horizontal.
e.	Relative coordinates	v.	Located at 0', 0' usually the lower-left corner of the drawing.

3. The hyphen (-) prefix in front of a command, such as -Array, forces AutoCAD to display the command's dialog box.

   True / False

4. Match the term with its best meaning:

a.	Absolute coordinates	i.	Invisible rectangle that encompasses the entire drawing.
b.	Aperture	ii.	Snaps to a geometric feature, such as the center of a circle.
c.	Direct Distance Entry	iii.	Measurement made relative to the origin.
d.	Extents	iv.	Square cursor that indicates AutoCAD's search area for object snaps.
e.	Object Snap	v.	Point in a direction, then type a distance.

5. Entering the coordinates 20,30 means:

   a. x=20, y=30
   b. y=30, x=20
   c. Distance = 20, angle = 30
   d. Angle = 20, distance = 30

6. Entering the coordinates @20<30 means:

   a. x=20, y=30
   b. y=30, x=20
   c. Distance = 20, angle = 30
   d. Angle = 20, distance = 30

7. A zoom factor of 0.8x makes the drawing 80% smaller.

   True / False

8. A polyline can contain lines and arcs.

   True / False

9. Match the object snap abbreviation with its application:

a. cen	i. End of a polyline.
b. end	ii. Center of an arc.
c. int	iii. Perpendicular to a line.
d. qua	iv. Intersection of a line and a circle.
e. per	v. 90-degree point on an arc.

10. After a command ends, you can press the F1 key to repeat the command.

    True / False

Chapter 4

Adding Details to a Drawing

In the last chapter, you created the outlines of the lot, the house, and the roadways. In this chapter, you add details to the yard, such as the lawn, trees, and a pool.

Key Concepts

- Editing a polyline
- Using non-modal editing commands
- Zooming by picking two corners of a window
- Applying a hatch pattern to an area
- Drawing circles and ellipses
- Real-time zooming and panning, and Aerial View
- Creating an array of objects and a parallel offset of objects
- Creating a symbol (block)

Key Terms

Drag—hold down the left mouse button while moving the cursor or object

Grips—editing points on an object

Hatch—a repeating pattern that indicates the material of an object

Modal editing—start the editing command, then select the object

Non-modal editing—pick the object, then select the editing command

Palette—a dialog box-like graphical selection of options

Abbreviations	
L	last (object selection)
W	window (object selection and zooming)

Command	Shortcut	Menu Selection
Array	ar	Modify | Array
BHatch	h	Draw | Hatch
Block	b	Draw | Block | Make
Circle	c	Draw | Circle
Insert	i	Insert | Block
DsViewer	ds	View | Aerial View
Ellipse	el	Draw | Ellipse
Offset	o	Modify | Offset
Pan	p	View | Pan | Real Time
PEdit	pe	Modify | Object | Polyline
Zoom Window	z w	View | Zoom | Window

Dividing the Lot

The yard has a lawn and a garden area. In this section, you draw the boundary between the two areas with a polyline, then smooth it with the PEdit (polyline edit) command.

1. Before starting the PLine command, take these steps.
   - Turn off ortho mode by clicking on the word **ORTHO** on the status line.
   - Make sure the working layer is set to Lawn by selecting layer **Lawn** from the toolbar; the color should be magenta (pink).
2. Select **Draw | Polyline** from the menu bar.

```
Command: _pline
Specify first point:
```

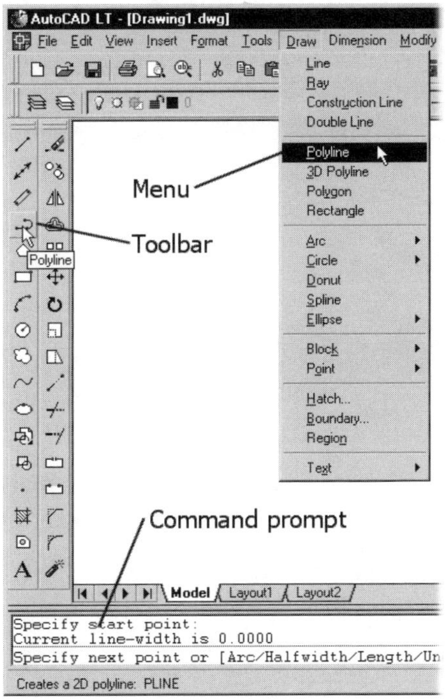

Selecting the Polyline command from the toolbar (at left) and menu (at right).

3. Hold down the **Shift** key on the keyboard, press the right mouse button, then let go of the **Shift** key. A new menu pops up. This shortcut menu lists all AutoCAD's object snap modes. Move the cursor down to **Midpoint**. Notice that AutoCAD LT prints a brief description of the Midpoint object snap on the status bar: "Snaps the midpoint of an arc or line: mid."

   Click the mouse button on **Midpoint**.

   Specify first point: _mid
   of [pick diagonal lot line]

4. Move the cursor to anywhere on the diagonal lot line and pick it. AutoCAD snaps the start of the polyline to the precise midpoint of the diagonal line.

5. Pick a few more points at roughly 10' to 20' intervals, moving your way down to the bottom yard line.

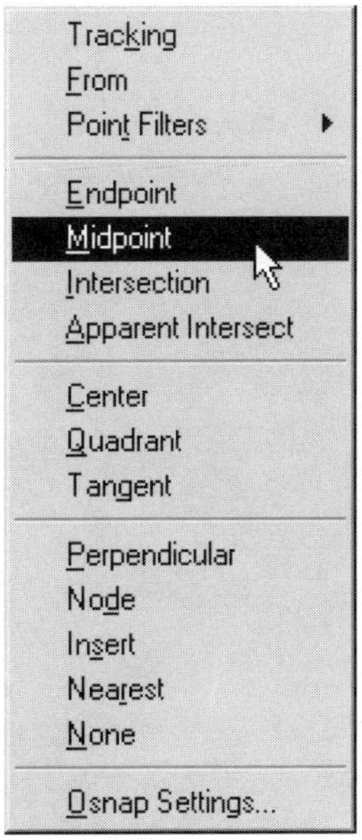

Selecting Midpoint object snap from the shortcut menu.

Tip: If you cannot tell how far ten feet is, move the cursor to the coordinate display and double-click on it until you see the display change to relative coordinates—similar to 7'-10"< 261. This means the cursor is currently 7'-10" from the last-picked point, at an angle of 261 degrees (measured counterclockwise from the East).

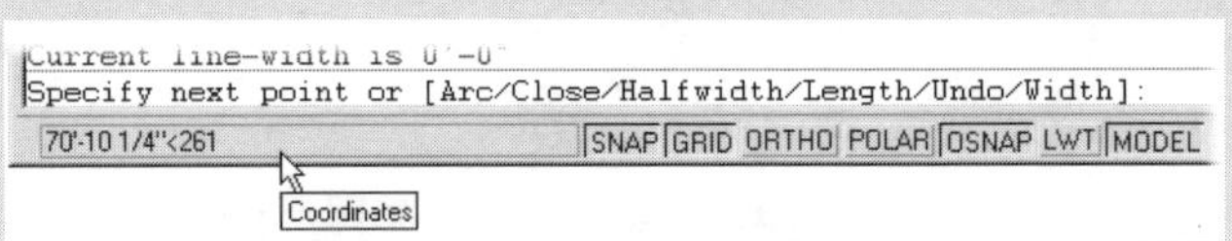

Click the coordinate display to toggle display modes.

6. When you get to the bottom lot line, press **Shift+Right mouse button** and select **Nearest** from the cursor menu that appears.
7. Pick anywhere the line crosses the cursor's aperture box. Again, AutoCAD snaps the polyline precisely on the lot line.

Using object snap to end the polyline.

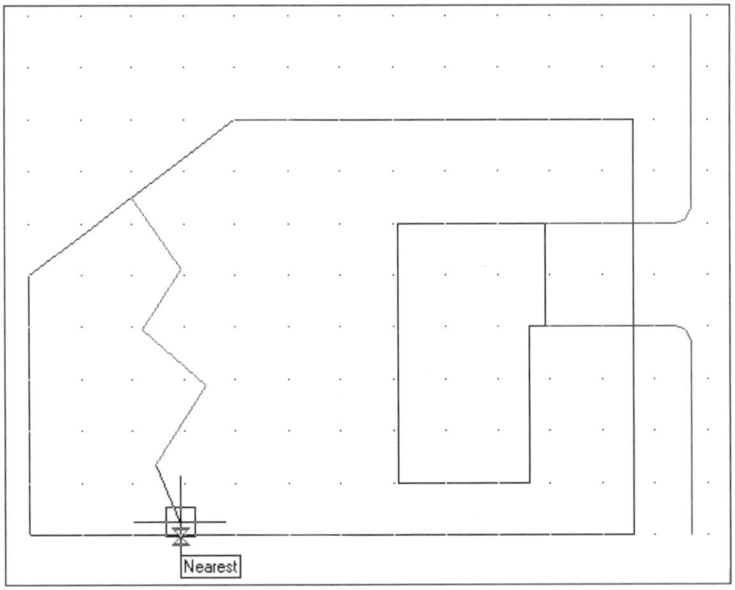

8. Press **Esc** to end the PLine command.

Note: The PLine command has the following options for drawing polylines:

Endpoint of line: Specifies the location of a polyline *vertex* (the endpoint of a polyline segment).

Arc: Draws a *polyarc;* you can switch between lines and arcs when drawing a single polyline.

Close: Joins the last endpoint with the starting point.

Halfwidth: Specifies the width of the polyline by the distance from the centerline to the outside of the polyline.

Length: Draws the polyline a specific distance, in the manner of direct distance entry.

Undo: Undoes the last polyline drawing operation.

Width: Specifies the width of the polyline; allows independent starting and ending width for tapered polylines.

Esc: Exits the command.

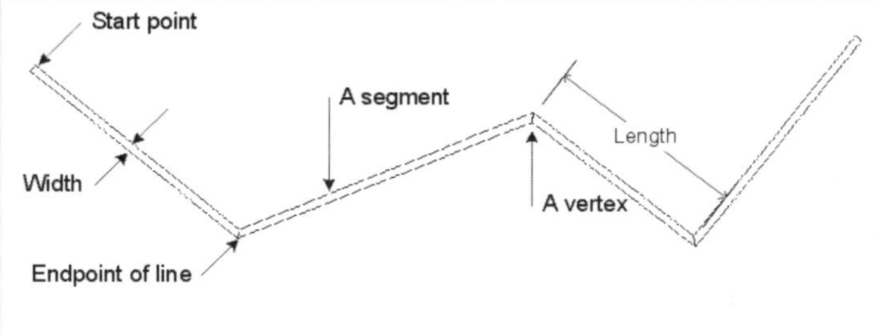

The elements of a polyline.

Smoothing the Polyline

■ Modify | Objects | Edit Polyline

Here is the reason you created the boundary as a polyline: you now use the PEdit command (short for "polyline edit") to smooth the crooked polyline segments into a flowing curve. If you had used the Line command, you couldn't smooth the lines.

1. To edit the polyline, begin the PEdit command by selecting **Modify | Object | Polyline** from the menu. AutoCAD responds as follows:

   Command: _pedit
   Select polyline or [Multiple]:

2. Pick the polyline.
3. The PEdit command has many options. Its purpose is to let you change the look of a polyline. You use the Spline option to smooth the straight lines into a flowing curve, as follows:

   Enter an option [Close/Join/Width/Edit vertex/Fit/Spline/Decurve/Ltype gen/Undo]: s

   The straight lines disappear and are replaced by a smooth curve. (Technically, AutoCAD redrew the line segments as a cubic Bezier curve based on the polyline frame.)

The polyline converted into a spline.

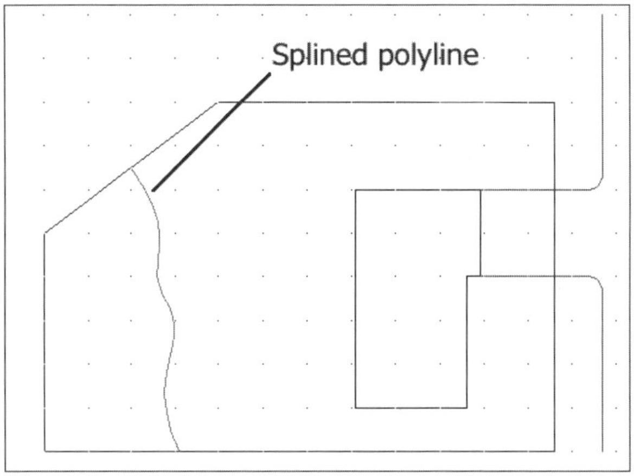

4. To exit the PEdit command, press **x**, as follows:

   Enter an option [...]: x
   Command:

Note: The PEdit command performs the following polyline editing functions:

Close: Closes an open polyline; a segment is drawn between the starting and ending vertices of the polyline (this prompt is only displayed when an open polyline is selected).

Open: Opens a closed polyline; the last segment drawn is erased (this prompt is only displayed when a closed polyline is selected).

Join: Joins another polyline, line, or arc connected to this polyline to form a single polyline. This option fails when there is even a slight gap between the objects to be joined.

Width: Changes the width of the polyline. This option applies a uniform width to all segments making up the polyline; variable width is not possible.

Edit vertex: Edits the width and position of individual vertices; also allows you to insert and remove vertices.

Fit: Applies a curve fit to the polyline.

Spline: Applies a Bezier spline to the polyline.

Decurve: Reverts a curve fit or splined polyline to its original form.

Ltype gen: Toggles linetype generation. When on, the linetype is stretched from the starting to ending point of the polyline. When off, the linetype stretches from vertex to vertex.

Undo: Undoes the last polyline editing operation.

eXit: Exits the command.

Non-Modal Editing

Earlier, you began the PEdit command, then selected the polyline to edit. This is called *verb-noun* editing or *modal* editing. The *verb* is the command (PEdit), followed by the *noun*, or the object (the polyline). Modal editing means that you first enter a mode (the PEdit command mode) before performing the action, such as selecting the polyline and the editing option.

In AutoCAD, you have the option to first pick the object, then move or stretch it into place. This is called *noun-verb* editing or *non-modal* editing. Non-modal editing is a faster way to edit. Here we use non-modal editing to change the shape of the splined polyline separating the garden from the lawn.

1. First though, use the Zoom command to enlarge the view of the splined polyline. Select the Zoom Window icon (looks like a magnifying glass with a square in the center) from the toolbar. AutoCAD launches the Zoom command with the Window option:

```
Command: '_zoom
Specify corner of window, enter a scale factor (nX or nXP), or
[All/Center/Dynamic/Extents/Previous/Scale/Window] <real time>:
_w
Specify first corner: [pick]
```

2. The Window option lets you pick a rectangular area on the screen to magnify. When you pick the two corners, you specify the two opposite corners of the rectangle (see illustration).

Make two picks to specify a window.

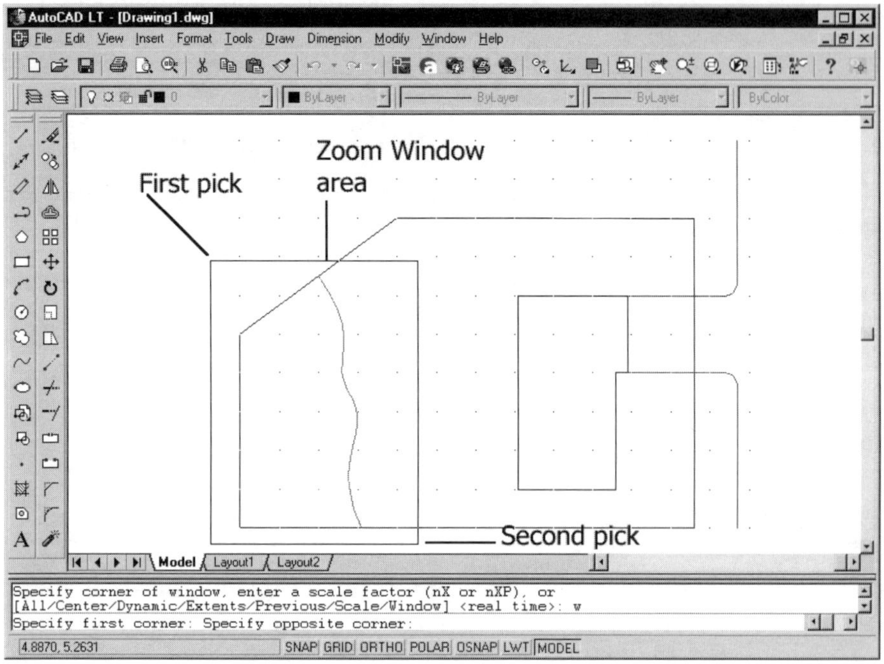

Specify opposite corner: [pick]

Tip: If floating toolbars obscure the polyline, use the scroll bar to move the polyline into view. Grab the horizontal scroll bar and drag it until you clearly see the polyline.

3. In Chapter 1, I pointed out the small square at the center of the cross-hair cursor. It is called the *pickbox*. When AutoCAD displays the pickbox, you are allowed to pick an object without first starting an editing command.

   Pick the polyline. The polyline changes from solid to dashed. Notice the small blue squares that appear on the polyline, including at both ends. The blue squares are called *grips*, since they let you grip an object.

The squares are called grips.

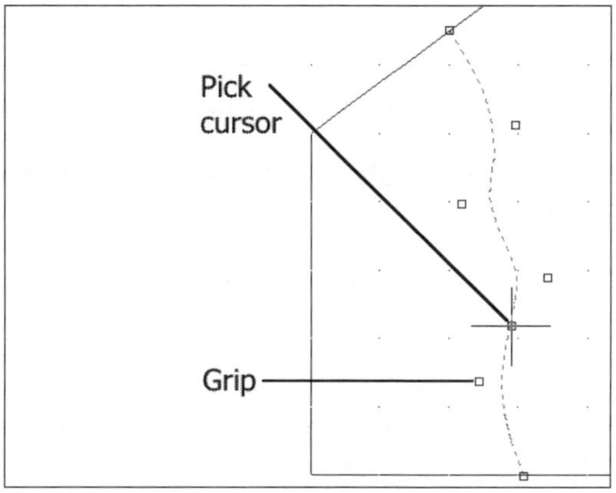

4. Pick the blue square at the bottom end of the polyline. It turns solid red. The red square is called a *hot grip* since editing commands affect it (not the *cold* blue squares). In addition to the solitary red square, a new prompt appears in the command area:

   ** STRETCH **
   <Stretch to point>/Base point/Copy/Undo/eXit: nea
   to [pick lot line]

Moving the splined polyline.

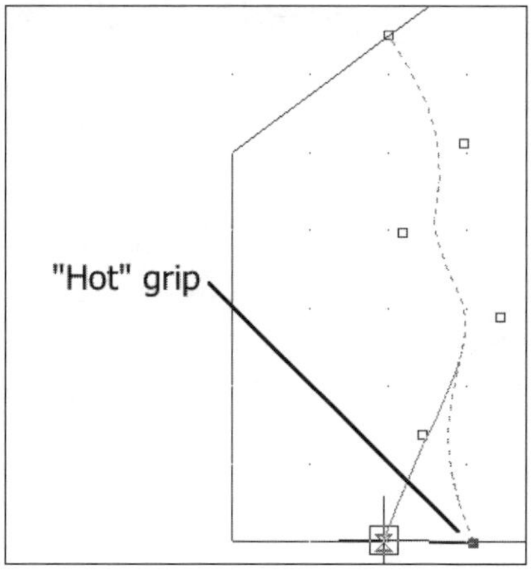

5. Move the cursor. As you do, the last segment of the polyline curves and arcs to follow you. The NEAR object snap ensures the polyline ends precisely at the lot line.

6. Click along the lot line where you want the end of the polyline to move to.

7. Now go ahead and interactively reshape the rest of the polyline, segment by segment. The blue grips indicate the vertices of the straight polyline segments you originally drew, before splining it with PEdit.
8. When you are finished reshaping the polyline, press **Esc** twice to exit modeless editing.
9. Select the **View | Zoom | Previous** command from the menu bar to see the entire drawing again.

Note: AutoCAD lets you employ six editing commands during non-modal editing: Stretch, Copy, Move, Rotate, Scale, and Mirror. To see all the commands, press the Spacebar. Each time you do, AutoCAD displays a different set of prompts:

```
** STRETCH **
<Stretch to point>/Base point/Copy/Undo/eXit: [Spacebar]
** MOVE **
<Move to point>/Base point/Copy/Undo/eXit: [Spacebar]
** ROTATE **
<Rotation angle>/Base point/Copy/Undo/Reference/eXit: [Spacebar]
** SCALE **
<Scale factor>/Base point/Copy/Undo/Reference/eXit: [Spacebar]
** MIRROR **
<Second point>/Base point/Copy/Undo/eXit: [Spacebar]
```

After the ** MIRROR ** command, AutoCAD repeats the cycle with the ** STRETCH ** command. The other non-modal editing options are:

Base point: Specifies a base point other than the hot grip.

Reference: Specifies a reference point other than the hot grip.

Undo: Undoes the last editing operation.

eXit or press Esc: Exits non-modal editing mode.

Hatching the Lawn

■ Draw | Hatch

You created a boundary between the lawn and garden but how do you show the difference between the grassy area and the dirt area? One way is to add a symbol to the lawn area. The symbol should indicate grass. In

AutoCAD LT, this is done with the BHatch command (short for "boundary hatch").

1. First, switch back to the Lawn layer via the toolbar.
2. Select **Draw | Hatch** to start the BHatch command and AutoCAD displays the Boundary Hatch dialog box:

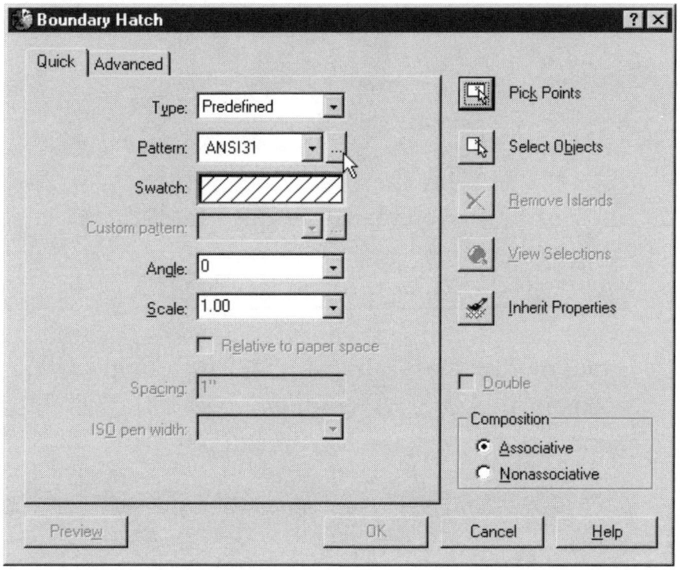

The Boundary Hatch dialog box.

3. Since you don't see the grass hatch pattern, click the ellipsis (...) button (next to the Pattern text box).
4. AutoCAD displays the Hatch Pattern Palette. Click the **Other Predefined** tab.

   A *palette* looks like a dialog box but presents a graphical palette of options. (Palettes were known as *icon menus* in earlier versions of AutoCAD.) Scroll down the list and select **Grass**.
5. Click **OK** to return to the Boundary Hatch dialog box.

The Hatch Pattern palette.

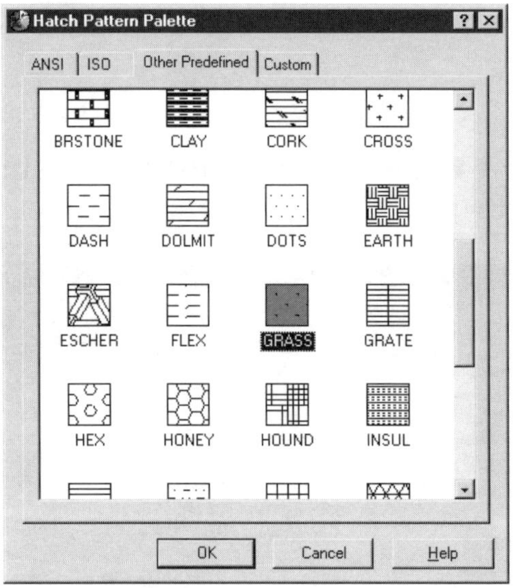

6. Look down at the Scale factor. Change Scale to **10'**. (If the scale factor is too small, AutoCAD refuses to draw the hatch pattern and complains, "Hatch spacing too dense, or dash size too small.")

Dialog box after selecting Grass, and changing the scale.

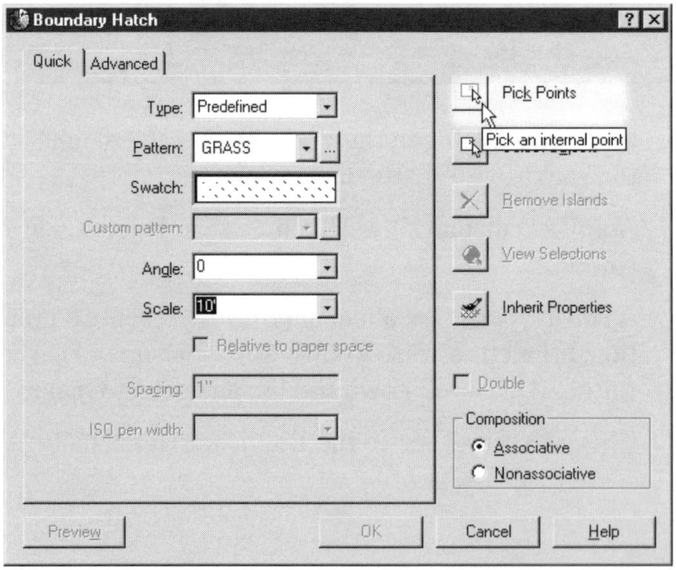

7. AutoCAD has a useful feature called *boundary* that searches a contiguous area, no matter how many different borders the area has. To employ this feature, click the **Pick Points** button.

8. The dialog box disappears and AutoCAD prompts:

   Select internal points: [pick]

9. Pick a point anywhere in the lawn area. AutoCAD spends a few moments reflecting upon your selection:

   Select internal point: Selecting everything...
   Selecting everything visible...
   Analyzing the selected data...
   Analyzing internal islands...

   and then highlights the object(s) that make up the boundary.

10. Press **Enter** to return to the dialog box:

    Select internal points: [Enter]

    AutoCAD draws the boundary out of a polyline, although you cannot see it.

11. After you press **Enter**, the Boundary Hatch dialog box reappears. A hatch pattern applied at the wrong scale can cause a lot of grief:

    - Scale too large: You can't see the hatch pattern.
    - Scale too small: Hatch pattern looks black and takes a longer time to display.

    To help you out, AutoCAD includes a preview feature. Click the **Preview** button.

Previewing the hatch pattern—looks good!

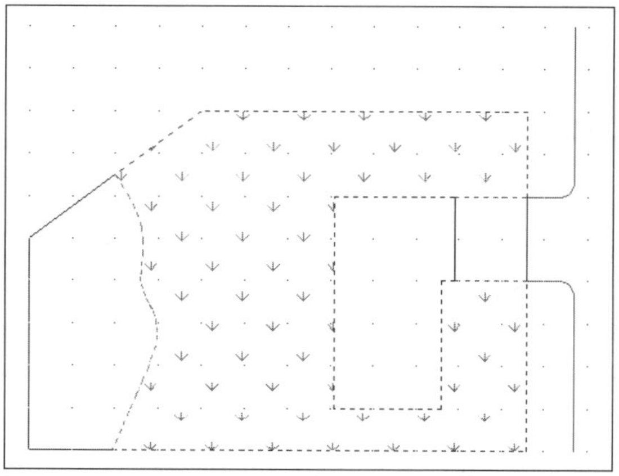

AutoCAD quickly hatches the lawn area. It looks like the scale factor is good enough. Note how precisely AutoCAD applied the hatch pattern, and how it automatically clips the pattern along boundaries. Try doing a hatch pattern that neatly and that quickly by hand!

12. Press **Enter** to end the preview and return to the dialog box:

    <Hit enter or right-click to return to the dialog> [Enter]

13. The Boundary Hatch dialog box reappears. Click on the **OK** button. The dialog box disappears for the last time, and the hatch pattern is in place.

    With all this hard work on your drawing, it's a good idea to save the drawing to the computer's hard disk with the Save command—right now!

Tip: Once a hatch is in place, it is not cast in stone. You can change it in several ways:

- New to 2000i: double-click the hatch entity and AutoCAD brings up the Edit Hatch dialog box. This works with many other objects as well.
- When you move (or stretch) any object making up the boundary of the hatch, the hatch automatically updates itself.
- The hatch does not have to stay in place. You can move it, copy it, erase it, etc.
- The HatchEdit command displays a dialog box that allows you to change the pattern, scale factor, rotation angle, etc.

Creating a Symbol

You've given the lawn area its grass. Let's now add some trees and shrubs to the garden area. In this section, you create a symbol for a tree. The symbol is commonly used by landscape architects and consists of a circle and radiating lines. After you create one tree symbol, you learn how to add many tree symbols with just a single command.

Drawing a Circle

- Draw | Circle
- View | Regen

1. Before starting the first tree, make sure the working layer is set to **Plants** (select layer **Plants** from the toolbar).
2. Select **Draw | Circle | Center, Radius** to draw a six-inch radius circle, as follows:

   Command: _circle Specify center point for circle or [3P/2P/Ttr (tan tan radius)]: [pick in garden area]
   Specify radius of circle or [Diameter]: 6

   With a radius of 6, AutoCAD draws the circle with a one-foot diameter. AutoCAD LT has several ways to draw a circle. Above, you used the most common method: pick the circle's center point and specify the radius of the circle.

Drawing a circle by specifying its center and radius.

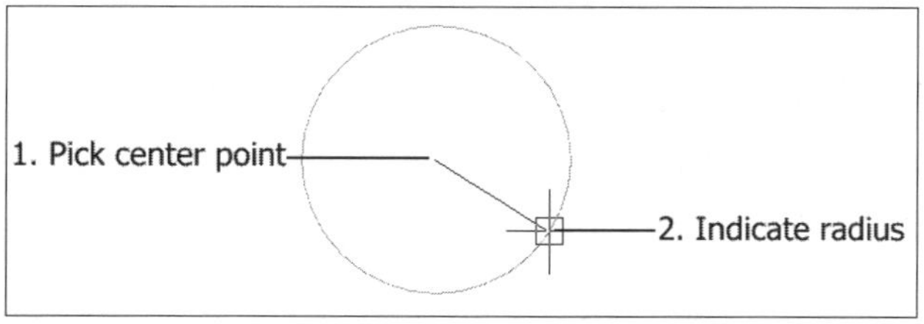

Note: AutoCAD LT 2000i has several ways to specify the parameters of a circle:

Center, Radius: The center point and the radius distance.

Center, Diameter: The center point and the diameter distance.

2 Points (2P): Two points on the circle's circumference and diameter.

3 Points (3P): Three points on the circle's circumference.

Tan,Tan,Radius (TTR): Two points of tangent (to other objects) and the circle's radius.

Tan,Tan,Tan: Three points of tangent to other objects.

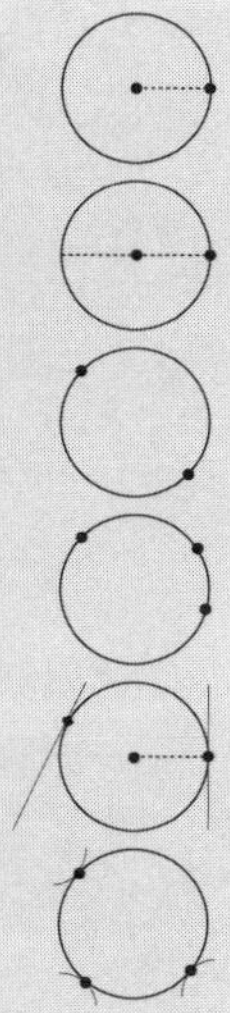

3. The one-foot circle looks very small on the screen. The Zoom command lets you see your work more clearly. Select **View | Zoom | Realtime**.

   Command: '_zoom
   All/Center/Extents/Previous/Scale(X/XP)/Window/<Realtime>:
   Press Esc or Enter to exit, or right-click to activate pop-up menu.

   The cursor changes to a magnifying glass. Drag the cursor over the circle. As you move the mouse forward and backward, AutoCAD dynamically increases and decreases the zoom. (As an alternative, if you have a wheelmouse, roll the wheel forward to zoom in.)

Note: To *drag* means to hold down the left mouse button, then move the mouse. Finally, let go of the mouse button.

4. If you find the circle moving off the edge of the window, right-click the mouse. AutoCAD displays this shortcut menu:

The shortcut menu for the Zoom command.

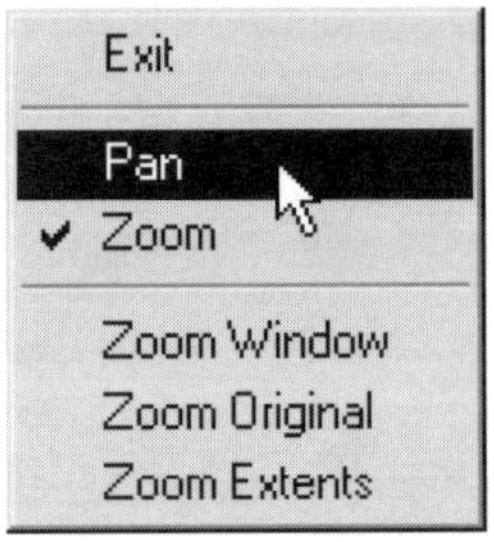

5. Click **Pan**. AutoCAD switches to real-time pan mode. (As an alternative, click the scroll bars to pan the drawing.) Move the tree circle to the center of the drawing area. If necessary, right-click again and switch back to Zoom mode.
6. When the tree circle is at a satisfactory size, press **Esc** to get out of real-time pan-and-zoom mode:

   Press Esc or Enter to exit, or right-click to activate pop-up menu. [Esc]

7. Oops! What happened to the circle? If the circle looks like an octagon, select **View | Regen** to clean it up, as follows:

   Command: _regen
   Regenerating drawing.

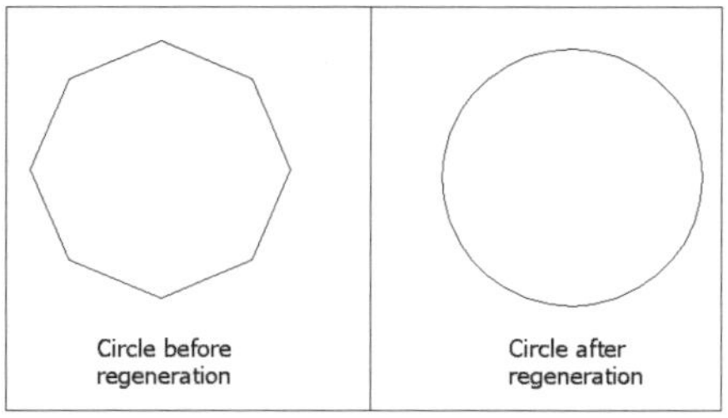

Zoomed-in circle before (at left) and after regeneration (at right).

Now that the circle looks rounder and larger, it is easier to work with.

Creating an Array

■ Modify | Array

To draw the array of lines (which represent the branches), you draw one line, then use the Array command to create the radiating lines.

Note: The Array command creates three types of arrays:

- Rectangular array (left, in the figure below).
- Polar array with rotated objects (center, below).
- Polar array with not-rotated objects (right, below).

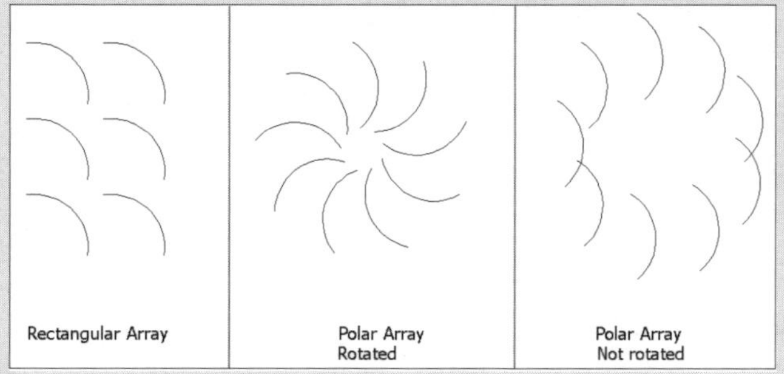

AutoCAD creates three kinds of arrays.

1. Select the **Line** icon from the Draw menu, and proceed as follows:

   Command: _line
   Specify first point: cen
   of [pick circle]
   Specify next point or [Undo]: [pick point anywhere outside of circle]
   Specify next point or [Undo]: [Enter]

   Here you used the CENter object snap to begin the line at the precise center of the circle. The other end of the line goes beyond the edge of the circle.

Chapter 4

Drawing a line from the precise center of a circle.

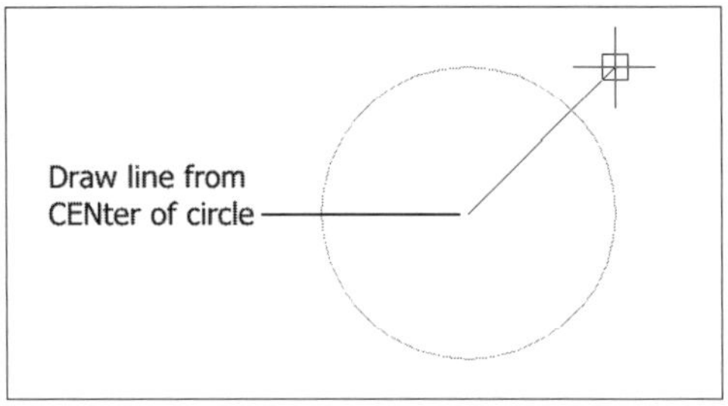

2. From the Modify menu, select **Array.** Notice that AutoCAD displays the Array dialog box.
3. Click the **Polar Array** radio button. (The Rectangular Array option creates linear, square, and rectangular arrays of objects.)

The Array dialog box.

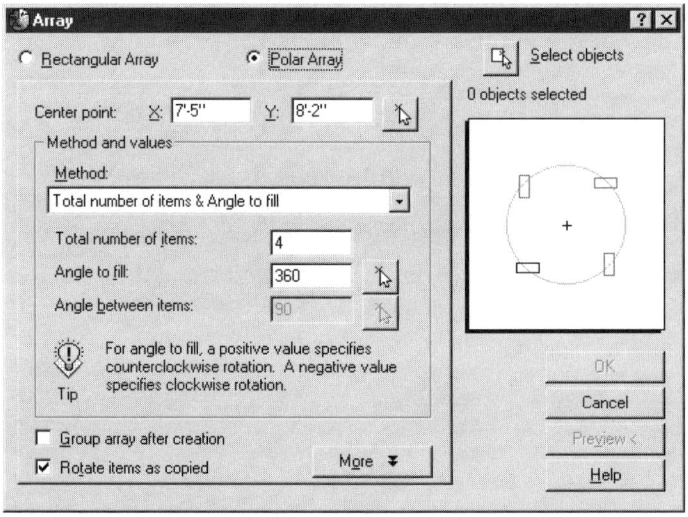

4. The dialog box can look intimidating, what with all those buttons and fields. The process you go through is: (1) select objects to array; (2) specify the centerpoint of the array; (3) define the array method; (4) take care of any other options; (5) preview to make sure it looks right; and (6) commit!

   Select the line that you want to array. Click **Select Objects**. The dialog box disappears so that you can see the screen. Notice the prompt that AutoCAD displays:

Select objects: L
1 found

5. Type **L** to select the line, then press **Enter**.

Note: The letter "L" (short for last) is shorthand notation for selecting the last object drawn still visible on the screen. After you select one or more objects, AutoCAD reports the number selected: "1 found."

6. Press **Enter** to end object selection and return to the dialog box:

   Select objects: [Enter]

7. Click the **Pick Center Point** button, located across from the Center point fields. (The button is labeled with a small cross and an arrowhead; when you pause your cursor over the button, it displays the Pick Center Point tooltip.) Notice that the dialog box disappears (again) so that you can see the screen.

8. Use endpoint object snap to select the end of the line at the center of the circle, as follows: hold down the **Shift** key and press the left mouse button. When the cursor menu appears, select **Endpoint**. Select the line:

   Specify center point of array: _endp
   of [pick line at center of circle]

   Although you are using a dialog box to construct this array, notice that you still need to watch the Command prompt area from time to time! After picking the center of rotation, the Array dialog box immediately returns.

9. Now you need to tell AutoCAD how you want the array constructed. You use the default method "Total number of items" and "Angle to fill." In the Total number of items field, erase **4** and type **25**. The Angle to fill field reads 360, which is what you want (items are arrayed around a full circle).

10. At the bottom of the dialog box are two more options:

    Group array after creation: No, since you will create a block later.

    Rotate items as copied: Yes, since you want the lines to radiate.

11. With all the parameters (hopefully) set up correctly, click **Preview.** AutoCAD quickly draws 24 more lines around the circle, completing the tree symbol.

The preview of the arrayed lines.

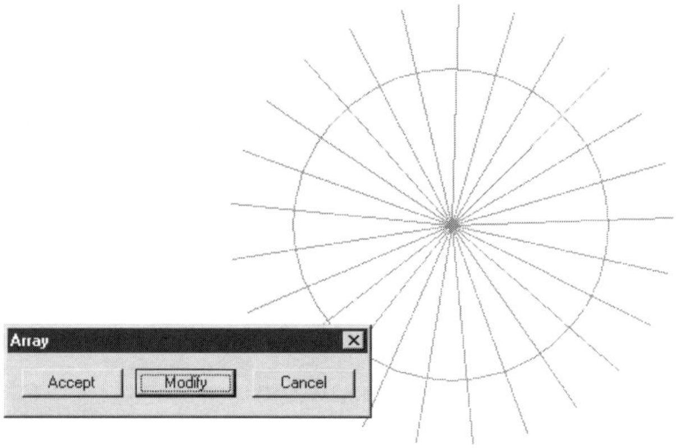

12. Since the array looks correct, click **Accept**. The array is left as is; unlike the BHatch command, the Array dialog box does not reappear.

Making a Block

■ Draw | Block | Make

Although CAD draws a symbol more quickly and accurately than you can by hand, there are many keystrokes involved. The way to reduce the number of keystrokes is to turn the symbol into a block, and then insert the block into the drawing. In this section, you add several more trees to the garden area.

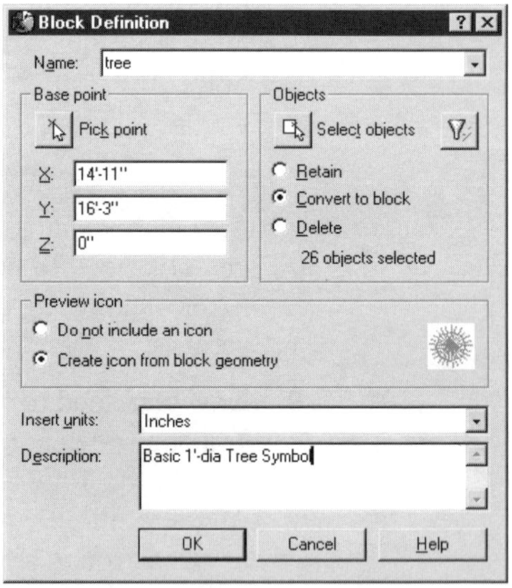

The Block Definition dialog box.

1. To turn the tree symbol into a block, use AutoCAD LT's Block command (changed from BMake in the previous versions of AutoCAD LT). Select **Draw | Block | Make** from the menu bar. AutoCAD displays the Block Definition dialog box.

2. Type **Tree** in the field next to Name. You can give the block any name you like, up to 255 characters long.

3. You need to tell AutoCAD which objects to turn into a block. AutoCAD lets you select objects several different ways. So far, you have picked them (one at a time) with your mouse or with the L option. Just as you windowed the zoomed-in view, you can window the objects you want to select with the W option (short for *window*).

   Click the Select Objects button. The dialog box disappears and AutoCAD prompts:

   ```
   Select objects: w
   Specify first corner: [pick corner]
   Specify opposite corner: [pick corner]
   26 found Select objects: [Enter]
   ```

   You pick the two corners of a rectangle that encompasses the circle and 25 lines making up the tree symbol. Make sure that all 26 objects are inside the selection rectangle, otherwise AutoCAD will not include them. The dialog box reappears and reports:

   ```
   26 found.
   ```

4. Click **Delete Selected Objects after Block Creation** to turn on the feature.

5. The Pick Point fields are used later by the Insert command. It is also known as the *insertion point*, the point where the block is placed in the drawing. The center of the tree symbol is a logical insertion point.

   Click the **Pick Point** button. The dialog box disappears and AutoCAD prompts:

   ```
   Specify insertion base point: cen
   of [pick circle]
   ```

   Using the **CENter** object snap ensures a precise selection. The dialog box reappears with the x,y,z-coordinates of the base point filled in.

6. For the Description field, type anything you like up to 255 characters long:

   ```
   Basic 1'-dia Tree Symbol
   ```

7. Click on the **OK** button. AutoCAD removes the dialog box, erases the old tree symbol, and records the tree symbol as a block.

Drawing Overview

■ View | Aerial View

Now that the tree symbol is stored as a block, you can insert it into the drawing. First, though, you need to see more of the garden area. You could use the Zoom command to enlarge the view. AutoCAD LT, however, includes a handy drawing viewer named Aerial View that lets you easily zoom and pan about the drawing.

1. Click the **Aerial View** icon on the toolbar (looks like an airplane).

   Or, select **View | Aerial View** from the menu bar.

   Or, type **Alt+VW** (press the **Alt** key, press **V**, and press **W**).

   Or, type **DsViewer** at the Command: prompt, as follows:

   Command: dsviewer

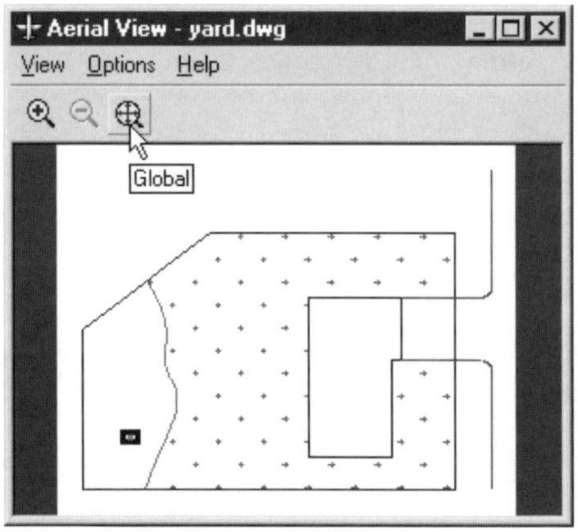

The Aerial View window.

   Any of these actions causes AutoCAD to launch the Aerial View window. Like floating toolboxes, Aerial View is an independent window that floats anywhere on the Windows desktop.

2. Click on the last icon on the Aerial View's toolbar; the tooltip identifies it as Global. That way you always see the full extent of the drawing, no matter how far in or out you zoom. Look closely to see a small black square where you drew the tree; that is Aerial View's way of showing where AutoCAD's current zoomed-in view is located.

3. To change the zoom level, click the second button on the Aerial View toolbar (Zoom). Then click two points in the Aerial View, just like doing a Zoom Window.
4. As you move to the second corner, AutoCAD immediately zooms to the new view and a resized rectangle shows the new view in the Aerial View window.

Note: If you find Aerial View frustrating, do a Zoom All followed by a Zoom Window, as follows:

```
Command: zoom
All/.../ Window/<Scale(X/XP)>: a
Regenerating drawing.
Command: [Enter]
All/.../ Window/<Scale(X/XP)>: w
First corner: [pick]
Other corner: [pick]
```

If you find Aerial View getting in the way, simply grab it by the title bar and slide it off the side of the screen.

Adding Many More Trees

- Insert | Block

With the larger view in place, let's see how to insert the tree blocks into the drawing.

1. Select **Insert | Block** from the menu bar. AutoCAD displays the Insert dialog box.
2. The Tree block name should be shown in the Name field.

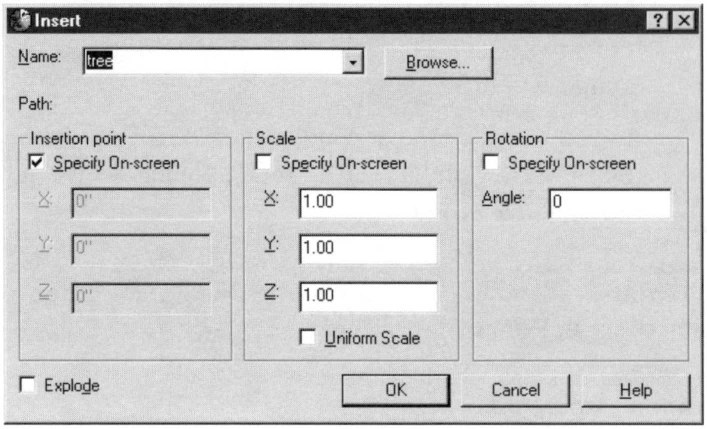

The Insert dialog box.

3. Click **OK**. You can ignore everything else in the Insert dialog box, other than to ensure that **Specify On Screen** is turned on for **Insertion point** and **Scale** (a check mark should be shown).

4. Click on the **OK** button. AutoCAD prompts you to pick a spot to place the block:

   Specify insertion point or [Scale/X/Y/Z/Rotate/PScale/PX/PY/PZ/PRotate]: [pick a point in the garden area]

5. When you supply an X-scale factor, AutoCAD draws the block larger or smaller than the original symbol.

   Enter X scale factor, specify opposite corner, or [Corner/XYZ] <1>: 5

   By specifying an X-scale factor of 5, AutoCAD draws the block five times larger. Since you drew the original symbol 1' in diameter, the newly inserted tree is five feet in diameter. You can see that it makes sense to draw a symbol to *unit size* (to the nearest inch or foot); that makes it easier to scale the block during insertion.

6. If you want to stretch or squeeze the block, the DdInsert command allows you to specify a different Y-scale factor. This is useful for inserting rectangles (such as different sized lumber) based on a unit square.

   Enter Y scale factor <use X scale factor>: [Enter]

7. Try a different way of repeating the Insert command; right-click and AutoCAD LT displays a new cursor menu.

Select **Repeat INSERT** and AutoCAD redisplays the prompts of the Insert command.

The shortcut menu repeats the last command.

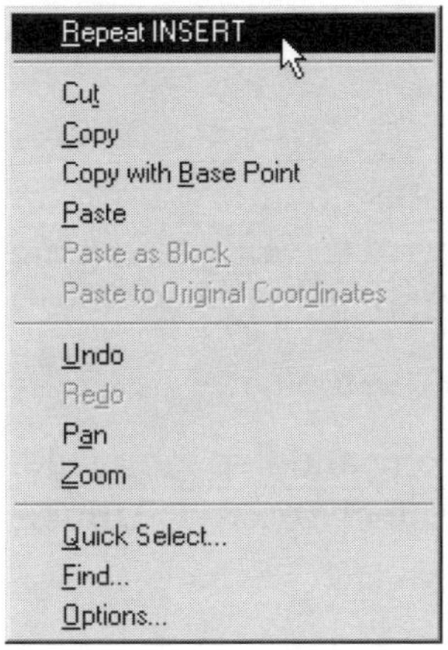

8. Add several more trees around the garden area using the Insert command and different X-scale factors, such as 6.0, 4.0, 2.0, and 1.5. Use the Mirror command to double the number of trees.

Trees placed in the garden area.

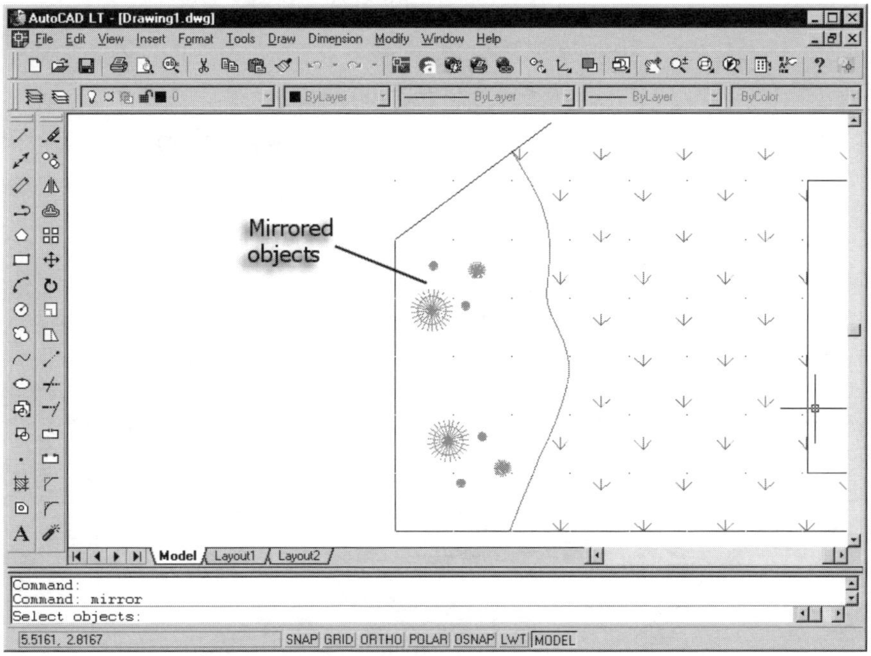

Drawing the Pond

- Draw | Ellipse
- Modify | Offset

Drawing the garden pond illustrates another pair of AutoCAD commands. You draw the oval-shaped pond with the Ellipse command, and add the pond's edging with the Offset command.

1. Switch to the Pond layer via the toolbar.
2. The pond is an oval 15 feet long and five feet wide. Draw the pond with the Ellipse command. Select **Draw | Ellipse | Center** from the menu.

   Command: _ellipse
   Specify axis endpoint of ellipse or [Arc/Center]: [pick]
   Pick the starting point of the ellipse anywhere in the garden area.

3. The other end of the pond is 15 feet away. Use direct distance entry by moving the cursor down and right:

   Specify endpoint of axis: [move cursor] 15'

4. The pond is five feet wide:

   Specify distance to other axis or [Rotation]:5'

Drawing the pond as an ellipse.

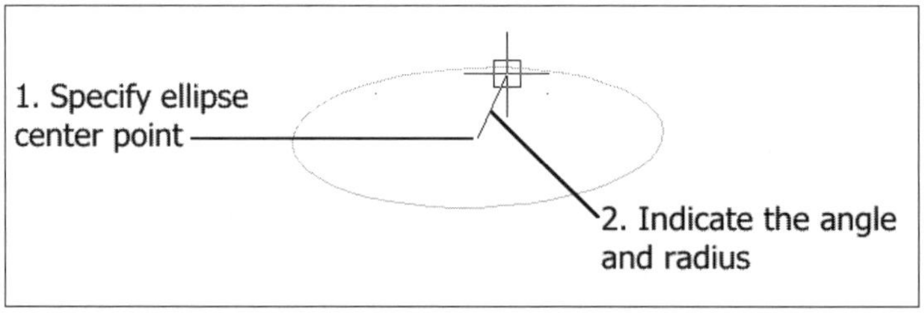

Note: AutoCAD lets you draw an ellipse by three different methods:

Center: Specifies the ellipse's center point, the endpoint of the major axis, and the "width" (minor axis).

Axis,End: Specifies the endpoints of the ellipse's major and minor axes.

Arc: Draws an elliptical arc.

In addition, the Ellipse command draws *isocircles*, which are isometric circles. This option is available only when you turn on Iso mode via the Snap or DdRModes commands.

5. You could draw the rock edging of the pond by repeating the Ellipse command but the ellipses probably wouldn't be centered precisely. Instead, the Offset command helps you creates concentric ellipses. Select **Modify | Offset** from the menu bar, as follows:

   Specify offset distance or [Through] <Through>: 1'
   Select object to offset or <exit>: [pick ellipse]
   Specify point on side to offset: [pick outside ellipse]
   Select object to offset or <exit>: [Enter]

   The Offset command also creates parallel lines, parallel polylines, and concentric circles and arcs.

Adding a second ellipse with the Offset command.

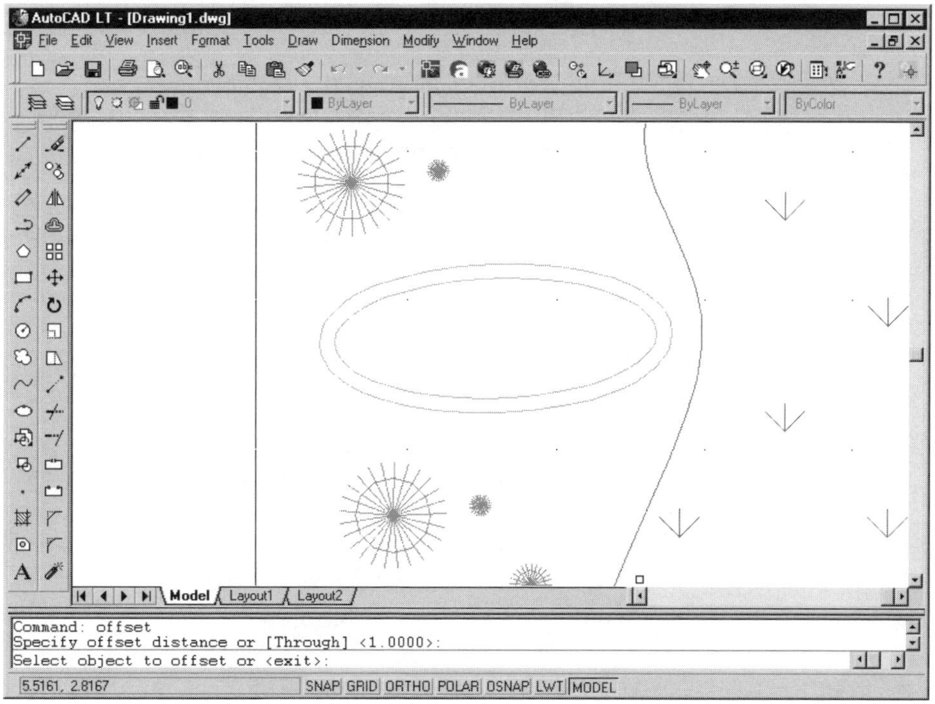

Remember to save your work with the Save command. To see the progress you are making in learning AutoCAD, plot your drawing with the Plot command by clicking on the printer icon on the toolbar.

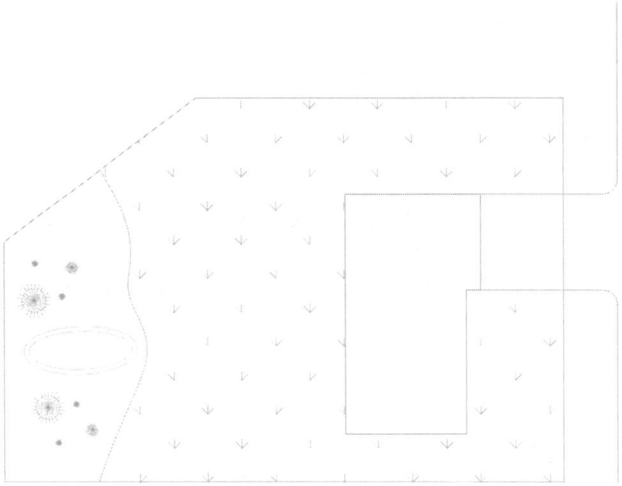

Review Questions

1. When drawing a circle with the 3P option, AutoCAD requires:
   a. Three points of tangent to other objects.
   b. The center point and two points on the circumference.
   c. Three points on the circle's circumference.
   d. Two points of tangent and the circle's radius.
2. The purpose of the PEdit command's Spline option is to:
   a. Replace the polyline segments with a spline object.
   b. Redraw the polyline segments as a cubic Bezier curve based on the polyline's frame.
   c. Reduce the polyline segments to a single circular arc.
   d. Regenerate the polyline segments to show the polyline's frame.
3. Match the command name with its best meaning:

a.	Array	i.	Copies an object parallel to an existing object.
b.	PEdit	ii.	Adds an associative hatch pattern to a closed area.
c.	Offset	iii.	Displays an overview of the entire drawing.
d.	BHatch	iv.	Creates a repeating rectangular or circular pattern of objects.
e.	DsViewer	v.	Edits the look of a polyline.

4. In which menu would you find the following commands:

a.	Block	i.	Modify
b.	Insert	ii.	File
c.	Pan	iii.	View
d.	Array	iv.	Draw
e.	Save	v.	Insert

5. A polyline is restricted to a single width.

   True / False

6. Match the command name or option with its expanded name:

a.	PEdit	i.	Linetype
b.	Pline	ii.	Tangent, tangent, radius
c.	BHatch	iii.	Polyline Edit
d.	Ltype	iv.	Polyline
e.	TTR	v.	Boundary Hatch

7. The PEdit command can "close" an open polyline.

   True / False

8. The small square at the center of the crosshair cursor is called the:

   a. Aperture
   b. Crosshair
   c. Object snap
   d. Pickbox

9. Match the term with its best meaning:

a.	Hot grip	i.	Noun-verb editing
b.	Cold grip	ii.	Command
c.	Verb	iii.	Object
d.	Non-modal	iv.	Red square
e.	Noun	v.	Blue square

10. The Aerial View window helps you zoom and pan the drawing independent of the drawing itself.

    True / False

Chapter 5

Making Changes to the Drawing

In the last chapter, you added details to the drawing, such as the lawn, some trees, and a pond. In this chapter, you learn how to change parts of the yard, and how to get information out of the drawing.

Key Concepts

- Modifying the properties of objects
- Applying a linetype and linetype scaling to selected objects
- Using "filters" to create a selection set of objects
- Changing the length of open and closed objects with the Lengthen and Stretch commands
- Finding out information about objects in the drawing
- Positioning the cursor with tracking

Key Terms

Extend—extends an open object to a *boundary*, which is defined by another object

Trim—cuts an object at a *cutting edge*, which is defined by another object

Properties—a listing of all properties of an object

Tracking—move during a drawing or editing command

Linetypes—a line pattern made of dashes, dots, and gaps

Filters—creates a specified subset of objects

Abbreviations	
ISO	International Organization for Standardization
.LIN	Linetype definition file

Command	Shortcut	Menu Selection
Dist	di	Tools | Inquiry | Distance
Lengthen	len	Modify | Lengthen
Linetype	Alt+FN	Format | Linetype
List	li	Tools | Inquiry | List
Properties	Ctrl+1	Modify | Properties
QSelect	Alt+TQ	Tools | Quick Select
Stretch	s	Modify | Stretch
Tracking	tk	...

Changing the Look of Lines

- Format | Linetype
- Modify | Properties

When you drew the lot lines, back in Chapter 3, they showed up on the screen as solid lines. Lot lines, however, are usually shown by a dashed pattern. Just as AutoCAD comes with several hatch patterns, it also includes a number of line patterns called *linetypes*.

AutoCAD allows you to change lines from solid to patterned. It takes two steps to change a line from solid (called "continuous" by AutoCAD) to a dashed linetype: (1) load the linetype, and (2) change the line to the new linetype.

1. First, use the Zoom command so that you see the entire drawing on the screen, as follows:

```
Command: zoom
Specify corner of window, enter a scale factor (nX or nXP), or
[All/.../Scale/Window] <real time>: a
```

Then do an 80% zoom to add some breathing space around the drawing:

Command: zoom
Specify corner of window, enter a scale factor (nX or nXP), or
[All/.../Scale/Window] <real time>: 0.8x

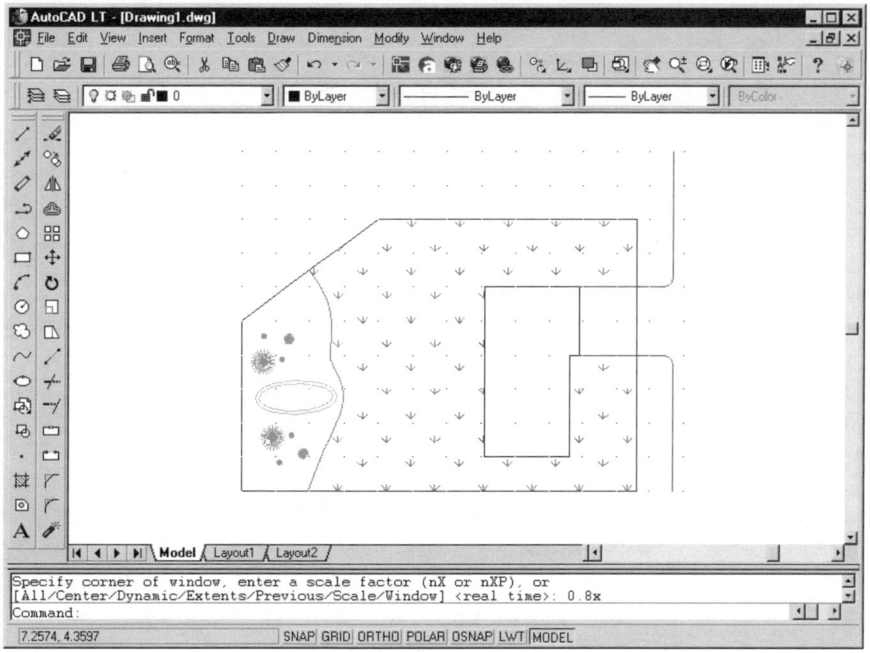

Viewing the entire drawing with the Zoom command.

2. The linetype definitions are stored in a file separate from AutoCAD LT and the drawing. You can recognize the file by its .LIN extension. Before you can use a linetype, you must load it's definition into the drawing.

   From the menu bar, select **Format | Linetype.** Notice that AutoCAD displays the Linetype Manager dialog box. The dialog box displays the three linetypes found in all new drawings:

- Bylayer: lines take on the linetype assigned to the layer.
- Byblock: lines take on the linetype of the block.
- Continuous: lines are solid.

The Linetype Manager dialog box displays the linetypes loaded into the current drawing.

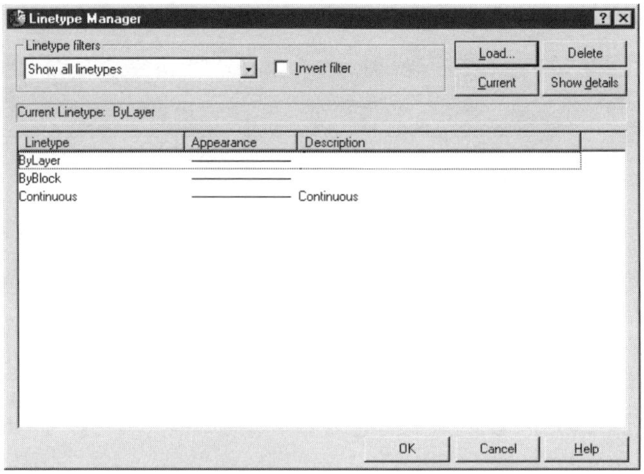

3. Click the **Load** button. Notice that AutoCAD displays the Load or Reload Linetypes dialog box. This dialog box lists in alphabetical order the names of all linetype definitions stored in the Aclt.Lin file. Scroll through the list to see what is available:
   - Linetype names prefixed with ACAD_ISO are compliant with ISO (International Organization for Standardization).
   - Linetype names postfixed with X2 are double scale.
   - Linetype names postfixed with 2 are halfsize (0.5x scale).

The Load or Reload Linetypes dialog box.

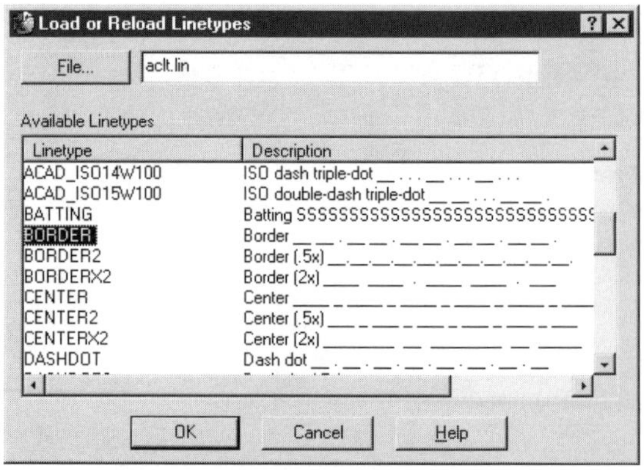

4. To load the Border linetype, select it, then click **OK**.
5. Back in the Linetype Manager dialog box, ensure that the **Border** linetype is listed. Click **OK** to dismiss the dialog box.

The Border linetype is loaded into the current drawing.

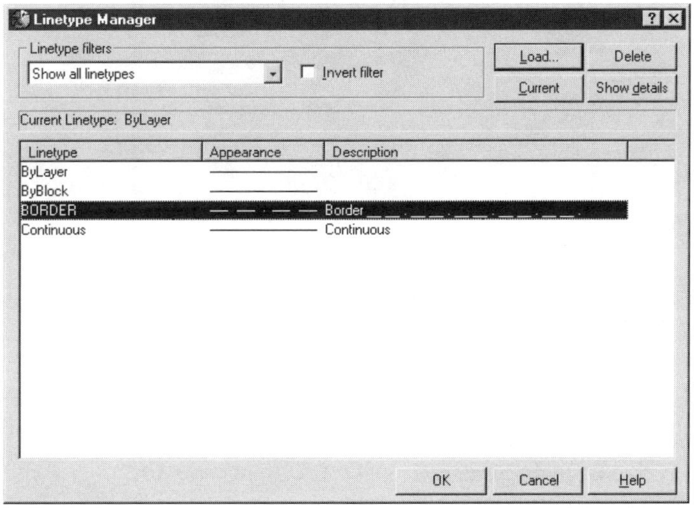

6. The Properties command allows you to change several properties of one or more objects, including the linetype. You use it now to change the lot lines from Continuous to Border. From the menu bar, select **Modify | Properties**. Notice that the Properties window opens. (As an alternative, you can press Ctrl+1 to open and close the window.) A window is different from a dialog box; you can leave the window open while you continue to work on the drawing.

Note: The Properties window was added to AutoCAD LT 2000. It replaces two commands found in older versions of LT: DdModify and DdChProp.

A *property* describes the object. Properties include color, linetype, starting and ending coordinates, and thickness. The Properties window displays information in several different ways:

- When no objects are selected, the Properties window displays the current settings, repeating some of the information displayed by the Object Properties toolbar with nothing selected.
- When a single object is selected, the Properties window displays everything AutoCAD knows about the object. The list of properties varies, depending on the object selected.

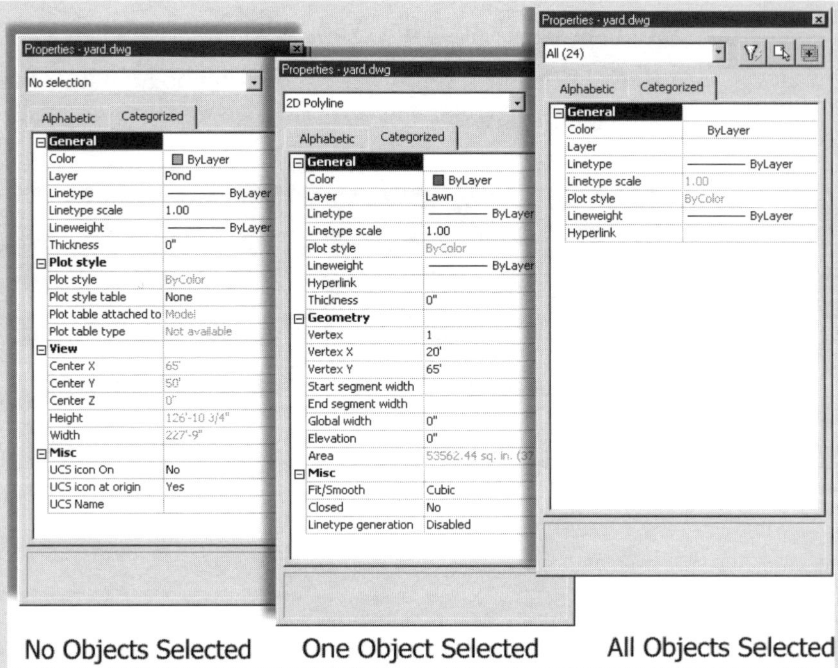

- When two or more different objects are selected, the Properties window displays all properties in common. Naturally, the list of properties displayed varies, depending on which kinds of objects are selected. The more different objects are selected, the fewer properties are displayed.

When two or more objects are selected, you can have the Properties window display a subset. Notice that the list box reports the number of objects in brackets.

The Properties window does more than display the properties of selected objects(s). It allows you to change the value of individual properties:

- Values that can be changed are displayed in black text.
- Values that cannot be changed are displayed by gray text.
- Values that have not yet been assigned are shown blank.

There are three ways you can change a property value: type a new value, select a preset value from a list box (if available), or click a small button (when available) and select the value directly from the drawing or from a dialog box.

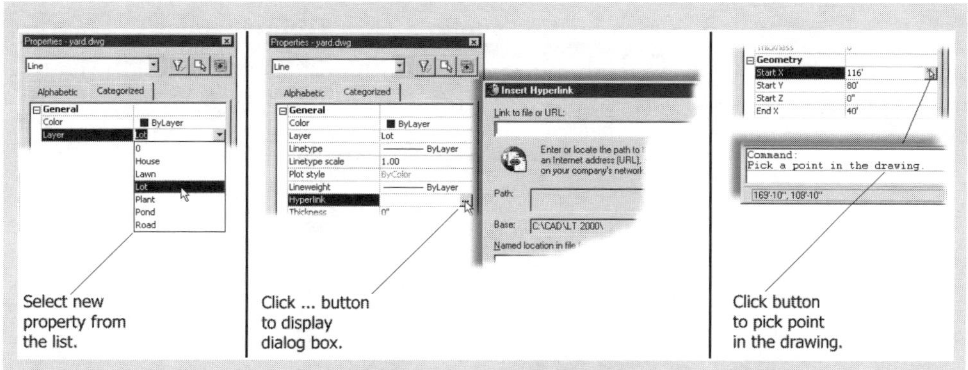

Property values can be changed by a variety of methods.

7. At the Command: prompt, AutoCAD does not ask you to select the objects you want to change (as it would with the Change and ChProp commands). As an alternative, click the **Select Objects** button (new in 2000i) that supports all selection options, such as Previous and All. Go ahead and select the lot lines, as follows:

   Command: [pick all five lot lines]

   As you pick the lot lines, notice that AutoCAD highlights them: they are shown as dashed lines, with blue grip squares appearing on each line. Over in the Properties window, notice that the list box reports Line (5), indicating that five lines have been selected.

The properties of the five selected lines.

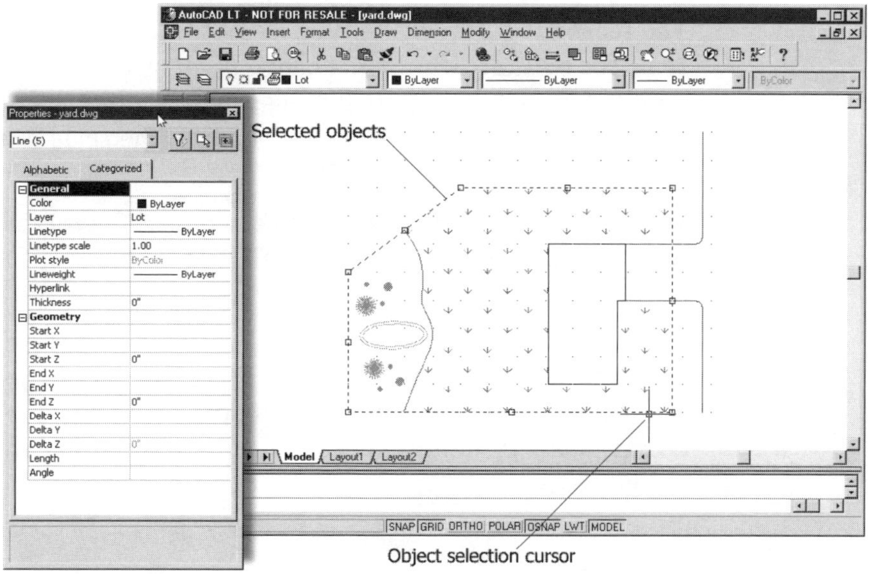

8. In the Properties window, click **Bylayer** next to Linetype. Notice that AutoCAD displays a list of the linetypes loaded in the drawing: Bylayer, Byblock, Continuous, and Border.

Selecting the Border linetype for the selected lines.

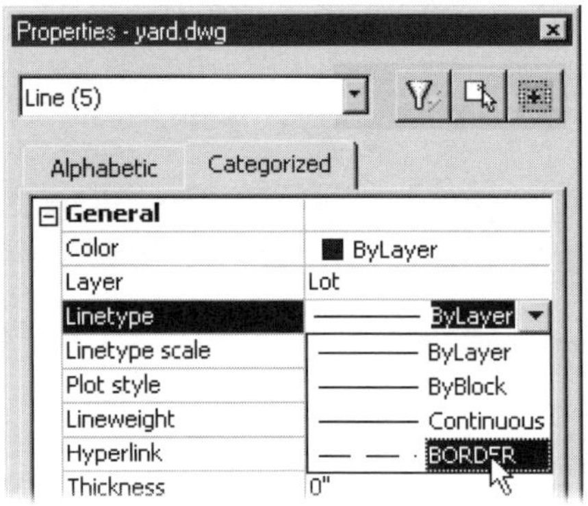

9. Select the Border linetype by clicking its name.
10. Looking over at the drawing, you cannot tell that anything has changed. Press **Esc** twice to remove the highlighting. Still no change; the lines still look continuous. In the next section, you find out how to fix this problem.

Changing the Linetype Scale

■ Tools | Quick Select

The reason the Border linetype looks continuous is that linetypes are sensitive to scale. Until you change it, all linetypes start with a scale factor of 1.0, which is too small to display the pattern of dots and dashes. What we need is a way to change the linetype scale factor.

> **Warning:** Setting linetype scale is one of the trickiest aspects of AutoCAD. Linetypes look continuous when the scale is too high and too low.

One method of setting the scale is to use the older LtScale command. Another method is to reuse the Properties window and its Linetype scale option; this'll give me a chance to show you another way to select objects, called *filtering*.

In an automobile engine, the oil filter removes unwanted grime from the oil. In a similar way, AutoCAD's filter removes unwanted objects from the selection set. To see the filter in action, we'll use the QSelect command (short for "quick select").

1. You can access quick select from the Properties window. Click the **Quick Select** button, as shown by the illustration. Notice the Quick Select dialog box.

Making a quick selection of objects.

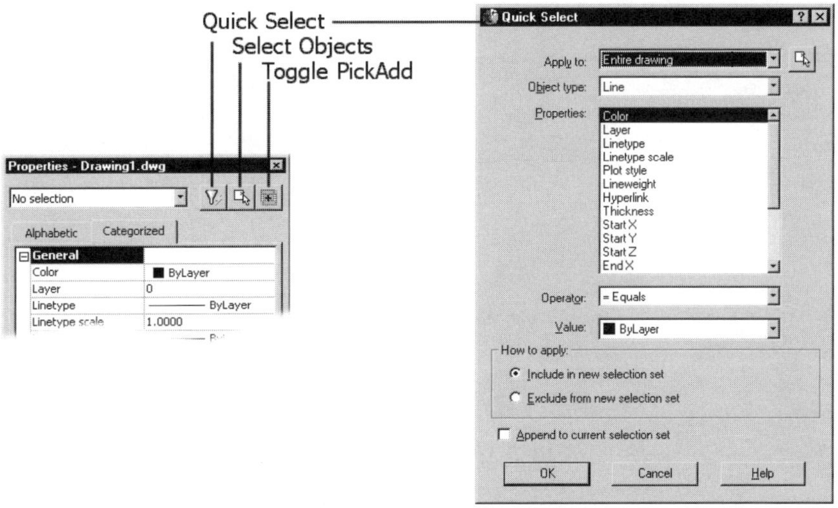

2. In the Quick Select dialog box, make the following choices; you are limited to a single choice in each section:

   - Apply to: **Entire drawing**; the choices here are typically the entire drawing or the selection set.
   - Object type: **Line**; the choices here are Multiple (meaning more than one type of object) or a specific object.
   - Properties: **Layer**; the choices here are any property valid for the selected object. If object type is Multiple, then properties in common are listed.
   - Operator: **= (Equals)**; the choices here are equals, not equal, greater than, and less than.
   - Value: **Lot**; the choices here are whatever values are valid for the selected property.

   You have selected all lines on layer Lot.

3. Click **OK**. Notice that AutoCAD highlights the five blue lot lines.
4. In the Properties window, click the text entry field next to **Linetype scale.**
5. Change 1.0 to **10**, and press **Enter.**
6. To see the linetypes more clearly, press **Esc** twice.

The lot lines have the Dashed linetype.

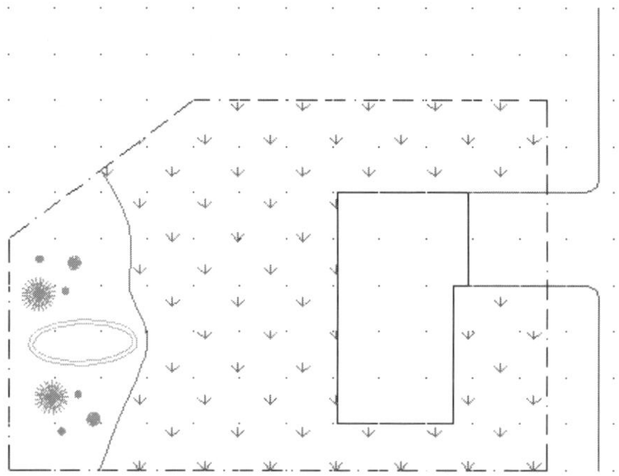

> **Tip:** As of 2000i, the Properties window sports two additional buttons:
>
> Select Objects. Allows you to select objects just like the Select command, or the Select objects prompt of any editing command.
>
> Toggle PickAdd. Toggles the value of the PickAdd system variable. When on (the default), each object you select is added to the selection set. When off (the icon shows a 1), you have to hold down the Shift key to add objects to the selection set, which mimics the Windows selection style.

Changing Line Length

■ Modify | Lengthen

As an example of how the Lengthen command changes the length of an object, extend the edge of the street to the bottom of the screen. You may want to zoom into the area around the pond, first.

1. Select **Modify | Lengthen** from the menu bar.
2. Pick the lower road line, as follows:

   Command: _lengthen
   Select an object or [DElta/Percent/Total/DYnamic]: [pick line]
   Current length: 37'-0"

3. AutoCAD reports that the line is currently 37 feet long. Select the **DYnamic** option:

   Select an object or [DElta/Percent/Total/DYnamic]: dy

4. When prompted, pick the line again:

   Specify new end point.
   Select an object to change or [Undo]: [pick line]

5. As you move the cursor up and down, AutoCAD lengthens and shortens the line. Move the cursor to the bottom of the screen, and click.

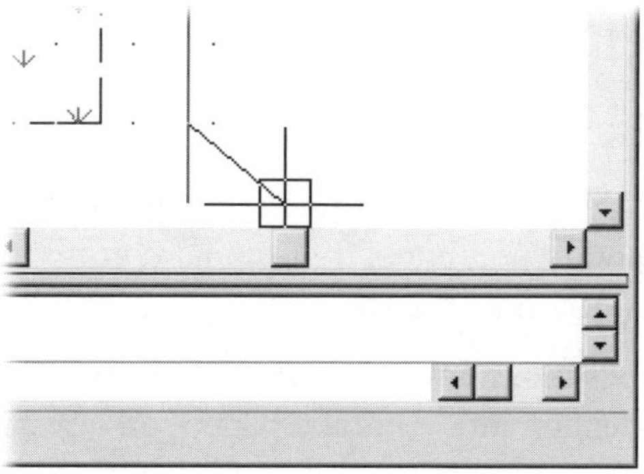

Dragging the line longer, via the Lengthen command.

6. Press **Enter** to end the command.

   Select an object to change or [Undo]: **[Enter]**

Note: The Lengthen command provides you with the following options for changing the length (longer or shorter) of an *open* object, such as a line, arc, or polyline:

Angle: Changes the arc's angle (available only for arcs); default = 0 degrees.

DElta: Changes the length to an absolute amount (*delta* is the Greek word used by mathematicians to indicate change) where you show the delta by picking a point the required distance from the endpoint; default = 0 units.

DYnamic: Changes the length by interactive dragging.

Percent: Changes the length relative to 100%; less than 100%, such as 50%, shortens the object; more than 100%, such as 200%, lengthens the object; default = 100%.

Total: Changes the length to an absolute amount; default = 1 unit.

Undo: Undoes the last change.

There are other commands that allow you to change the length of lines:

- Change: changes the length of a group of lines to a common endpoint.
- Extend: extends a line to a boundary object.
- Trim: cuts back a line to a cutting edge object.

Changing the Look of the Pond

■ Modify | Stretch

So far, you have used several editing commands to change objects. You've used PEdit to modify polylines, non-modal editing to stretch the sketch line, and the Lengthen command to alter the length of lines.

One of AutoCAD's most powerful editing commands is called Stretch. The Stretch command lets you take part of an object and stretch it wider or thinner. Here you apply it to the pond to change its shape. You may find it helpful to zoom into the area around the pond.

1. Select **Modify | Stretch** from the menu bar.

   ```
   Command: _stretch
   Select objects to stretch by crossing-window or crossing-polygon...
   ```

2. Select objects with a crossing-window selection mode, as follows:

   ```
   Select objects: _c
   Specify first corner: [pick]
   Specify other corner: [pick]
   ```

   C is short for "crossing," an object selection mode similar to the window mode you used earlier with the Zoom command. In this case, AutoCAD selects all objects within the selection rectangle and all objects crossing or touching the rectangle (see following figure). You may find crossing mode somewhat faster than window mode since you don't have to draw as large a rectangle.

Pick two points to specify the crossing selection mode.

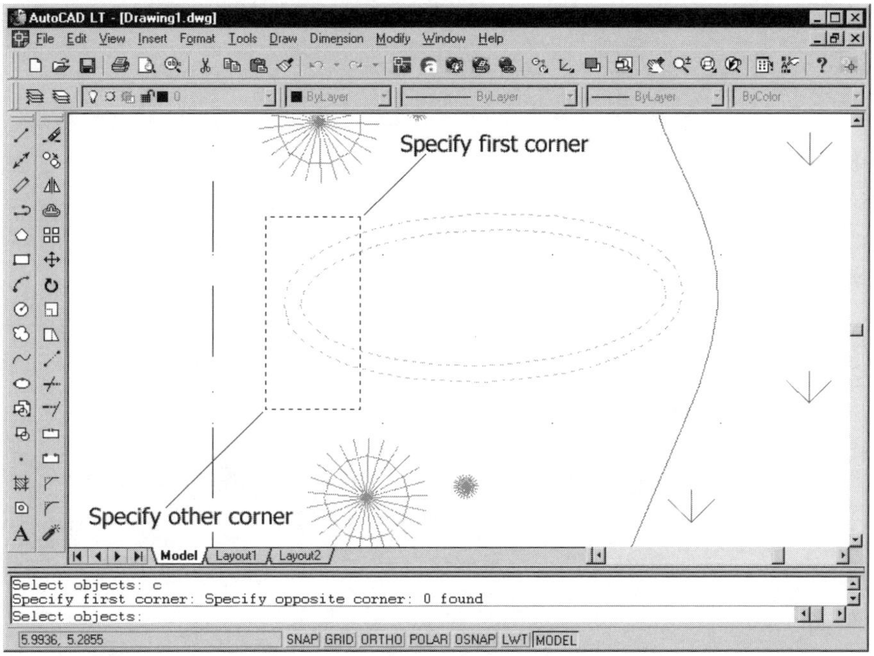

3. After picking two corners of a rectangle that covers part of the pond, press **Enter**:

   2 found Select objects: [Enter]

   If the pond were entirely inside the object selection rectangle, the Stretch command would only move the pond, not stretch it.

   If you accidentally select objects other than the pond, type the **R** option (short for "remove from selection set") and select the objects to remove.

   You may find it helpful to zoom in for a closer look.

4. To tell AutoCAD how much you want the pond stretched, pick two points that indicate the distance:

   Base point: [pick near pond]
   New point: [pick away from pond]

Stretching the pond longer.

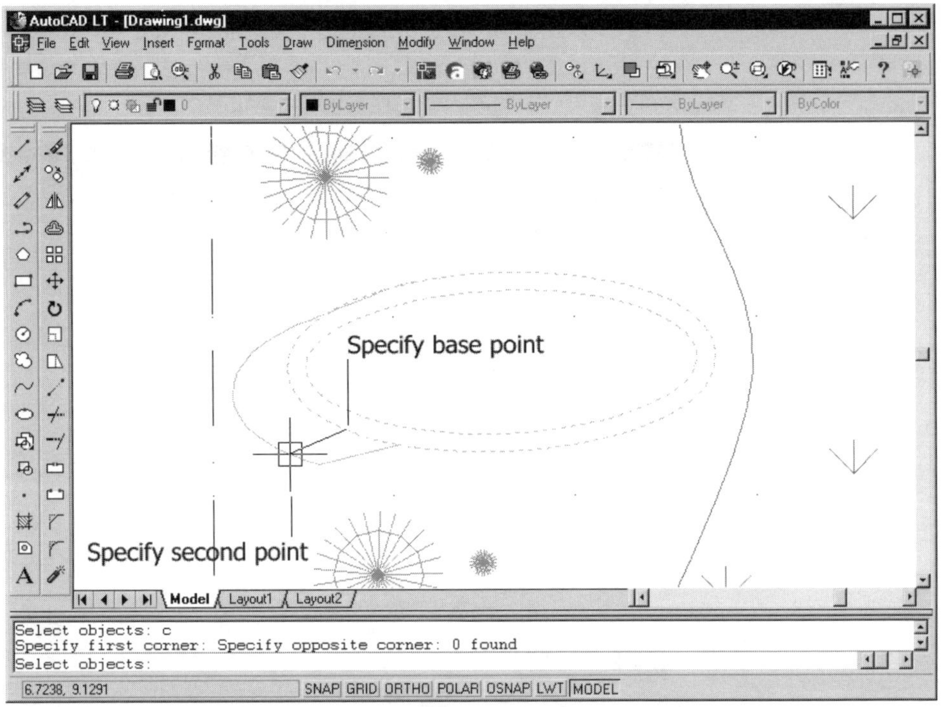

5. You have now created a whole new look to your pond! If you don't like it, you undo the stretch with the U command, as follows:

   Command: u
   U: STRETCH

   . . . and try stretching the pond again.

Note: AutoCAD has 17 options for you to use during object selection, as shown in the following table. I find that I use only a few of the options, such as pick, W, C, L, P and Enter.

Mode	Abbreviation	Meaning
Object	[pick]	Selects a single object.
Window	W	Selects all objects within a rectangular window.
Window Polygon	WP	Selects all objects within a polygonal window.
Crossing	C	Selects object crossing and within a rectangular window.

Mode	Abbreviation	Meaning
Crossing Polygon	CP	Selects all objects crossing and within a polygon.
Fence	F	Selects objects along a fence polyline.
Box	B	Switches to W or C mode, depending on cursor movement.
Automatic	AU	Allows a single pick, or switches to W or C mode, depending on the pick point and cursor movement.
Single	SI	Selects first object encountered.
Last	L	Selects the most recently created object visible on the screen.
Previous	P	Selects most recently selected object.
Multiple	M	Delays database scanning.
Undo	U	Removes most recent selection group.
Remove	R	Enters remove-objects mode.
Add	A	Enters add-objects mode.
End	Enter	Ends object selection.
Cancel	Esc	Cancels object selection.

6. If the pond isn't exactly where you can want it, you can relocate it with the Move command. Move the pond, as follows:

   Command: m

   Here you used another of AutoCAD's shortcuts. M is the abbreviation for the Move command.

7. Continue the Move command, as follows:

   MOVE Select objects: c
   Specify first corner: [pick]
   Specify opposite corner: [pick]
   2 found Select objects: [Enter]
   Specify base point or displacement: [pick edge of pond]
   Specify second point of displacement or <use first point as displacement>:[pick new location]

8. Clean up the screen with the Redraw command, as follows:

   Command: r
   REDRAW

> **Note:** AutoCAD lets you specify a command by typing just one or two letters at the Command: prompt. Shortcut keystrokes are listed at the start of each chapter. The complete list of command name abbreviations (called *aliases*) is stored in the file AcLt.Pgp, (found in the \AutoCAD LT 2000i\Support folder) and is listed in Appendix C, "AutoCAD LT 2000i Commands."

Adding the Fence

- Tracking
- List
- Dist

You've decided to add a fence to the backyard. You can use AutoCAD to help you plan the materials you need. After you draw the fence as a polyline, you find out from AutoCAD how long that fence is.

1. First, switch the working layer to **House** by selecting its name from the toolbar.

   Use the Zoom Previous command to see the full drawing.

2. If you turned off the INTersection object snap, turn it back on. Right-click **OSNAP** on the status bar, then select **Settings** from the shortcut menu. In the Drafting Settings dialog box's Object Snap tab, click **Clear All**, then select **Intersection.** Click **OK**.

The Object Snap tab of the Drafting Settings dialog box.

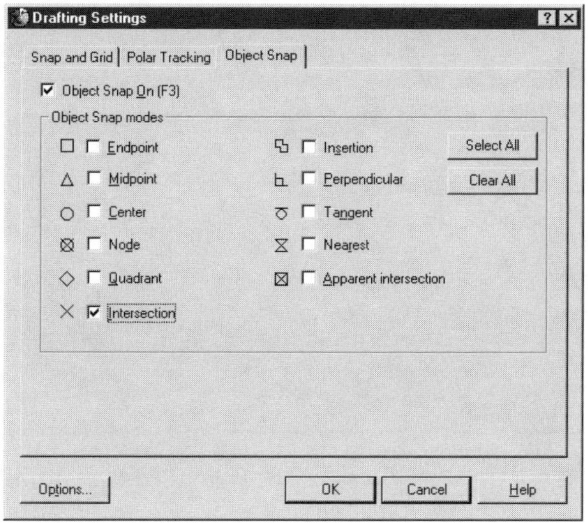

3. The fence is drawn as a 4"-wide polyline. Along the way, you'll use a variety of object snap modes and tracking mode. Begin the PLine command by selecting **Draw | Polyline**.
4. Start the polyline at the middle of the upper house line (1 in the figure). Use the MIDdle object snap override to precisely locate the polyline's starting point, as follows:

```
Command: _pline
Specify start point: mid
of [pick upper house line]
Current line-width is 0'-0"
```

5. To change the width of the polyline from zero to four inches, use the w option, as follows:

```
Specify next point or [Arc/Close/Halfwidth/Length/Undo/Width]: w
Specify starting width <0'-0">: 4"
Specify ending width <0'-4">: [Enter]
```

(You can specify a different starting and ending width to produce tapered polylines.)

6. Now that the starting point and width are set, continue drawing the fence. Follow the path shown by the circled numbers in the following figure.

The fence line is shown by the 4″-wide polyline.

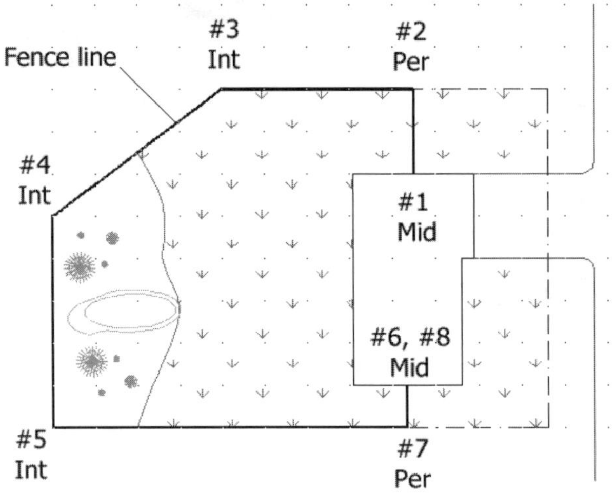

Specify next point...: per
to [pick upper lot line, #2]
Specify next point...: [pick upper right of diagonal, #3]
Specify next point...: [pick lower left of diagonal. #4]
Specify next point...: [pick lower left corner of lot, #5]
Pause the picking action here.

7. When you get to the bottom of the house, you get into a bit of tricky geometry. You want the fence to end at the same relative location as its starting point. You're not sure of the x-coordinate, which is located somewhere along the bottom line of the lot. Fortunately, you can find that point with *tracking*.

   Enter tk (the abbreviation for tracking), as follows:
   Specify next point...: .tk
   First tracking point: per
   of [pick lower house line, #6]

   Tracking lets you position the next endpoint of a line. When in tracking mode, you move the cursor (to indicate direction), then pick a point (to specify a distance). AutoCAD constrains cursor movement by turning on ortho mode automatically. This means you move the cursor in the vertical or horizontal direction only.

8. Indicate the other tentative point, then press **Enter** to exit tracking mode:

   Next point (Press ENTER to end tracking): per
   to [pick lot line below house, #7]

Next point (Press ENTER to end tracking): [Enter]

9. Complete the fence by drawing the last polyline segment PERpendicular to the lower house line (#7), and exit the PLine command, as follows:

   Specify next point...: per
   to [pick lower house line, #8]
   Specify next point...:: [Enter]

10. Now that you've drawn the fence, you can use the List command to tell you the length. Select the **List** icon from the toolbar:

    Command: _list
    Select object: [pick fence polyline]
    1 selected 1 found
    Select objects: [Enter]

The fence is approximately 258 feet long.

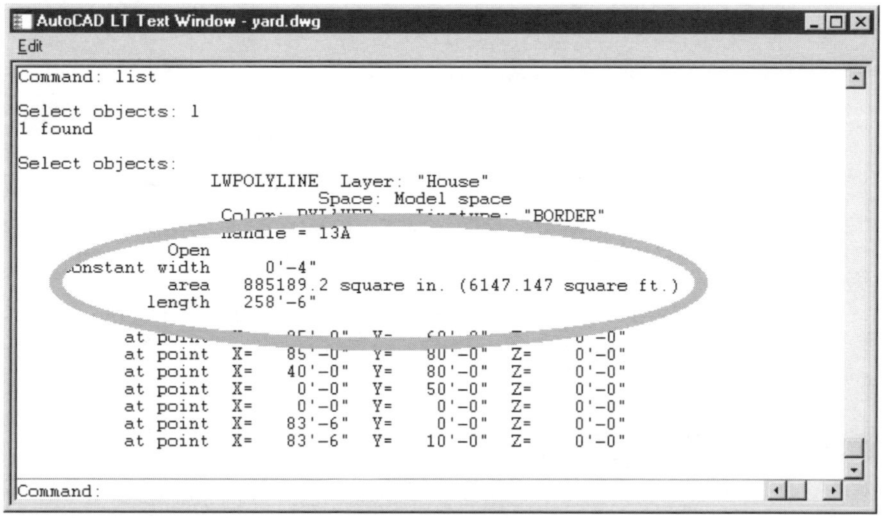

AutoCAD flips to the Text window and lists line after line of information. Whenever the listing pauses, press **Enter** until the Command: prompt reappears.

The List command tells you every piece of information about the polyline (technically called a *lwpolyline*) that AutoCAD has stored in its database. Most of the information is about the vertices ("at point"). The total length of the polyline is shown at the end of the listing: roughly 258 feet. You now know how much fencing you need.

Press function key **F2** to flip back to the graphics window.

11. You measure distances directly on the drawing with the Dist command (short for *distance*). Find the shortest distance from the house to the pond with the Dist command. From the menu bar, select **Tools | Inquiry | Distance**:

    Command: _dist
    Specify first point: nea
    to [pick inside edge of pond]
    Specify second point: per
    to [pick house wall]
    Distance = 41'-3", Angle in XY Plane = 0, Angle from XY Plane = 0
    Delta X = 41'-3", Delta Y = 0'-0", Delta Z = 0'-0"

The pond is approximately 41 feet from the house.

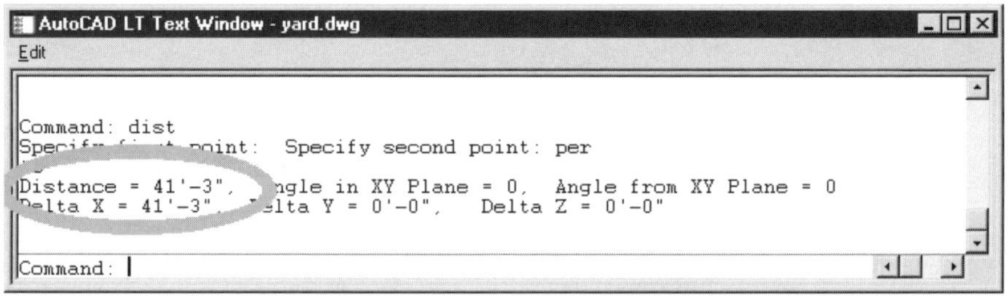

The beeline distance from house to pond is just over 41 feet. The value on your drawing may differ, depending on where you located the pond.

Remember to save the work you have done on the drawing. You may also want to plot out the drawing.

Review Questions

1. Once drawn, the linetype of a line cannot be changed.

   True / False

2. To make a drawing relatively small in the AutoCAD window, use the Zoom command with the following zoom factor:

   a. 1.2x
   b. 0.8x
   c. 8x
   d. 1

3. Match the command name with its function:

a.	Dist	i.	Locates an x or y coordinate relative to the previous point.
b.	List	ii.	Measures the 3D distance and angle between two points.
c.	Lengthen	iii.	Lists object properties in the Text window.
d.	Stretch	iv.	Makes open objects longer and shorter.
e.	Tracking	v.	Moves and stretches objects.

4. The first time you use a linetype pattern in an AutoCAD drawing, you must:

   a. Select the linetype from the Object Properties toolbar.
   b. Define the linetype using an ASCII text editor.
   c. Load the linetype pattern from Aclt.Lin file into the drawing.
   d. Nothing; AutoCAD LT does not support linetypes.

5. When an object is assigned the Bylayer linetype, the object:

   a. Takes on the Continuous linetype.
   b. Is displayed only when its layer is frozen.
   c. Takes on the linetype assigned to its layer.
   d. No longer appears in the drawing.

6. A linetype name that ends with "X2" is scaled half-size.

   True / False

7. The Properties window changes the properties of only one object at a time.

   True / False

8. When the Properties window shows a value in gray, this means:

   a. The value is the default value.
   b. The value has been changed from its default.
   c. The value can be changed.
   d. The value cannot be changed.

9. The QSelect command allows you to select objects based on their properties.

   True / False

10. When you cannot see the linetype pattern, the problem is:
    a. The linetype scale is too large.
    b. The linetype scale is too small.
    c. Both of the above.
    d. Neither of the above.

Chapter 6

Adding Notes and Dimensions

In the last chapter, you changed parts of the yard and learned how to get some information out of the drawing. In this chapter, you add the finishing touches by drawing callouts on the drawing, and dimensioning the yard.

Key Concepts

- Placing text in the drawing
- Defining styles and justification for text
- Changing text and placing rotated text
- Turning text into quick text
- Attaching a variety of dimensions to objects in the drawing

Key Terms

Associative—an object that updates automatically when an associated object is updated; in AutoCAD LT, applies to hatch patterns and dimensions only

Callouts—text that describes parts of the drawing

Dimension—indicates a measured distance; usually consists of a line, a pair of arrowheads, and text

Justification—determines the relative positioning of text, such as right, left, and centered

Layout—shows the drawing as it will appear plotted; much like an editable print preview

Text Boundary—an invisible rectangle that determines the word wrap of paragraph text

Abbreviations	
Dim	Dimension
Dimvars	Dimensioning variables

Command	Shortcut	Menu Selection
DdEdit	ed	Modify | Object | Text
MText	t	Draw | Multiline Text
QText	qt	...
Rotate	ro	Modify | Rotate
Style	st	Format | Text Style
Text	dt	Draw | Single Line Text

Dimensioning Commands		
DimAligned	dal	Dimension | Aligned
DimBaseline	dba	Dimension | Baseline
DimContinue	dco	Dimension | Continue
DimLinear	dli	Dimension | Linear
DimRadius	dra	Dimension | Radius

Before starting the exercises in this chapter:

- In AutoCAD LT, open the Yard.dwg file.
- Do a Zoom All to make the full drawing visible.
- Create a new working layer called Text.
- Assign the color White to the layer.
- Make the layer current by clicking the Current button.

Adding Notes to the Drawing

- Draw | Text | Multiline Text

With the yard plan largely complete, you now add callouts (text) to describe the different parts of the yard.

1. In AutoCAD LT, you add callouts with the MText command (short for "multiline text"). Select **Draw | Multiline Text** from the menu bar.

   Command: _mtext
   Current text style: "Standard" Text height: 1/4"

   AutoCAD displays the status of MText, i.e., style and height. The current style is called "Standard." A *text style* defines the properties of text, such as font, size, and orientation—just like a word processing style. You learn more about text styles later in this chapter.

   The height of the text should be large enough to be legible when you plot the drawing. For a drawing of 130 feet to fit a sheet of paper 11 inches wide, text at 1/4" height would appear about 0.0002" tall. Clearly, that's not going to work. The text needs to be taller; we'll work that out later.

2. Pick a starting point for the note within the house outline.

   Specify first corner: [pick]

3. AutoCAD displays a long prompt. Type **H** to specify a different height of the text:

   Specify opposite corner or [Height/Justify/Line spacing/Rotation/Style/Width]: H
   Specify height <1/4">:

   Here we have to do some multiplication and division to work out the height of the text. We want the text to be legible when printed, which means it must be *scaled up* (made larger). To figure out how much larger, we work with these facts:

   - The drawing is 130 feet long, and will be scaled down to fit the 11"-wide paper.
   - We want the text to be legible when the drawing is printed, which means it should be printed about 3/8" tall.

Here's how to calculate the size:

Step 1. Set up the distances to determine the scale factor:

```
Distance in drawing        130ft.
-------------------   =    ------
Distance on paper          11in.
```

Step 2. Convert the feet to inches to make the units consistent:

```
130ft x 12in/ft        1,560in.
---------------   =    --------- = 142
11in.                  11in.
```

The scale factor is 142. That means the text in the drawing should be 142 times larger to plot big enough to be legible on the page.

Step 3. Multiple the printed text size by the scale factor:

3/8″ x 142 = 53″

We need to specify a text height of 53".

Specify height <1/4″>: 53″

4. Pick a second corner for the text boundary:

   Specify opposite corner or [Height/Justify/Line spacing/Rotation/Style/Width]: [pick]

   The *text boundary* is an invisible rectangle; it is neither displayed nor printed. AutoCAD uses the top of the boundary to determine where the text starts. The two sides of the boundary determine where to wrap the text. The bottom of the boundary is ignored; text will flow down as far as necessary. This is very similar to using a frame in desktop publishing software.

Picking the corners of the text boundary.

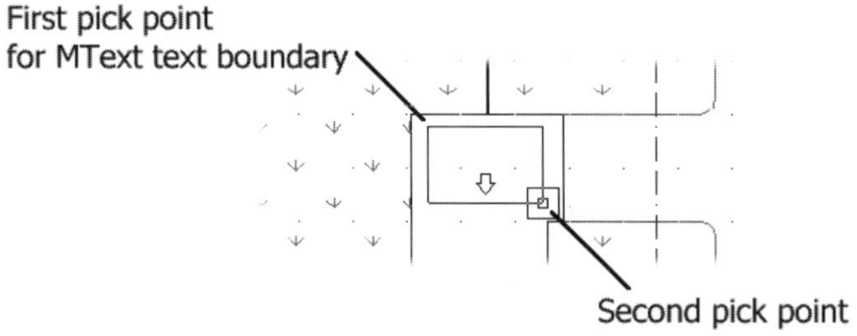

Tip: To change the size of the text boundary, select it. Four handles appear, one at each corner of the boundary. You can drag the boundary wider and narrower to change the word wrap.

Dragging the text boundary to change the word wrap.

5. After picking the second point, AutoCAD displays the Multiline Text Editor dialog box, where you enter the text. Although the dialog box has many options, for now just type **House**.

The Multiline Text Editor dialog box.

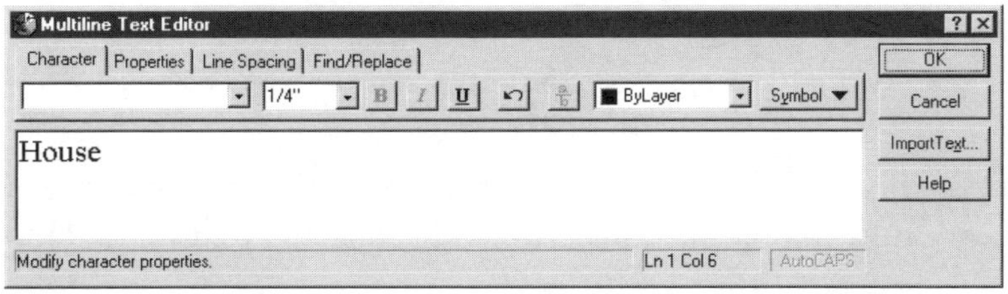

Note: The Multiline Text Editor has an Import Text button. If you want to place text in the drawing from an external file, created by a word processor or spreadsheet, click the Import button to load the file. The file must be in ASCII format and must be 16KB or smaller.

6. Click **OK**. AutoCAD places the word "House" in the drawing.
7. Zoom in on the word "House" to get a closer look, as follows:

```
Command: zoom
All/.../<Scale(X/XP)>: [pick]
Other corner: [pick]
```

The text looks kind of ugly—like a computer created it! This is the basic font, called TXT, found in every new AutoCAD drawing. Fortunately, AutoCAD comes with a large number of TrueType fonts that look better than this one. In the next section, you change the font.

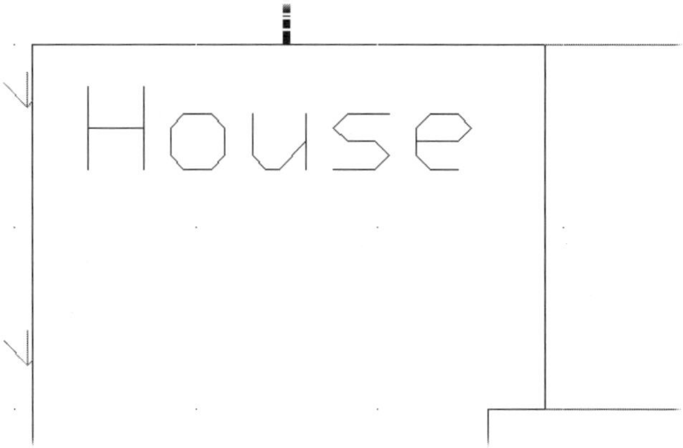

Creating a Text Style

■ Format | Text Style

Just like you have to load hatch patterns and linetypes into an AutoCAD drawing file, you have to load additional text fonts. Here's how:

1. Select **Format | Text Style** from the menu bar. The Text Style dialog box appears on the screen. AutoCAD LT comes with many different fonts and can read any TrueType (TTF) font on your computer system.

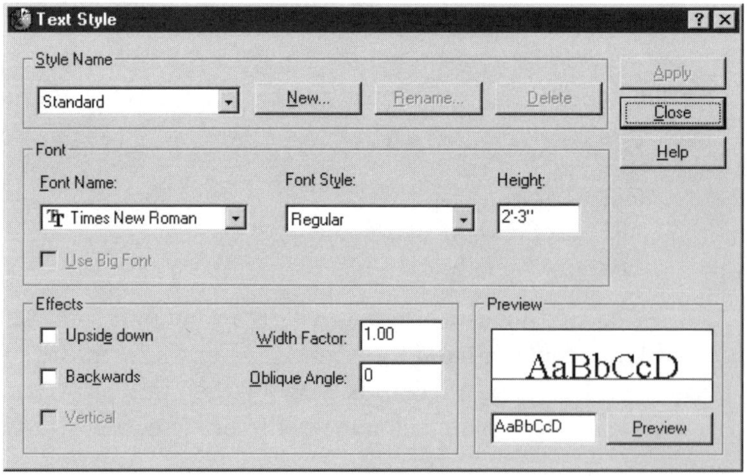

The Text Style dialog box.

2. Make the following changes to the dialog box:
   - Font Name: **Times New Roman**
   - Height: **2' 3"**

   Ignore the other settings, since you don't want the text to print backwards, upside-down, or vertically. From now on, all text you place will have use the Times New Roman font, 2'3" tall—text that is half as tall as before.

3. Click **Apply**, then **Close**. Notice that the word "House" is updated automatically for the new font.

The text with the new font.

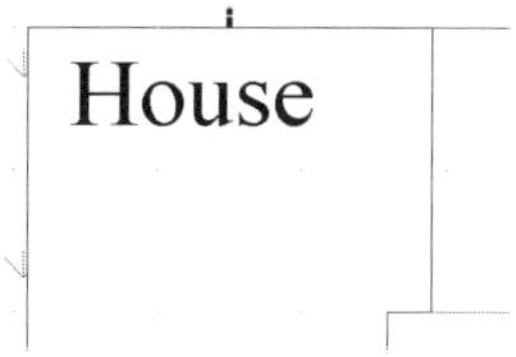

Text Justification Modes

■ Draw | Single Line Text

AutoCAD has a second command for placing text. The Text command places a single line of text at a time. In some ways, it is easier to use than the MText command. To try out the Text command, first zoom in to the pond.

1. Now that you see the pond area more clearly, start the Text command by selecting **Draw | Single Line Text**. This time, use one of AutoCAD LT's 15 justification modes, as follows:

```
Command: _text
Current text style: "Standard" Text height: 2'-3"
Specify start point of text or [Justify/Style]: j
Enter an option [Align/Fit/Center/Mid-
dle/Right/TL/TC/TR/ML/MC/MR/BL/BC/BR]: a
Specify first endpoint of text baseline: [pick one end of the pond]
Specify second endpoint of text baseline: [pick other end of pond]
```

   Here you selected the Align justification, which draws text fitted between two points. All of AutoCAD LT's text justification modes are listed in the following table:

Justification	Meaning
Start point	Baseline left
Align	Fitted between two points
Fit	Fitted with constant text height
Center	Baseline center
Middle	Exact center of text
Right	Baseline right
TL	Top left
TC	Top center
TR	Top right
ML	Middle left
MC	Middle center (equivalent to Middle)
MR	Middle right
BL	Baseline left (equivalent to Start point)
BC	Baseline center (equivalent to Center)
BR	Baseline right (equivalent to Right)

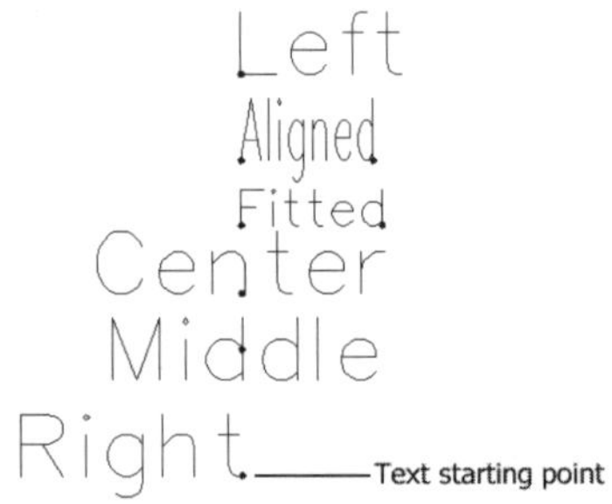

AutoCAD's most commonly used text justification modes.

2. Since the Align option's pick points define the width and angle of the text, AutoCAD doesn't ask for the height or rotation angle. Instead, the Text command goes straight to the prompt:

   Text: Pool

   Notice that the word "Pool" is drawn with the Times New Roman text font.

3. Press **Enter** to exit the Text command.

   Text: [Enter]

Placing text with the Text command.

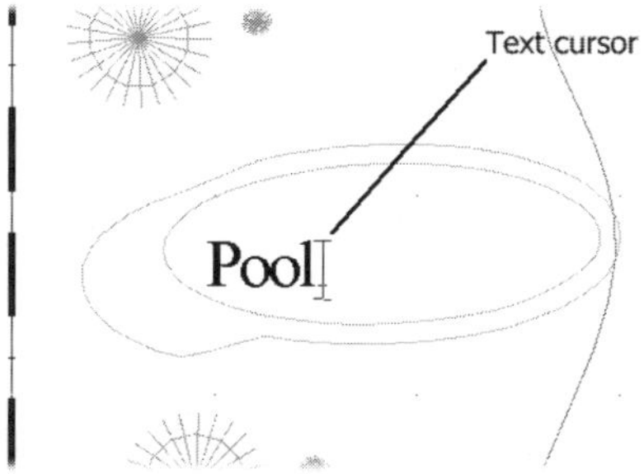

Changing Existing Text

■ Modify | Object | Text

Oops! "Pool" should read "Pond." When you need to change the wording of text, you use the DdEdit command.

1. Select **Modify | Object | Text** and follow the prompts:

   Command: _ddedit
   Select an annotation object or [Undo]: [pick Pool word]

   The Edit Text dialog box pops up on the screen.

2. Press the **Backspace** key, and type **Pond**.
3. Click **OK**. AutoCAD instantly replaces "Pool" with "Pond."

Editing text with the Edit Text dialog box.

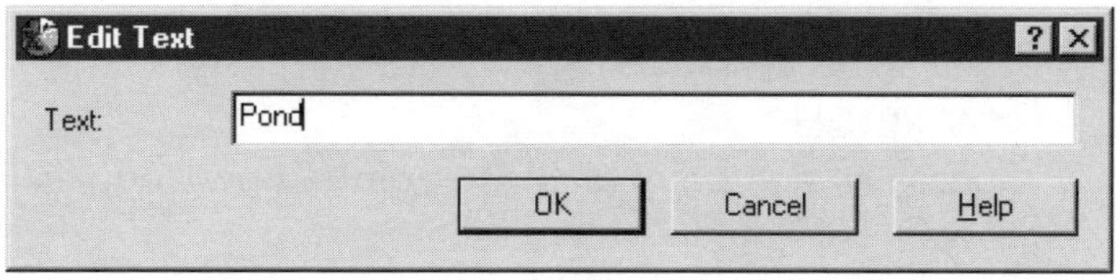

4. Exit the DdEdit command by pressing **Enter** again:

   Select an annotation object or [Undo]: [Enter]

> **Tip:** New in 2000i: double-click the text, and the appropriate text editor is displayed.

Placing Rotated Text

Now add some more text, such as the street name. Here we see how to write text sideways. First, perform a Zoom All to see the entire drawing.

1. Select the **Single Line Text** item from the Draw menu:

   Command: _text
   Current text style: "Standard" Text height: 2'-3"
   Specify start point of text or [Justify/Style]: [pick point on street]
   Specify rotation angle of text : 90
   Enter text: Donlyn Avenue
   Enter text: [Enter]

   By specifying a rotation angle of 90 degrees, AutoCAD draws the text sideways. You can, of course, place text at any angle—from 0 degrees through to 359 degrees.

2. If you picked the wrong rotation angle, you can rotate the text after the fact. Use the Zoom Window command to take a closer look. Pick the text, as follows:

   Command: [pick text]

   Notice that a small blue box, a handle, appears at the text's insertion point.

3. Click the blue box. Notice that it turns solid red.

4. Select the rotate option by pressing the spacebar until ** ROTATE ** shows up in the command prompt area. Then specify an angle of 180 degrees, as follows:

   ** STRETCH **
   Specify stretch point or [Base point/Copy/Undo/eXit]: [spacebar]
   ** MOVE **
   Specify move point or [Base point/Copy/Undo/eXit]:[spacebar]
   ** ROTATE **
   Specify rotation angle or [Base point/Copy/Undo/Reference/eXit]: 180

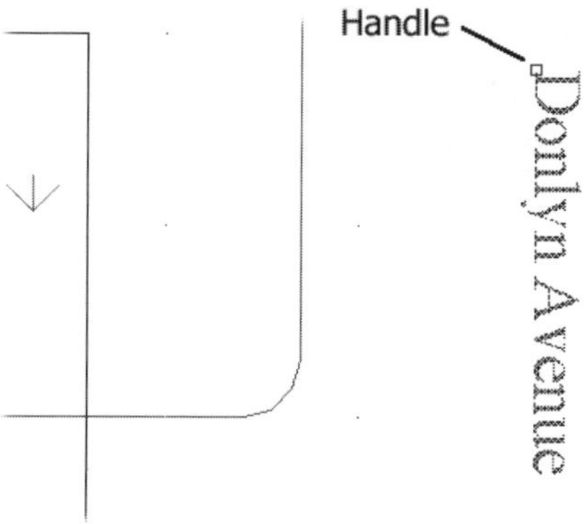

Rotating the text.

5. AutoCAD draws the text flipped over. Press **Esc** twice to exit non-modal editing:

   Command: [Esc]
   Command: [Esc]

A Fast Way to Place Lots of Text

An advantage to Text command is that it lets you place text all over the drawing. First, though, Zoom All to see the entire drawing.

1. Use the Text command to add more callouts to the drawing. Select **Draw | Text | Single Line Text**, as follows:

   Command: _text
   Current text style: "Standard" Text height: 2'-3"
   Specify start point of text or [Justify/Style]: [pick near bottom of drawing]
   Specify rotation angle of text Z: 0
   Text: 34486 Donlyn Avene

   Misspell "Avene" as shown; we correct it later. Notice the little square cursor displayed after each letter is drawn.

2. When you press **Enter** at the end of "Avene," the Text: prompt reappears. Now that you've noticed the spelling mistake in "Avene," you backspace with the Backspace key and type the correction, as follows:

   Text: [Backspace][Backspace]ue

3. Type the next line, as follows:

   Text: Abbotsford BC [Enter]

You don't need to start the Text command again; Text automatically jumps to the next line below.

4. Now move the cursor near to a clump of trees and press the pick button. Text is ready for you to type more words:

   Text: Birch [Enter]
   Text: trees [Enter]
   Text: [move cursor to other trees] Western [Enter]
   Text: Red Cedar [Enter]
   Text: [Enter]

Placing text all over the drawing with the Text command.

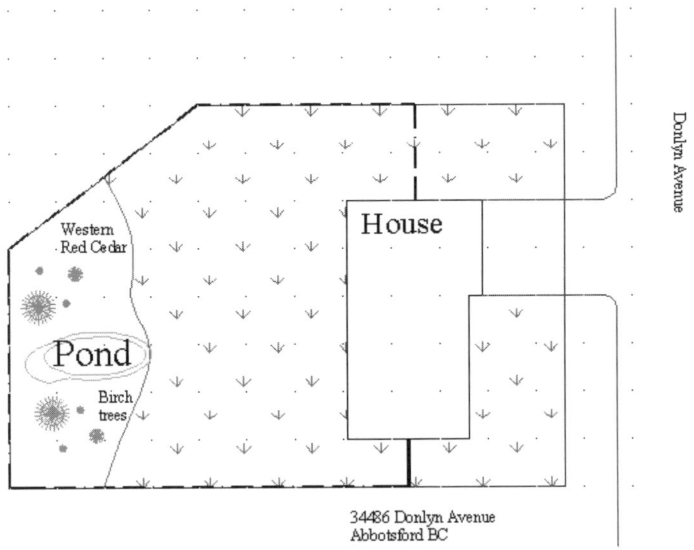

5. Press the **Enter** key one more time to end the Text command. When you do, AutoCAD rewrites all the text as left justified.

   At this point, it is a good idea to save your work using the Save command.

Reducing Text Display Time

■ QText

A lot of text in a drawing slows the display speed. AutoCAD has a special command, called QText (short for "Quick Text"), that changes text into rectangular outlines.

1. Change the text into outlines with the QText command:

   Command: qtext
   ON/OFF : on

   The callouts look no different! AutoCAD doesn't change the text until the next regeneration.

2. Force a screen regeneration with the Regen command. Select **View | Regen** and the text turns into rectangle.

QText displays text as rectangles for faster drawing regeneration.

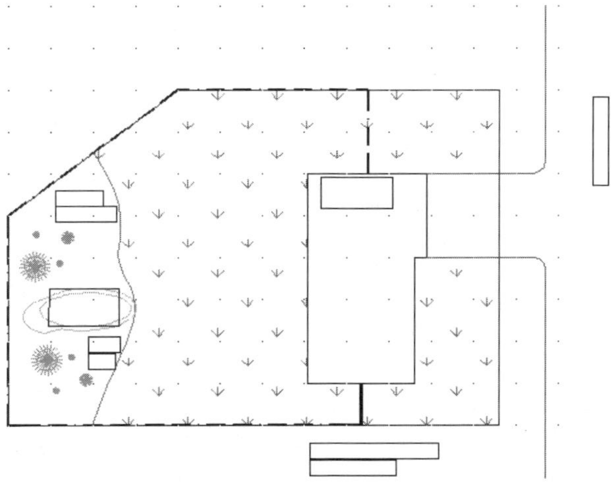

5. Change the outlines back to text, this time typing the QText command followed by the Regen command, as follows:

   Command: qtext
   Enter mode [ON/OFF] : off
   Command: regen
   Regenerating drawing.

Adding Dimensions

With several callouts placed on the drawing, you may want to dimension the lot. This is done with commands that start with Dim (short for "dimensioning").

The bad news is that there are dozens and dozens of settings for dimensions, which adhere to a variety of international standards. The good news is that AutoCAD comes preconfigured with most set to reasonable values. These settings are called *dimensioning variables*, or "dimvars" for short. (If you ever want to change dimvars, select **Dimension | Style** from the menu bar. When AutoCAD displays the Dimension Style Manager dialog box, click **Modify**.)

The Dimension Style Manager dialog box.

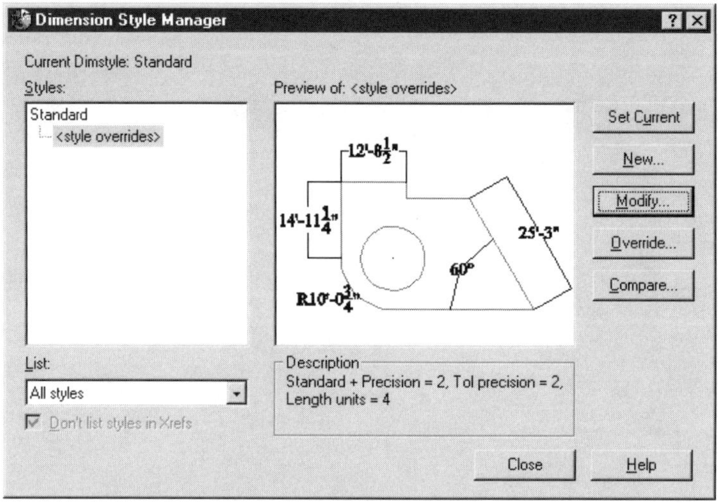

1. Like text, hatch patterns, and linetypes, the scale of a dimension is relative to the printed size. Before you start placing dimensions, you must set the dimension scale with the DimScale command. Use the same scale factor you calculated for text: 142.

   ```
   Command: dimscale
   Enter new value for DIMSCALE <0'-1">: 142
   ```

2. Since dimensioning often takes place at intersections, turn on INTersection object snap mode, as follows:

   ```
   Command: osnap
   Object snap modes: int
   ```

3. Ensure the linetype is set to Continuous. The dimensions will be placed on the Text layer, which should be current from the text exercises.

   Let's begin dimensioning!

Dimensioning the Yard

- DimLinear
- DimContinue

1. To begin dimensioning the drawing, select **Dimension | Linear** from the menu bar. The DimLinear command draws horizontal, vertical, and rotated linear dimensions—depending on how you move the cursor:
   - Select two points roughly horizontal, and AutoCAD automatically draws a horizontal dimension.
   - Select two points roughly vertical, and AutoCAD automatically draws a vertical dimension.
   - To create a rotated dimension, specify "R" at the prompt.

   Command: _dimlinear

2. Dimension the lower lot line, as follows:

Drawing a horizontal dimension.

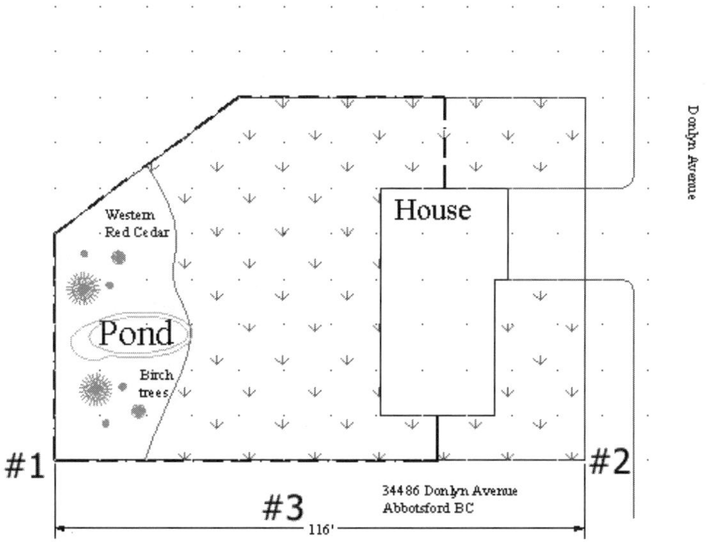

Specify first extension line origin or <select object>: [pick intersection near #1]
Specify second extension line origin: [pick #2]
Specify dimension line location or

[Mtext/Text/Angle/Horizontal/Vertical/Rotated] [pick below the lot line at #3]
Dimension text = 116'

AutoCAD knows the length of the line, and places it for you.

Note: AutoCAD automatically draws all the components of the dimension: the two extension lines, the dimension line, the arrowheads, and the dimension text.

5. Try another horizontal dimension. This time, we will use AutoCAD's object dimensioning. In this method, you pick the object (the top lot line), and AutoCAD dimensions it. This method uses just two picks, instead of three. Press the spacebar to repeat the DimLinear command:

   Command: [Spacebar] _dimlinear
   Specify first extension line origin or <select object>: [Enter]
   Select object to dimension: [pick line at #4]
   Specify dimension line location or
   [Mtext/Text/Angle/Horizontal/Vertical/Rotated]: [pick #5]
   Dimension text = 76': [Enter]

Placing a dimension by picking an object.

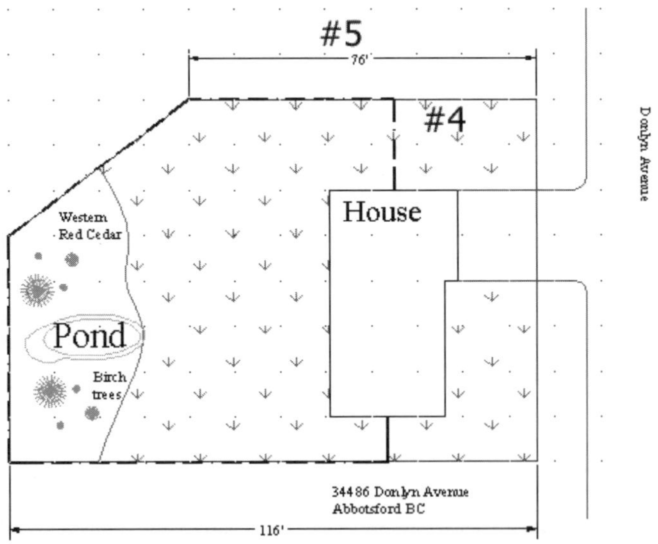

3. Continue the horizontal dimension with the DimContinue command. From the menu, select **Dimension | Continue**.

Command: _dimcontinue
Specify a second extension line origin or [Undo/Select] <Select>: [pick #6]
Dimension text = 40' [Enter]
Specify a second extension line origin or [Undo/Select] <Select>: [Enter]

4. Pick the point at #6 in the illustration. Since AutoCAD knows where the last extension line was, it only needs to know the location of the next extension line to draw in the second dimension.

Placing a dimension by continuing from an existing dimension.

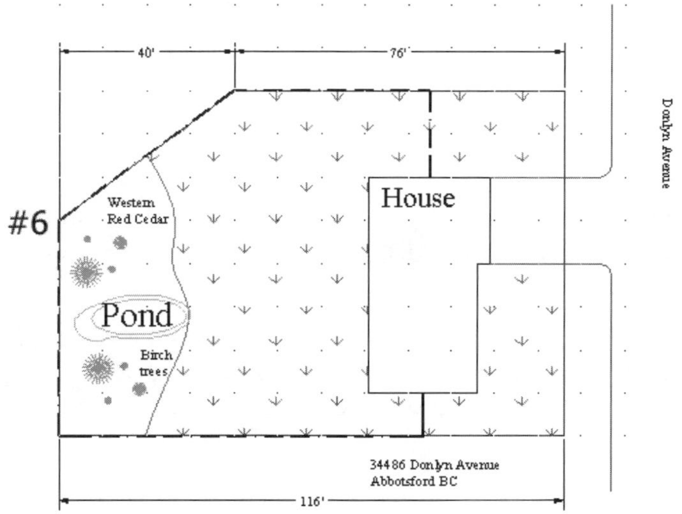

Vertical and Baseline Dimensions

■ DimBaseline

1. To draw a vertical dimension, select **Dimension | Linear**. As noted earlier, the DimLinear command draws vertical dimensions, in addition to horizontal dimensions. Since you will be moving the cursor in a vertical direction, AutoCAD knows from this to draw a vertical dimension.

Specify first extension line origin or <select object>: [pick #7]
Specify second extension line origin: [pick #8]
Specify dimension line location or [Mtext/Text/Angle]: [pick #9]
Dimension text = 20'

Chapter 6

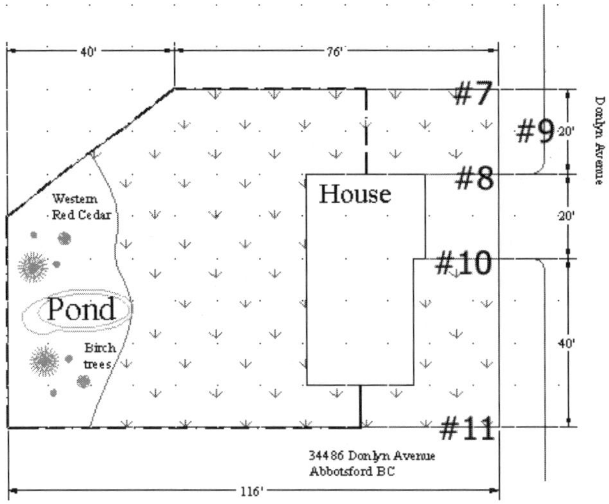

Placing vertical and continued dimensions.

> **Note:** The MText option lets you substitute a paragraph of text for the dimension text; it displays the same Multiline Text Editor dialog box we saw earlier in this chapter.
>
> The Text option lets you modify or replace the dimension text. AutoCAD uses the symbol (double angle brackets) to represent the measured dimension text. You could, for example, enter:
>
> Enter dimension text <20'>: Fence length =
>
> and AutoCAD would place Fence length = 20' as the dimension text.
>
> The Angle option allows you to rotate the dimension text. For example, enter:
>
> Specify angle of dimension text: 45
>
> and AutoCAD would rotate the dimension text by 45 degrees.

2. Use the DimContinue command to continue the vertical dimensions along the right side of the lot at points 3 and 4.

   Command: dimcontinue
   Specify a second extension line origin or [Undo/Select] <Select>: [pick #10]
   Dimension text = 20'
   Specify a second extension line origin or [Undo/Select] <Select>: [pick #11]
   Dimension text = 40'

Specify a second extension line origin or [Undo/Select] <Select>:
[Enter]

3. A variation on the DimContinue command is the DimBaseline command. Rather than continue a dimension from the previous extension line, DimBaseline dimensions from the original extension line. To see how it works, first place a vertical dimension:
   - Place the first extension line at #12.
   - Place the second extension line at #13.
   - Place the dimension line at #14.

Placing vertical and baseline dimensions.

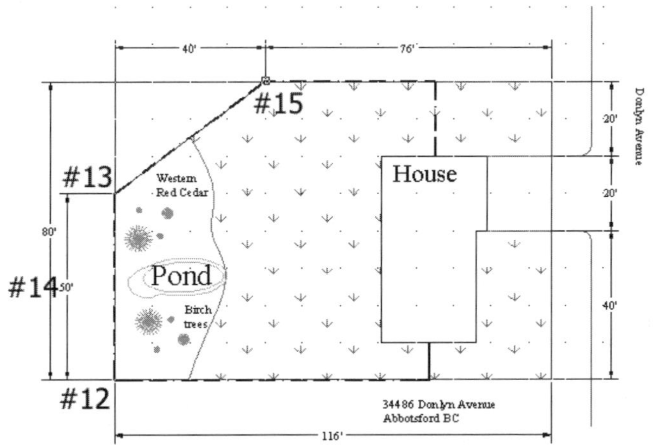

4. Try the DimBaseline command. From the menu, select **Dimension | Baseline.**

   Command: _dimbaseline
   Specify a second extension line origin or [Undo/Select] <Select>:
   [pick #15]
   Dimension text = 80'
   Specify a second extension line origin or [Undo/Select] <Select>:
   [Enter]

   Press **Enter** to exit the command. The DimBaseline and DimContinue commands work with all linear and angular dimensions.

Chapter 6

Aligned and Radial Dimensions

- DimAligned
- DimRadius

So far, you have dimensioned the straight and angled portions of the lot line with horizontal and vertical dimension commands. To dimension an angled line, you use the DimAligned command.

1. From the menu, select **Dimension | Aligned.**

```
Command: _dimaligned
Specify first extension line origin or <select object>: [pick #16]
Specify second extension line origin: [pick #17]
Specify dimension line location or
[Mtext/Text/Angle]: [pick #16]
Dimension text = 50'
```

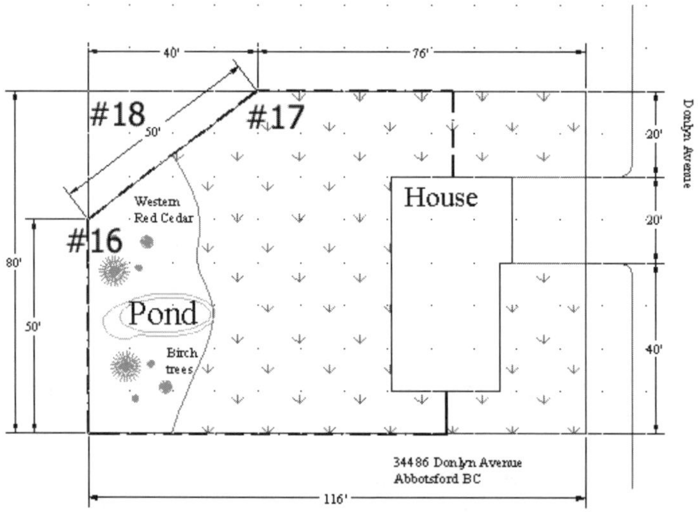

Placing an aligned dimension.

2. The dimensioning commands you have been using present much the same prompts to you. Now try some dimensioning commands that are a bit different. For example, the DimRadius command dimensions an arc or circle that you pick. From the menu, select **Dimension | Radius**, as follows (zoom in for a closer look, if necessary):

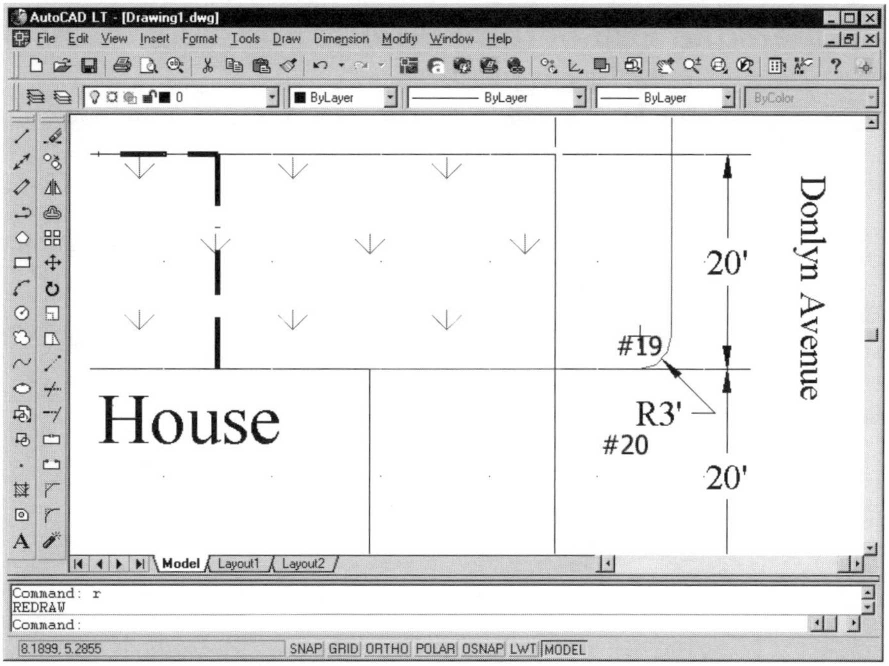

Placing a radial dimension.

Command: _dimradius
Select arc or circle: [pick #19]
Dimension text = 3'
Specify dimension line location or [Mtext/Text/Angle]: [pick #20]

The DimRadius command gives you some flexibility as to where you want to place the dimension text. As you move the cursor, AutoCAD ghosts in the leader and text.

3. Save your work, and print a copy.

Review Questions

1. Match the command name with its best purpose:

a.	Text	i.	Edits the text.
b.	MText	ii.	Places one line of text at a time.
c.	QText	iii.	Specifies the look of the text.
d.	Style	iv.	Changes text to rectangles.
e.	DdEdit	v.	Places a paragraph of text.

2. Which one of the following properties is not defined by a text style?
    a. Font
    b. Color
    c. Size
    d. Orientation

3. The text boundary of the MText command is similar to a frame in desktop publishing.

    True / False

4. AutoCAD LT's default text font is named:
    a. Arial
    b. Txt
    c. New Courier
    d. RomanS

5. Match the abbreviation with the best justification mode:

a.	F	i.	Bottom right
b.	R	ii.	Top left
c.	TL	iii.	Middle center
d.	MC	iv.	Right
e.	BR	v.	Fit

6. Which key do you press to exit the DdEdit command?
    a. Enter
    b. F2
    c. Alt+X
    d. Ctrl+E

7. When the QText command is turned on, AutoCAD displays faster because:
    a. The TXT font is substituted for the text.
    b. All the text is turned off (does not appear).
    c. Any layer with text is frozen.
    d. Rectangles are displayed in place of the text.

8. The DimLinear command draws:
    a. Horizontal dimensions.
    b. Vertical dimensions.
    c. Both of the above.
    d. None of the above.

9. When placing dimensions, to what do the < > characters refer?

   a. The dimension distance, as calculated by AutoCAD.
   b. Placeholder for metric distances.
   c. Nothing. This is a trick question.
   d. Placeholder for plus/minus tolerance.

10. You cannot override the dimensions calculated by AutoCAD.

    True / False

Chapter 7

The AutoCAD DesignCenter

Earlier in this book, you placed blocks with the Insert command, and hatch patterns with the BHatch command. In this chapter, you learn to use the AutoCAD DesignCenter, which is a visual, interactive method for placing blocks and other named objects. (Its predecessor, the Content Explorer for placing blocks and hatches, was introduced with AutoCAD LT 97.)

Key Concepts

- Understanding the AutoCAD DesignCenter
- Learning DesignCenter's user interface

Key Term

Symbol—another name for a block

Abbreviation	
xref	Externally referenced drawing

Command	Shortcut	Menu Selection
AdCenter	Ctrl+2	Tools | AutoCAD DesignCenter

Understanding AutoCAD DesignCenter

Until AutoCAD LT 2000, there was no single way to see named objects in drawings. Instead, you would use individual commands, such as Insert, Linetype, and Layer, to see the names. AutoCAD LT 97 introduced the Content Explorer, which provided access to blocks only. LT 98 added hatch patterns to the capabilities of the Content Explorer.

With LT 2000, Content Explorer was renamed the *AutoCAD DesignCenter*, and provided access to blocks, linetypes, dimension styles, layers, layouts, linetypes, text styles, and xrefs in every drawing—whether open in LT or in any other DWG file located on your computer, the network, and the Internet. In the case of blocks, you see not just the name but also a preview image.

Touring the DesignCenter GUI

Let's take a tour of DesignCenter to see what it looks like and how it works. Ensure AutoCAD LT 2000i is running. Open the AutoCAD DesignCenter by selecting **Tools | AutoCAD DesignCenter** from the menu bar. Or, press **Ctrl+2**. It may take a few seconds for the DesignCenter to regenerate its icons.

To close the DesignCenter, press **Ctrl+2** a second time.

> **Tip:** The DesignCenter is an independent window that you can move out from the AutoCAD LT window. Grab the window by its title bar (the double embossed line), and move it out of the AutoCAD window. As an alternative, double-click the title bar.

To dock it back inside LT, double-click the DesignCenter's title bar.

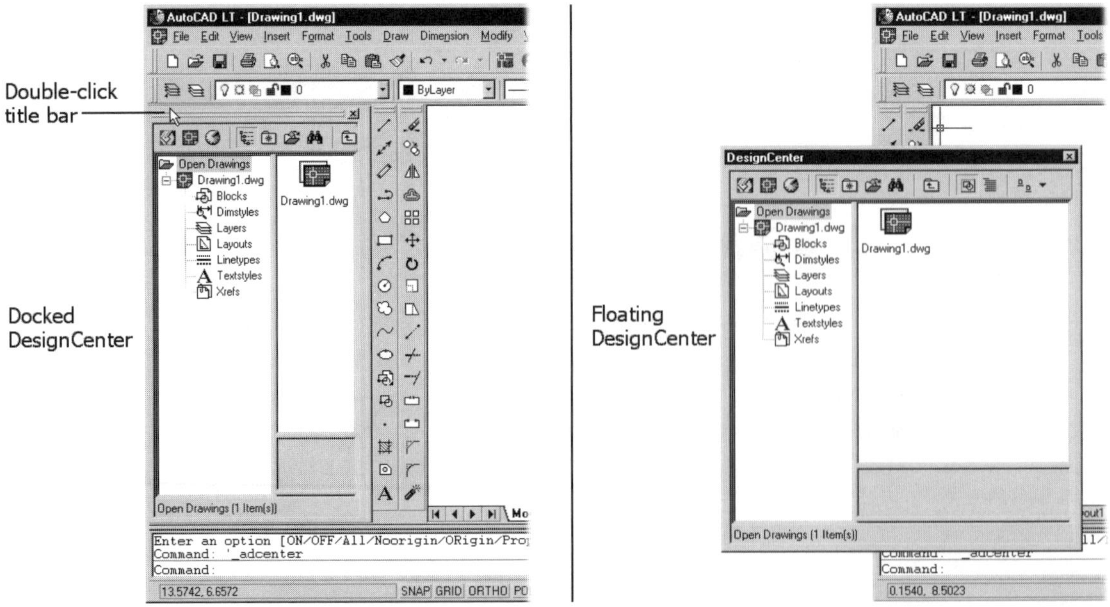

DesignCenter displays data from three (sometimes four) sources in a Windows Explorer-like tree view. Click one of the first three (or four) buttons on the toolbar to display different views, as follows:

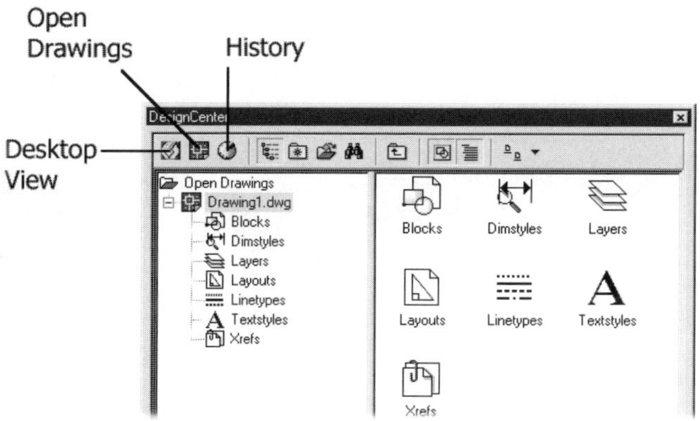

The first three toolbar buttons toggle the view.

Desktop displays all the drives, folders, and files on your computer. Just like Explorer, it can view the drives and files located on computers connected to yours via a local network and the Internet.

Open Drawings displays the drawings (and their contents) currently open in AutoCAD LT.

Custom Content displays proxy data. This button is available only when specific ObjectARX application is running; otherwise, you won't see this button.

History displays up to 120 of the documents you most recently viewed in DesignCenter.

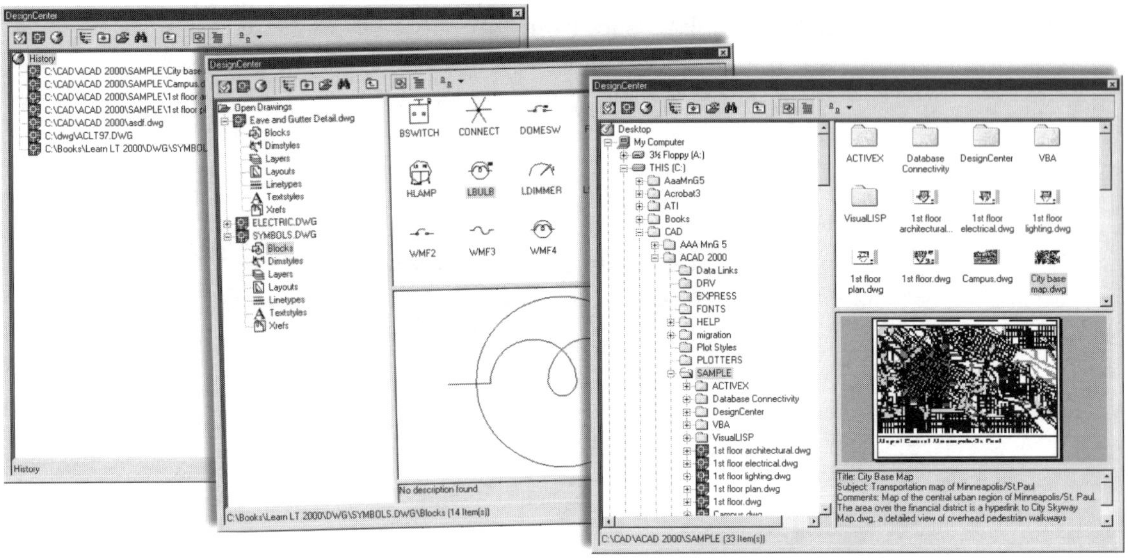

From left to right: History display, Open Drawings display, and Desktop display.

The middle set of buttons performs tasks, as follows:

The middle set of toolbar buttons performs tasks.

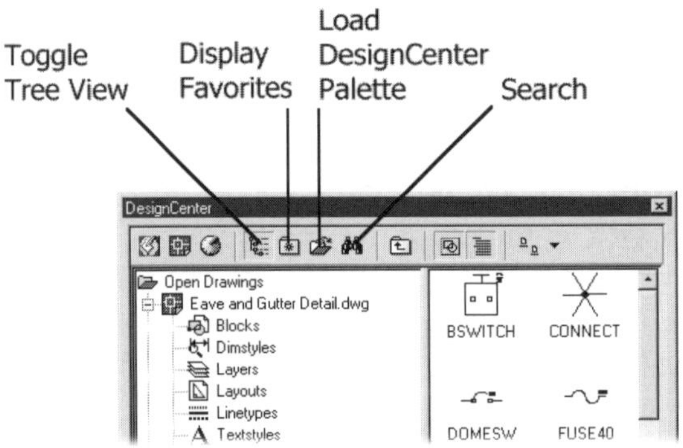

Toggle Tree View turns the tree view on and off.

Display Favorites allows you to store specific files (specifically, the files are stored in the \Windows\Favorites\Autodesk\DesignCenter folder).

Load DesignCenter Palette displays a file dialog box, which allows you to open drawings on your computer, your network, or from the Internet.

Search searches for filenames on your computer with the Find dialog.

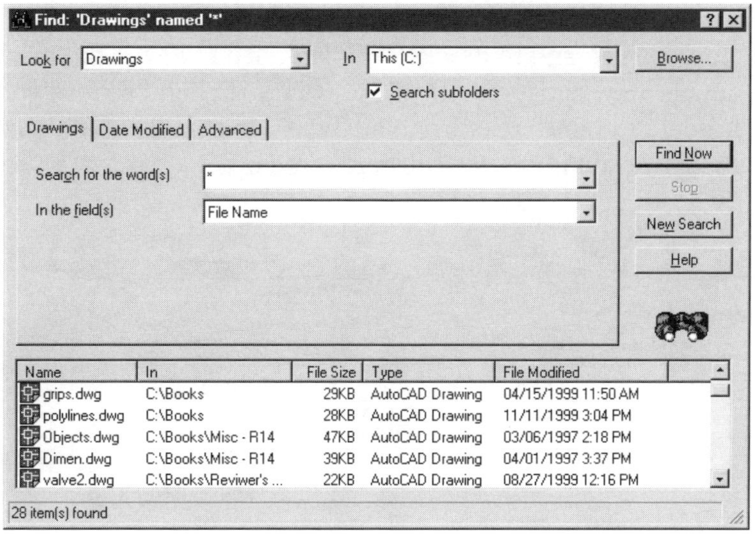

The Find dialog box.

The last set of buttons changes the display of the right pane:

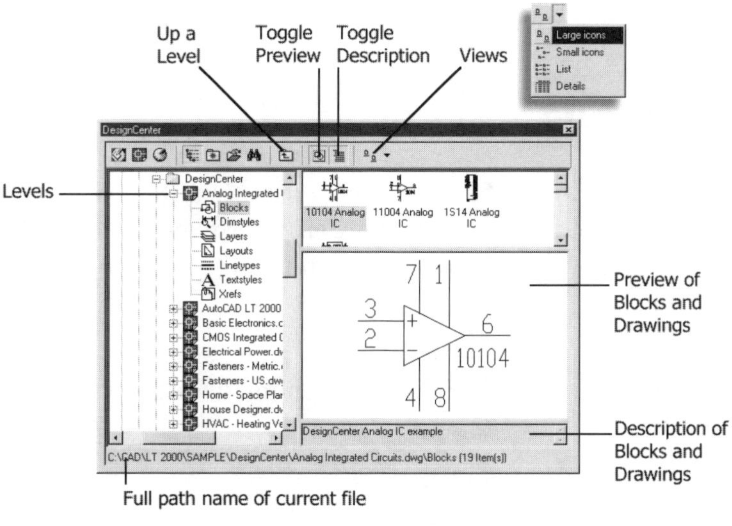

The right set of toolbar buttons changes the display of the pane.

Up moves up one level in the tree. For example, if you are viewing blocks in a drawing, click the Up button to view the drawing.

Preview toggles the display of blocks and drawings.

Description toggles the display of the description text for blocks and drawings.

Views changes the display of icons in the pane, from large icons, to small icons, to list, and to details. The large icon view is the most useful; the details view is useful only for drawings. In details view, you can click the headers—Name, File Size, and Type—to sort the list alphabetically; click them a second time to sort in reverse order.

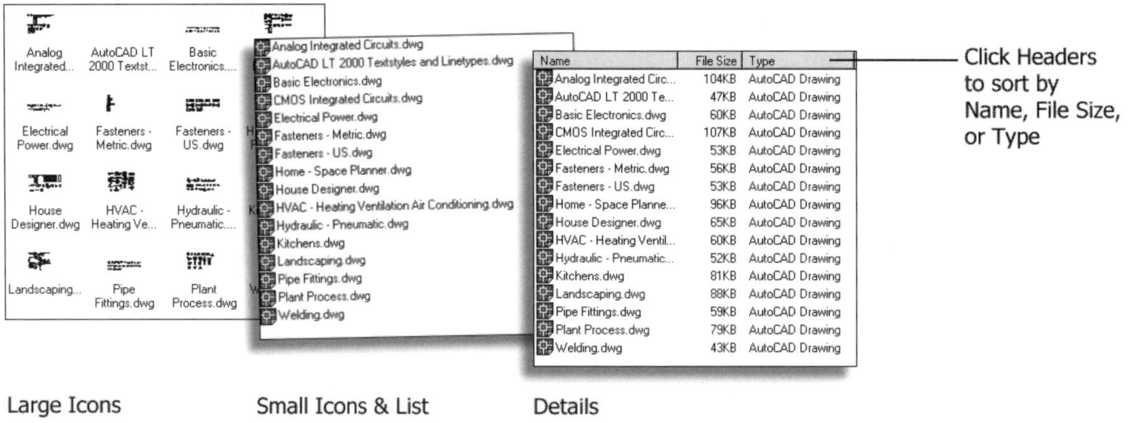

The Views button cycles through three views of filenames.

Right-Click Menus

The DesignCenter has a host of shortcut menus. When you right-click an element of the window, DesignCenter displays a shortcut menu with a variety of options. Some options mimic the toolbar buttons; others are unique.

Right-click an item in the left-hand pane, such as a filename or a folder. DesignCenter displays a shortcut menu with the same options as on the toolbar.

Right-click an item in the right-hand pane, such as the name of a block or text style. DesignCenter displays a shortcut menu with new options:

Add adds the selected dimension style, layer, layout, linetypes, and text styles to the current drawing. You may select more than one to add at a time by holding down the Ctrl key as you select.

Insert displays the Insert dialog box, which allows you to specify the block insertion parameters. This menu item appears only with blocks and xrefs.

Attach displays the External Reference dialog box, which allows you to specify options for attaching the xref to the drawing.

This shortcut menu mimics functions found on the toolbar menu.

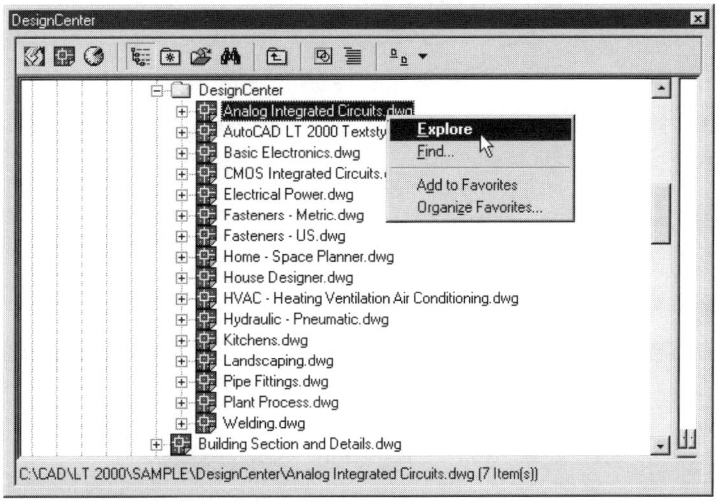

Copy copies the selected items to the Windows Clipboard. They can be pasted in another drawing with the Edit | Paste command.

This shortcut menu displays unique functions.

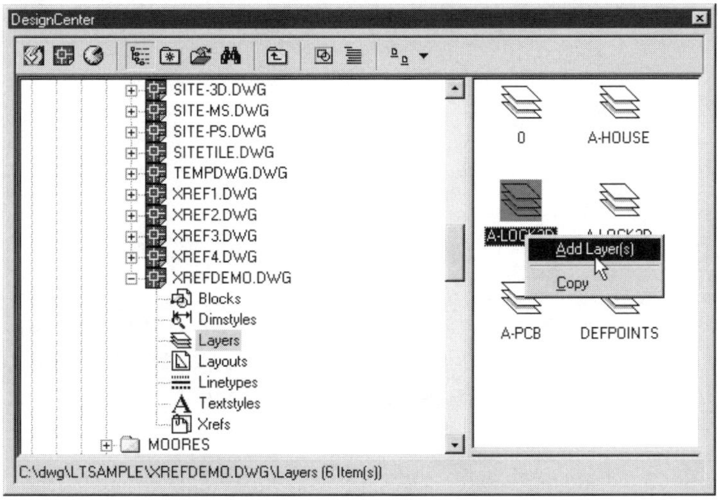

When right-clicking a drawing filename in the right-hand pane, you get a shortcut menu that combines the two above:

Explore opens the tree, giving you access to the drawing's blocks, dimstyles, layers, layouts, linetypes, text styles, and xrefs, if any.

Add to Favorites adds the drawing to the list of favorites.

Organize Favorites opens an Explorer-like window to the \Windows\Favorties\Autodesk folder, which allows you to move or delete shortcuts.

Insert as Block displays the Insert dialog box so that you can insert the drawing as a block in the current drawing.

Attach as Xref displays the **External Reference** dialog box so that you can attach the drawing as an xref to the current drawing.

Copy copies the drawing to the Windows Clipboard; you can then paste it into another AutoCAD drawing with the Edit | Paste command.

Open in Window opens the drawing in its own AutoCAD window.

The shortcut menu for drawings in the right-hand pane.

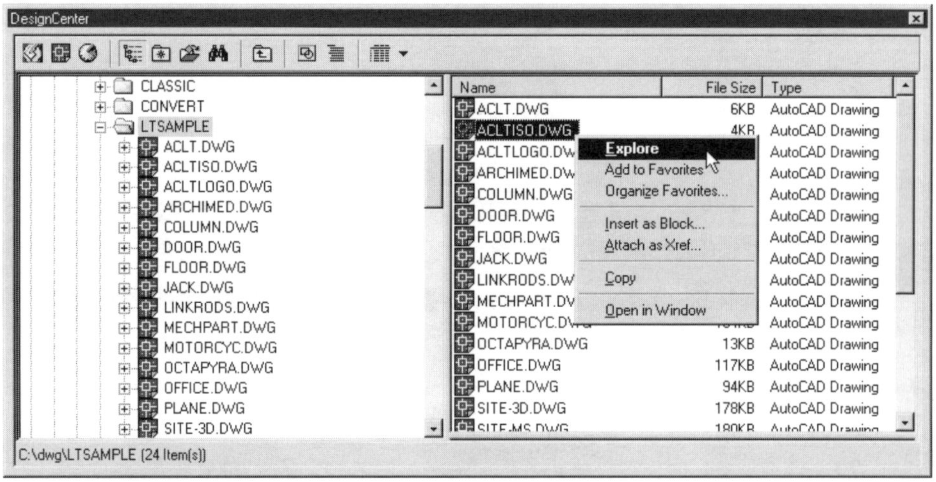

What DesignCenter Can and Can't Do

DesignCenter is a powerful tool that allows you to perform tasks by clicking and dragging, tasks that previously required entering commands and selecting options. Here is a summary of what DesignCenter can and cannot do for you:

- DesignCenter can preview blocks and DWG drawing files.
- DesignCenter allows you to share content between drawings: drag the item from DesignCenter into the drawing. Dragging a drawing from DesignCenter into the current drawing activates the -Insert command, with command-line prompts:

  ```
  -INSERT Enter block name or [?] <dcclpdata>:
  "C:\dwg\LTSAMPLE\ACLT.DWG"
  ```

Specify insertion point or
[Scale/X/Y/Z/Rotate/PScale/PX/PY/PZ/PRotate]: [pick point]
Enter X scale factor, specify opposite corner, or [Corner/XYZ] <1>:
[enter value or press Enter]
Enter Y scale factor <use X scale factor>: [enter value or press Enter]
Specify rotation angle <0>: [enter value or press Enter]

Dragging an xref into the drawing results in the following message:

Attach Xref "XREF2D": xref2

Dragging any other named object—text style, dimension style, linetype, layout, and layer—results in a message similar to the following:

Layer(s) added. Duplicate definitions will be ignored

This means that a layer with the same name is not added to the current drawing. This ensures you cannot accidentally overwrite the properties of an existing layer.

DesignCenter does not preview DXF files, nor any of AutoCAD support files, such as linetypes and text styles. DesignCenter doesn't allow you to edit properties; use the Properties command for that. To add the description to blocks displayed by DesignCenter, you must use the Block command.

Missing from DesignCenter are the tab sets found in Content Explorer; they have been replaced by Favorites. (You can migrate tab sets with the TabsetUpdate.Exe converter program found in the LT 2000i root folder.)

DesignCenter can display hatch patterns, but this is not obvious. To do this, (1) browse to the folder holding the PAT file; (2) drag the hatch pattern using the right mouse button; and (3) release the button to see the option menu.

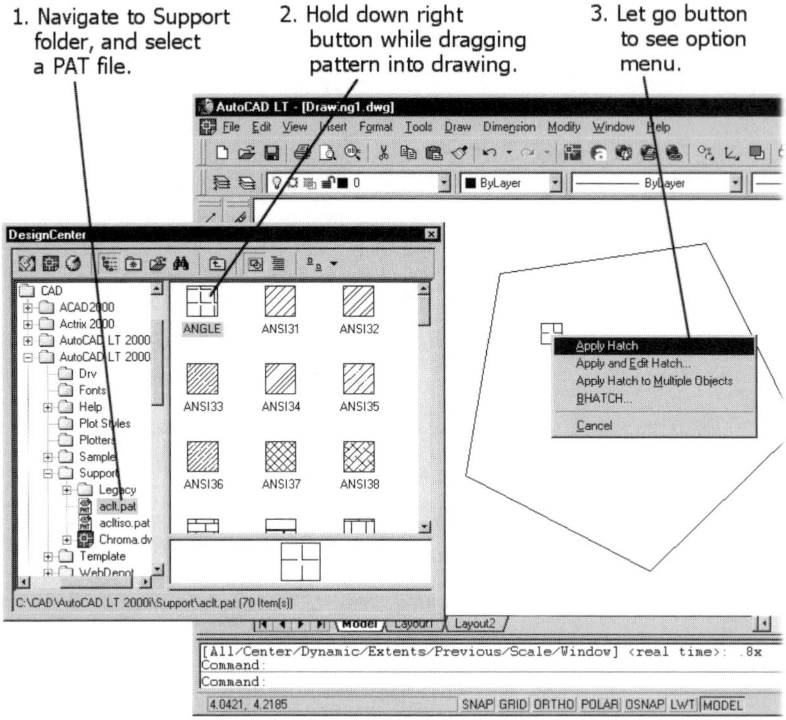

Review Questions

1. The Ctrl+2 shortcut opens and closes the AutoCAD DesignCenter.

   True / False

2. To switch the DesignCenter from docked to floating and back again:

   a. Press Ctrl+2.
   b. Double-click the title bar.
   c. Select Tools | Floating from the menu bar.
   d. This is not possible.

3. DesignCenter can display specific content of drawings found in any other computer connected to your computer:

   a. Via the Internet.
   b. Via the local area network.
   c. Both of the above.
   d. Neither of the above.

4. The purpose of the History button is to:
   a. Open the most recently accessed drawing.
   b. Add the current drawing to the Favorites list.
   c. Display proxy data.
   d. Display a list of previously opened files.
5. The purpose of the Open Drawings button is to:
   a. Display the drawings (and their content) currently open in AutoCAD LT.
   b. Add the current drawing to the Favorites list.
   c. Display proxy data.
   d. Display a list of previously opened files.
6. The purpose of the Desktop button is to:
   a. Display a list of previously opened files.
   b. Display all the drives, folders, and files on your computer, local network, and the Internet.
   c. Display the drawings (and their content) currently open in AutoCAD LT.
   d. Add the current drawing to the Favorites list.
7. The purpose of the Custom Content button is to:
   a. Add the current drawing to the Favorites list.
   b. Display a list of previously opened files.
   c. Display the drawings (and their content) currently open in AutoCAD LT.
   d. Display proxy data when specific ObjectARX applications are running.
8. DesignCenter allows you to share content between drawings.

   True / False
9. DesignCenter allows you to edit the properties of objects.

   True / False
10. DesignCenter displays hatch patterns in the drawing.

    True / False

Chapter 8

Creating Symbols and Attributes

Earlier in this book, you learned how to draw and edit an accurate drawing with AutoCAD LT. In this chapter, you learn how to use AutoCAD to create custom symbols with embedded information (called *attributes*). In addition, you put the AutoCAD DesignCenter to practical use.

Key Concepts

- Using symbols to speed up the drawing process
- Creating attributes and attaching them to symbols
- Creating a drawing made of (mostly) symbols

Key Terms

Block—the AutoCAD LT term for a symbol

Attribute—text data attached to a block

Insertion point—specifies where the block is inserted in the drawing

Unit size—a block drawn to fit a 1″x1″ area

Donut—a solid-filled circle

Tag—identifies the attribute

Prompt—attribute prompt displayed by AutoCAD during block insertion

Value—default value of the attribute

Command	Shortcut	Menu Selection
Arc	a	Draw | Arc
AttDef	at	Draw | Block | Define Attributes
DSettings	ds	Tools | Drafting Settings
Donut	do	Draw | Donut
Rectang	rec	Draw | Rectangle

There are two reasons for using symbols in a drawing: (1) symbols make you a faster drafter; (2) symbols can store attributes, which you can think of as customized information. As this chapter shows, it is much faster to add a symbol to a drawing, than it is to draw the symbol from scratch each time you need it.

You could use the Copy command to repeat a symbol, but it is more efficient to insert a symbol than simply copying it. A drawing that uses symbols takes up less disk space. Using symbols makes you a faster drafter; if you are a faster drafter, you complete more work in the same amount of time, or finish work sooner.

Later in this chapter, you learn how to create attributes, add them to the symbols, and store the symbols on disk. When it comes time to produce a bill of materials in the next chapter, AutoCAD produces automatically a list of all the symbols in the drawing. You can't do that with the Copy command!

Before You Begin

The example used for the design portion of this book creates a drawing of part of the electrical schematic of an antique automobile. If you have any sort of repair manual or other schematic drawing handy, you can reproduce your own schematic with AutoCAD LT.

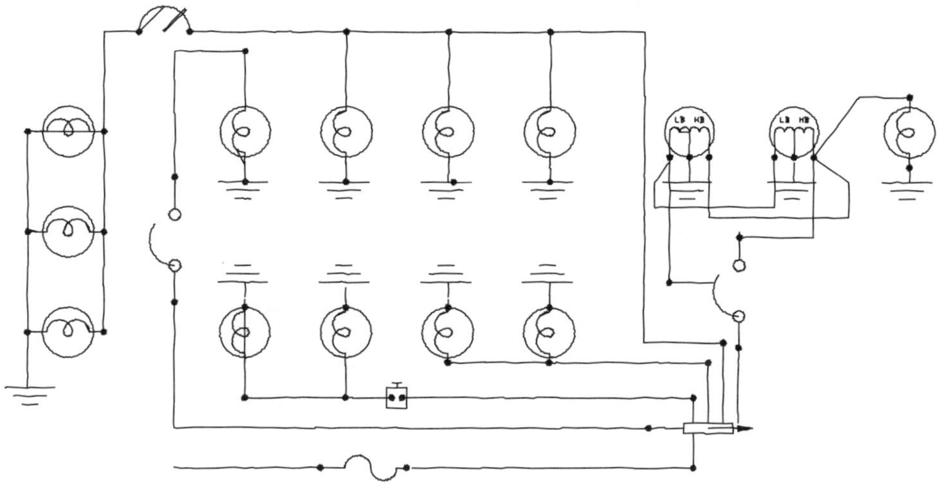

The sample drawing used for this chapter and the next.

If you'd rather not use your own, follow along with the schematic sketch above, the drawing used by this book.

Preparing for Drawing the Symbols

Creating a drawing with custom symbols takes two steps: (1) create the custom symbols; and (2) create the new drawing using the symbols.

AutoCAD calls symbols *blocks*, which is the name we use from now on. (Other drawing software calls the symbol a "component," "cell," "shape," or "part.")

1. Start AutoCAD LT from the Windows Start menu, or by double-clicking on its icon in the Windows Desktop.
2. When the Start Up dialog box comes up, click the **Start from Scratch** button, and click **OK.**
3. Since you will be drawing to an accuracy of 0.1 units, select **Tools | Drafting Settings** from the menu bar. AutoCAD displays the Drafting Settings dialog box.

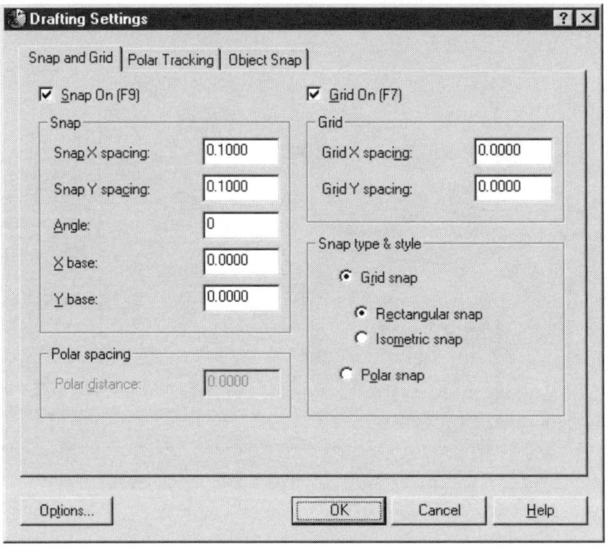

The Drafting Settings dialog box.

4. Make the changes in the Snap and Grid tab:
   - Snap On (F9) ☑ (on)
   - Snap X spacing **0.1**
   - Snap Y spacing **0.1**
   - Grid On (F7) ☑ (on)
   - Grid X spacing **0.0**
   - Grid X spacing **0.0**
5. Click **OK** when done. Notice that the screen fills with a fine grid of dots. Recall that the drawing limits determine the extents of the grid display.
6. Turn on ortho mode by clicking **ORTHO** on the status bar.
7. We don't need the UCS icon, so turn it off: **View |Display | UCS Icon |On.**
8. Perform a Zoom All to center the drawing limits.
9. Use the Save command to save the new drawing with the name "SYMBOLS."

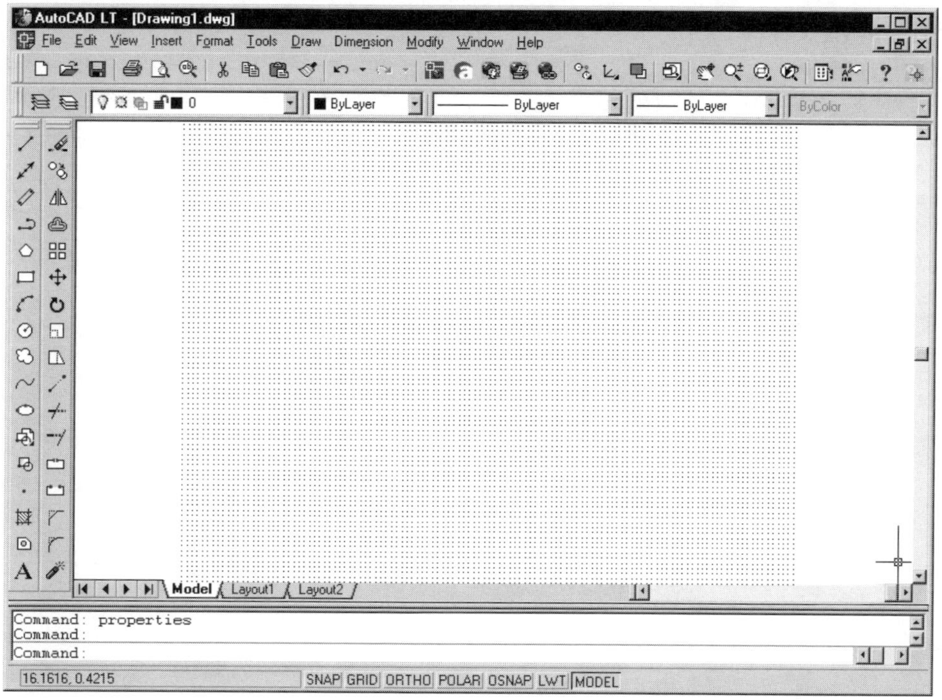

The drawing is prepared by setting snap, grid, and ortho.

Selecting the Parts

Before drawing the first block, look at the sketch of the schematic. Decide which parts should be converted into blocks. The light bulbs, for example, are likely candidates since there are so many of them. Other candidates are the headlamps and the ground symbols.

Here is the how a CAD drafter would think about the problem: create a block of anything you think should be listed in the bill of material, even if it appears just once. Thus, you should create a block of the dome light switch, the fuse, the instrument light dimmer, the light switch, the brake light switch, and even the solder connections.

The figure on the following page shows the blocks needed for the electrical schematic. The small black dot on each block shows its insertion point (more on this later in the chapter).

The parts that will become blocks.

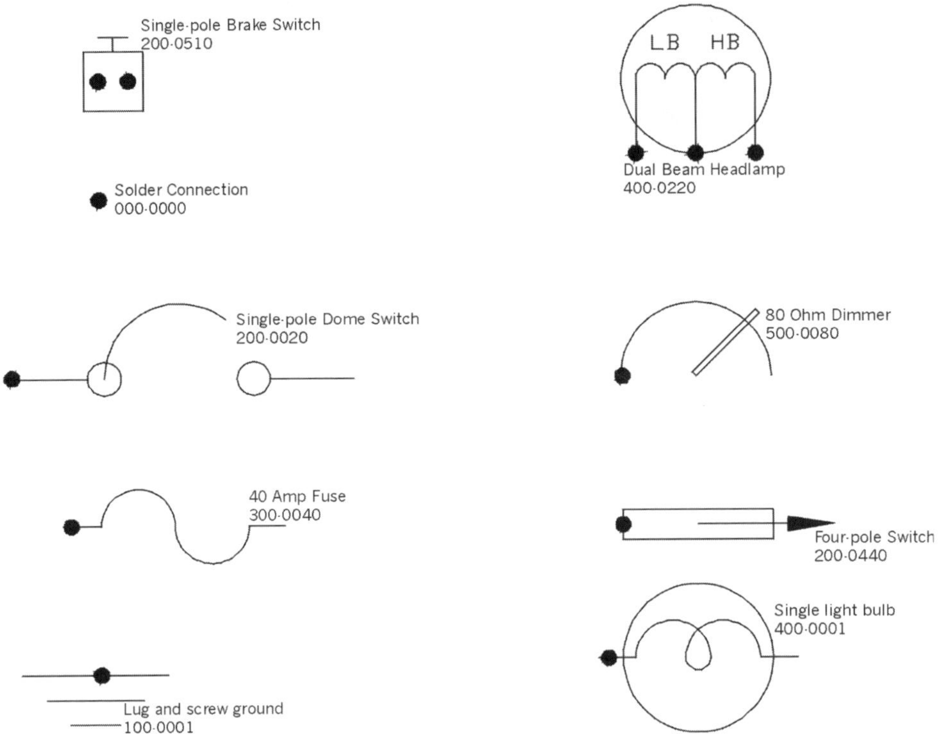

With the drawing set up, you begin drawing the blocks. Blocks are always drawn at unit size (with rare exceptions). *Unit size* means that the entire component is drawn within a one-unit square boundary. That makes it much easier to scale the component when it comes time to place it in the drawing.

Drawing the First Symbol

■ Draw | Donut

To start drawing components, begin with the easiest symbol, the symbol for the solder connection.

1. The dot for the connection is drawn with the Donut command, which draws solid-filled and thick-walled circles. Select **Draw | Donut** from the menu bar.

2. AutoCAD prompts you for the size of the donut, which will be 0.1 units in diameter. To make the donut solid filled, specify an inside diameter of zero, as follows:

    Command: _donut
    Specify inside diameter of donut <0.5000>: 0
    Specify outside diameter of donut <1.0000>: 0.1

3. AutoCAD prompts you to place the donut, as follows:

    Specify center of donut or <exit>: [pick anywhere in the screen]
    Specify center of donut or <exit>: [Esc]

    Notice that the Donut command automatically repeats itself until you end it by pressing **Esc**.

4. Enlarge the drawing area by selecting **View | Zoom | Window**. If the donut looks like an octagon, use the Regen command to turn the sides into a smooth circle.

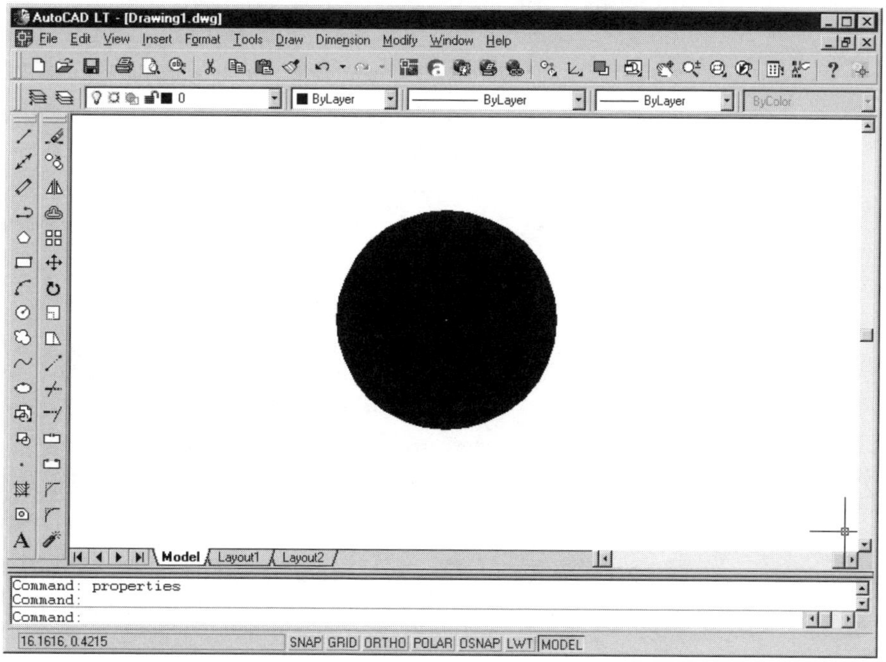

Drawing a donut (a solid filled circle).

Creating the Attribute Fields

■ Draw | Block | Define Attributes

With the solder connection block drawn as a donut, you now create the attribute fields. *Attributes* are customized descriptions of blocks; attributes only attach to blocks. While the drawing shows you how to assemble the electrical system, the attribute list tells you how many parts are needed for the assembly.

Just as the block is a graphical description of the ground symbol, the attribute is the textual description. The attribute information can describe the block's part number, the manufacturer, the price, and any other information you want to include.

A single block can contain one or more attributes. These attributes can have descriptive labels, such as "Product name," "Manufacturer," "Model number," "Stock number," "Serial number," and "Material." You could include "Price" as an attribute field but, since prices tend to change, it is better to deal with the price later in the spreadsheet program. Instead, use a price code.

Tip: You may be familiar with database software, where each record has one or more fields. In AutoCAD, an attribute (or field, in database terminology) has up to 256 characters. Each block (or record, in databases) can store up to 256 attributes (for a total of 64KB of information). You will probably find, however, that you rarely need to take advantage of that much space.

1. You create attributes with the AttDef command (short for "attribute definition"). From the menu bar, select **Draw | Block | Define Attributes.** AutoCAD displays the **Attribute Definition** dialog box.

   Although the dialog box has 15 fields to fill in, you work with just a few: the Attribute section, the Insertion Point, and Height.

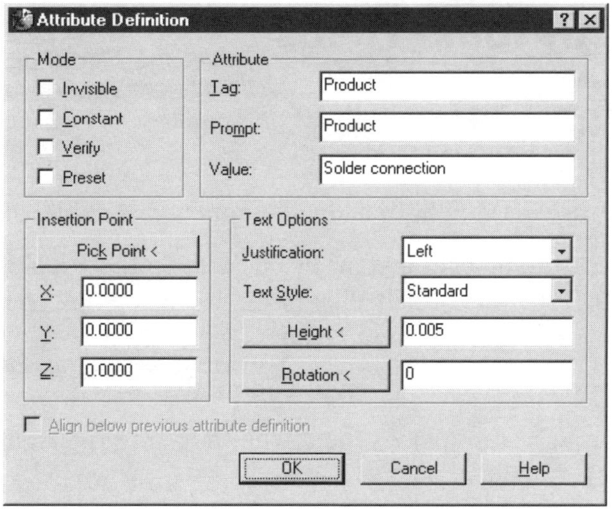

2. Type data in the **Attribute** section as follows:

   Tag: Product
   Prompt: Product
   Value: Solder connection

Note: AutoCAD uses these terms to define the three parts of an attribute: tag, prompt, and value.

Tag is an identifier used by AutoCAD (and you) to identify the attribute. The tag name can be up to 256 characters long but must contain no spaces.

Prompt is the prompt text AutoCAD displays when you later place the block in the drawing. The prompt can be up to 256 characters and may contain spaces for legibility. Do not include a colon (:) at the end; AutoCAD does that for you. In the example below, "Product" is the prompt:

Product <Solder connection>:

Value is the default value displayed when you later place the block in the drawing. This value is displayed by AutoCAD in angle brackets, as shown by "<Solder connection>" above. It can be up to 256 characters long. The default value can be anything, from a specific number to meaningless filler text. Here, you use a specific value, such as "Solder connection" because it is a common value. Using a default value can save you many keystrokes.

Alternatively, you could use a filler for the value. The filler reminds you (and other users) the size and type of the field. For example, "99999" suggests a five-digit number, while "AAA-999" suggests three characters, a dash, and three digits.

3. Set the text height to a small, unobtrusive value, such as 0.05.
4. Select the insertion point by clicking on the **Pick Point** button. AutoCAD temporarily dismisses the dialog box so that you can see the drawing and prompts you:

   Start point: [pick right edge of donut]

   Your pick point places the attribute data to the right of the connection block.
5. AutoCAD returns the dialog box. Click the **OK** button to dismiss it.

Adding a Second Attribute

You now add a second attribute directly below the first.

1. To do this, press the spacebar to repeat the AttDef command, as follows:

   Command: [Spacebar]
   _attdef
2. When the Attribute Definition dialog box appears, click the check box next to **Align below previous attribute definition.** This ensures the second attribute lines up cleanly below the first attribute.
3. Fill in the following data in the **Attribute** section:

   Tag: Stockno
   Prompt: Stock No.
   Value: 000-0000

   Ensure the tag name "Stockno" is a single word, with no spaces.

The next attribute definition.

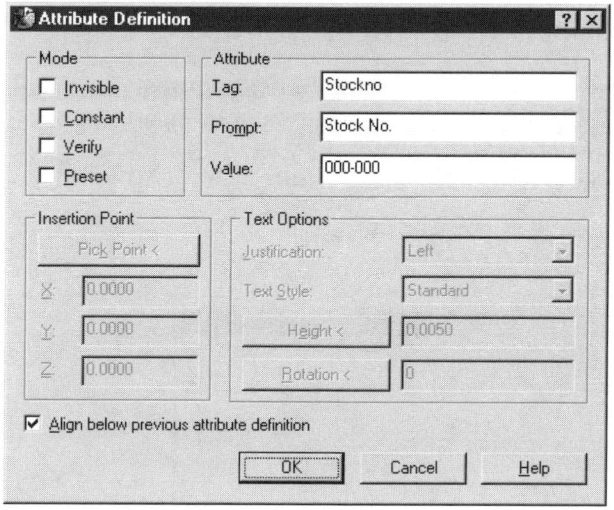

4. Click on the **OK** button and AutoCAD adds the second attribute. Notice that the attributes are identified by their tag names "PRODUCT" and "STOCKNO." This will change when the attributes are combined with the donut to create a block, as we see next.

The second attribute is aligned under the first.

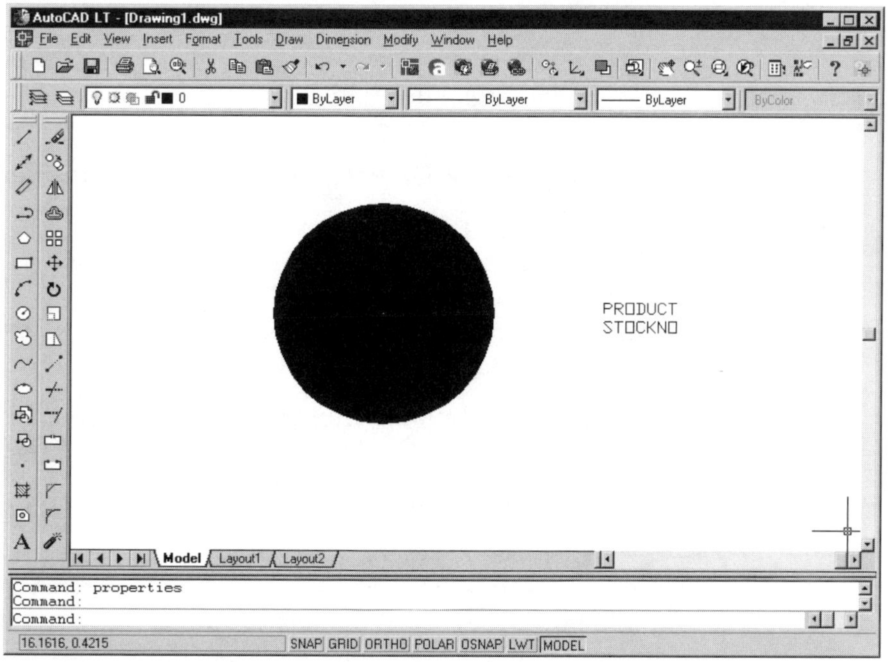

Combining the Object and Attribute into a Block

You now have a donut and the two attributes. The final step is to link the donut and attributes together in a single block.

1. Select **Draw | Block | Make** from the menu bar.
2. When the Block Definition dialog box appears, name the block "CONNECT."

The Block Definition dialog box.

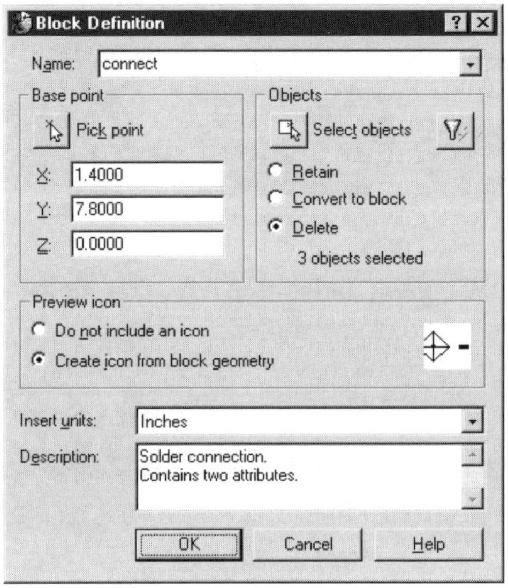

3. Click the **Select Objects** button, and select the donut and attribute, as follows:

   Select objects: c
   Specify first corner: [pick]
   Specify opposite corner: [pick]
   3 found Select objects: [Enter]

4. Change the base point from (0,0,0) to the center of the donut. Click the **Pick Point** button, and select the center of the donut, as follows:

   Specify insertion base point: [pick center of donut]

Note: The base point is also known as the *insertion point*, which is where the block is placed in the drawing. When deciding on the insertion point of a block, pick a convenient spot such as the center of the block or a connection point.

5. In the Objects section, turn on **Delete.**
6. Click **Close.** Notice that the donut and attribute text disappear. Don't worry! AutoCAD has stored the donut as a block definition, called CONNECT, in the drawing's database. You bring it back with the DesignCenter.

Inserting Blocks with DesignCenter

Use AutoCAD DesignCenter to insert the CONNECT block in the drawing. (If necessary, review Chapter 7 for the tutorial on using DesignCenter.)

1. Select **Tools | AutoCAD DesignCenter** from the menu bar.
2. Under Open Drawings, open Symbols.Dwg by clicking the plus sign (+). Click **Blocks.** Notice the Connect block.

The DesignCenter displaying the Connect block.

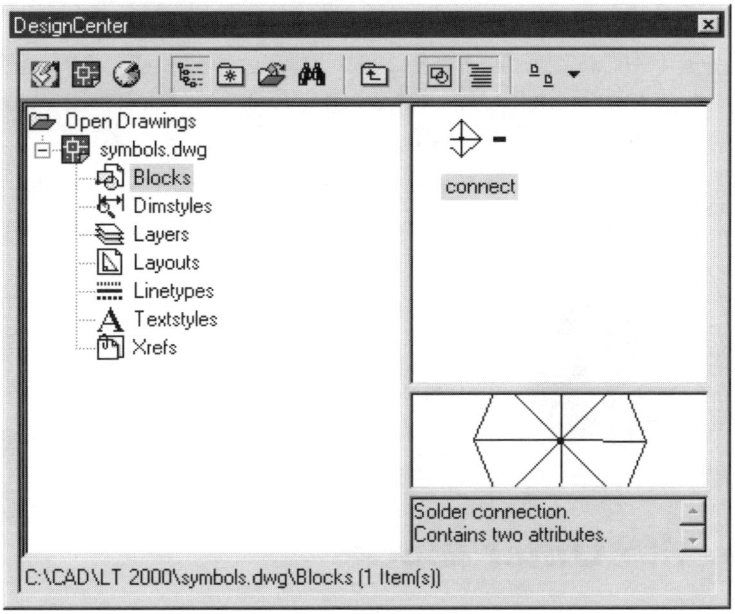

3. Drag the Connect block from the DesignCenter into the drawing. Pick any convenient point on the screen, then let go of the mouse button. Notice that AutoCAD displays the Edit Attributes dialog box.

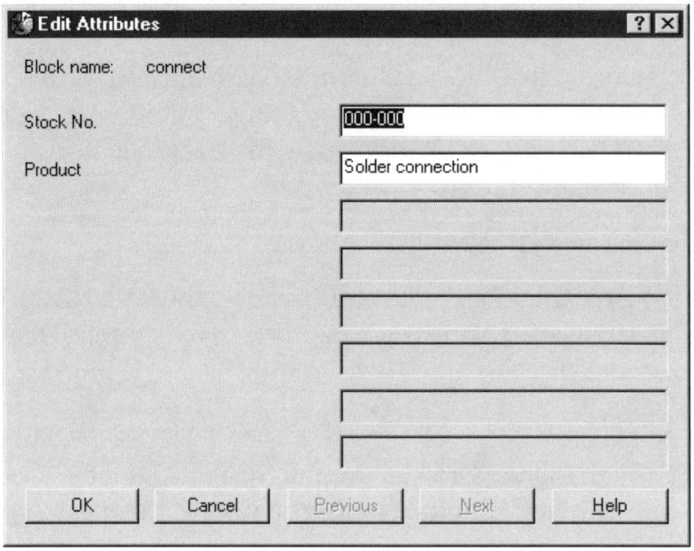

The Edit Attributes dialog box.

4. To accept the default attribute values—000-000 and Solder connection—click **OK**. Notice that AutoCAD draws the connection block accompanied by the words "Solder Connection" and "000-0000."

Congratulations! You've worked your way through the somewhat complicated process of creating a block and attaching attributes.

The inserted block and attribute text.

Creating the Other Blocks

There are nine more blocks that need attributes or need to be created for the electrical schematic. Here are the names and attributes you give them, along with tips for drawing the bits and pieces. The most complex is the headlamp, which is drawn with circles, arcs, lines, and donuts (see illustration).

The steps for drawing the headlamp block. The command is listed under each step.

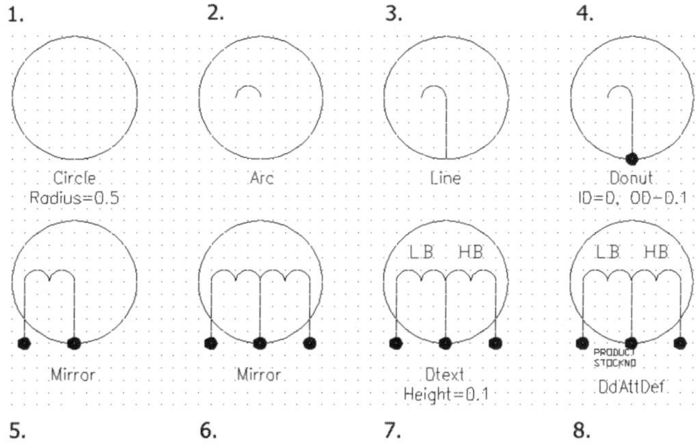

Chapter 8

Here is how I suggest you draw the headlamp symbol:

1. Turn on ortho mode; set snap and grid to 0.1. Draw a one-inch circle with the **Draw | Circle | Center,** Radius command, setting the radius to 0.5 inches.
2. The **Draw | Arc | Start, Center, End** command draws one filament arc.
3. Use the **Draw | Line** command to extend the center filament to the base of the lamp.
4. Place a 0.1 diameter donut at the end of the line with the **Draw | Donut** command.
5. Use the **Modify | Mirror** command to make a mirrored copy of the line, donut, and arc to the left.
6. Repeat the **Mirror** command to make a second copy of the filament to the left.
7. Use **Draw | Single Line Text** to add the "HB" (short for high beam) and "LB" (low beam) text. Use center justification and a text height of 0.1.
8. Use the **Draw | Block | Define Attributes** command to add the attributes:

Tag #1:	Product
Prompt:	Product
Value:	Dual beam headlamp

Tag #2: Stockno
Prompt: Stock No.
Value: 400 - 0220

Finally, use the **Draw | Block | Make** command to convert the drawing into a block called "HLamp." Select the middle donut as the insertion point.

The Remaining Symbols

1. Draw the four-pole switch with the Draw | Rectangle command, 1" x 0.2" in size. Add a line with the Line command. Draw the 0.3-long arrowhead with the PLine command, setting the starting width to 0.0, and the ending width to 0.2. The block name and attributes are:

   Block name: LSwitch
   Tag #1: Product
   Prompt: Product
   Value: Four-pole switch

   Tag #2: Stockno
   Prompt: Stock No.
   Value: 200-0440

2. Draw the single-pole brake switch with the Rectang command, with sides 0.4 units long. Finish with the Line and Donut commands. The block name and attributes are:

   Block name: BSwitch
   Tag #1: Product
   Prompt: Product
   Value: Single pole brake switch

   Tag #2: Stockno
   Prompt: Stock No.
   Value: 200-0510

3. Draw the 40-amp fuse with two arcs and two lines. The block name and attributes are:

   Block name: Fuse40
   Tag #1: Product
   Prompt: Product
   Value: 40 Amp fuse

Tag #2:	Stockno
Prompt:	Stock No.
Value:	300-0040

4. Draw the single-pole dome switch with an arc, donut, two circles, and lines. The block name and attributes are:

Block name:	DomeSw
Tag #1:	Product
Prompt:	Product
Value:	Single-pole switch
Tag #2:	Stockno
Prompt:	Stock No.
Value:	200-0020

5. Draw the single light bulb from a circle, arcs, lines, and donut. The block name and attributes are:

Block name:	LBulb
Tag #1:	Product
Prompt:	Product
Value:	Single
Tag #2:	Stockno
Prompt:	Stock No.
Value:	400-0001

6. Draw the 80-ohm dimmer with an arc, rectangle, and donut. The block name and attributes are:

Block name:	LDimmer
Tag #1:	Product
Prompt:	Product
Value:	80 Ohm dimmer
Tag #2:	Stockno
Prompt:	Stock No.
Value:	500-0080

7. Draw the lug and screw ground with three lines and a donut. The block name and attributes are:

Block name:	Ground
Tag #1:	Product
Prompt:	Product
Value:	Lug and Screw Ground

Tag #2:	Stockno
Prompt:	Stock No.
Value:	100-0001

8. Finally, save the Symbols drawing. For easy reference to your new symbol library, plot the drawing.

Drawing the Electrical Schematic

With the electrical blocks stored on disk, you now create the drawing of the automobile electrical schematic.

1. Select **File | New** to start a new drawing named "ELECTRIC." Make sure ortho mode is turned on; set snap to 0.1 and grid to 0.5.
2. Make sure the DesignCenter is open by selecting **Tools | AutoCAD DesignCenter**. Click the **Symbols.Dwg** icon (the drawing should still be open in AutoCAD LT).
3. Click **Blocks** to display the blocks in the Symbols drawings.

Displaying blocks in the DesignCenter.

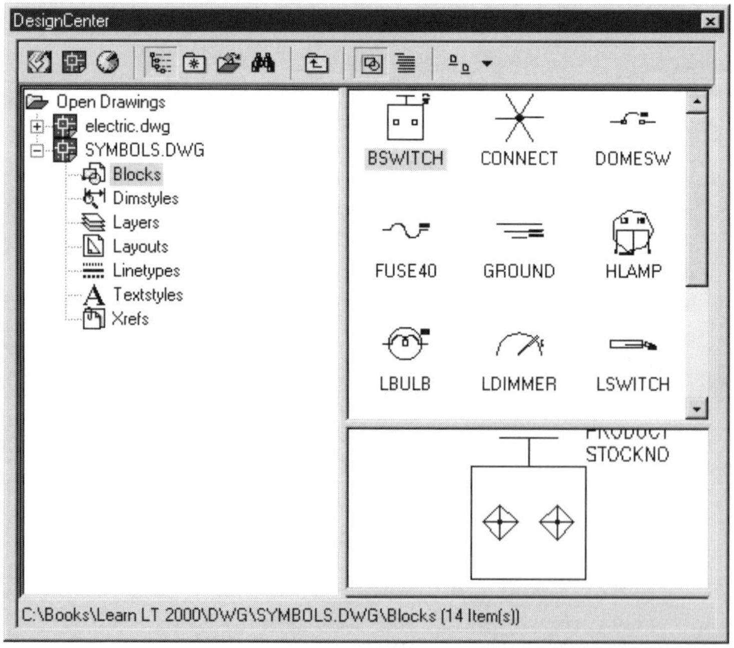

4. Drag the symbols into the drawing, placing them in roughly the correct location. (For now, do not drag the Connect block into the drawing.)

Placing blocks in the drawing.

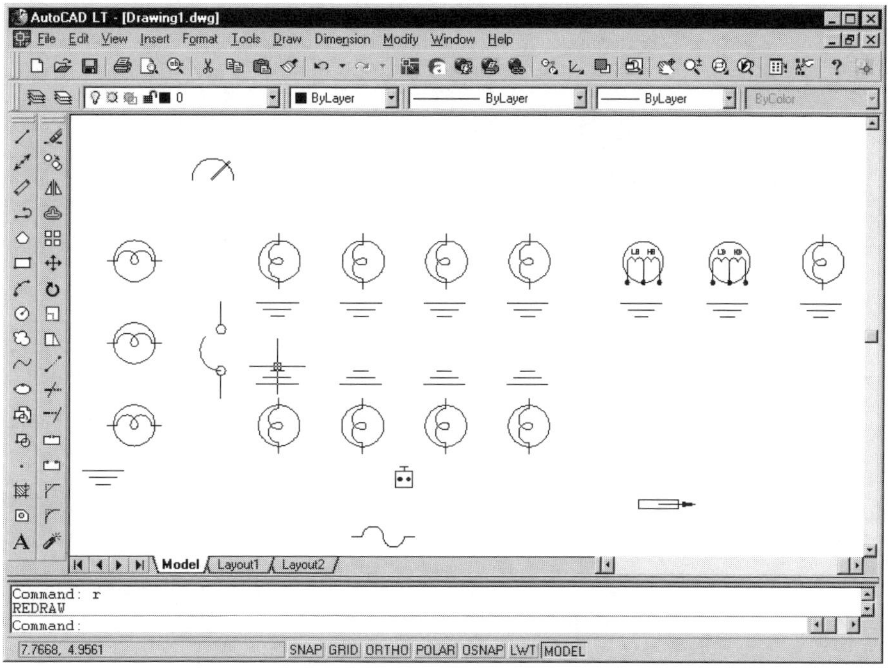

5. Once all the blocks are positioned, turn on ENDpoint object snap. On the status bar, right-click **OSNAP**. In the Object Snap tab of the Drafting Settings dialog box, turn on **Endpoint**, and click **OK**.

6. Use the Draw | Line command to connect the blocks, as shown in the following figure. If necessary, make liberal use of the Aerial View window to get a better view of your work.

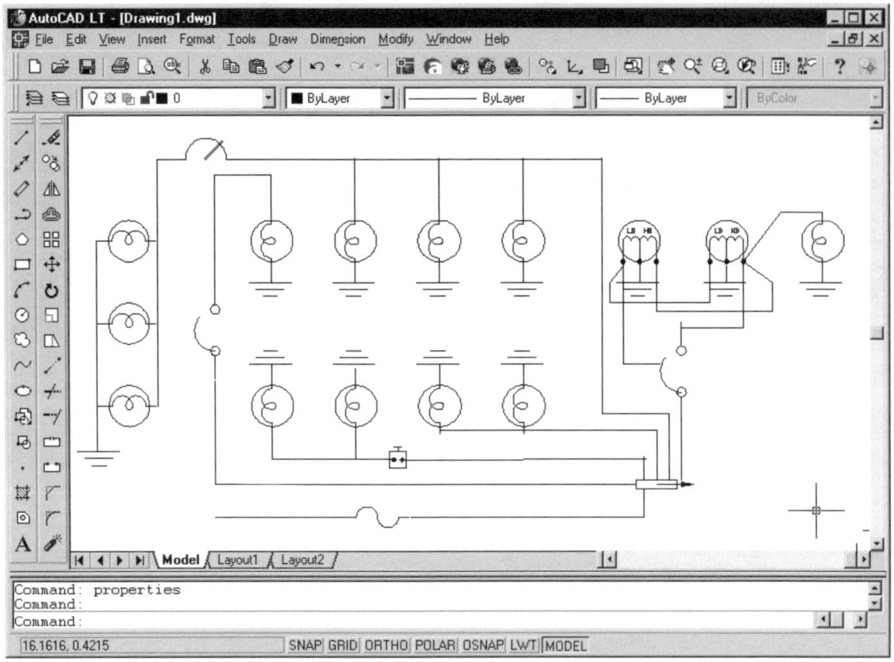

Connecting the blocks with lines (wiring).

7. Complete the electrical connections by placing the Connect block.
8. Add descriptive text to the schematic with the Text command.
9. Finally, save your work.

Review Questions

1. Using symbols helps speed up the drawing process.

   True / False

2. Match the term with its best meaning:

a.	Attribute	i.	Attribute prompt displayed by AutoCAD during block insertion.
b.	Insertion point	ii.	Identifies the attribute.
c.	Tag	iii.	Default value of the attribute.
d.	Prompt	iv.	Text data attached to a block.
e.	Value	v.	Specifies where the block is inserted in the drawing.

3. When referring to a block, the term unit *size* means:

   a. The block drawn to fit a 1 x 1 area.
   b. The block is drawn in metric units.
   c. The block is drawn in imperial units.
   d. The block is drawn with squares.

4. Which one of the following statements is true?

   a. Attributes can be attached to any object.
   b. Attributes can be attached only to blocks.
   c. Attributes can be attached only to groups.
   d. Attributes can be attached only to polylines.

5. It is more efficient to copy a symbol, than it is to insert it as a block.

   True / False

6. A block can store ______ attribute(s).

   a. 1 (one).
   b. 16
   c. 256
   d. Unlimited.

7. To attach an attribute to a block, use the ____________ command.

   a. Attach
   b. Block
   c. Attribute
   d. AttBlk

8. When objects are changed to a block, they:

   a. Disappear from the screen.
   b. Flash on and off.
   c. Turn red in color.
   d. Are placed on the BLOCKS layer.

9. Blocks that you create in a drawing can be accessed with the DesignCenter.

   True / False

10. Anything you draw can be turned into a block.

    True / False

Chapter 9

Bills of Material

In the last chapter, you learned how to create a custom symbol library with AutoCAD. In this chapter, you learn how to extract the attribute information with AutoCAD LT's attribute extraction command. It also provides a link to spreadsheet and database programs for further processing.

Key Concepts

- Exporting attributes from an AutoCAD drawing to a spreadsheet
- Creating a template file for formatting the attribute data
- Importing a spreadsheet into an AutoCAD drawing

Key Terms

Attribute extraction—exporting attribute data to a file

Template file—defines the format of the attribute export file

Abbreviations

CDF	comma-delimited format
DXF	drawing interchange format
SDF	space-delimited format
TXT	extension for an ASCII text file
XLS	excel spreadsheet file

Command	Shortcut	Menu Selection
AttDisp	ad	View | Display | Attribute | On
AttExt	Alt+TE	Tools | Attribute Extraction
PasteSpec	pa	Edit | Paste Special

Attribute Extraction

Last chapter, you created the drawing of an automobile electrical schematic. The drawing contains many components: a couple of headlamps, a fuse, several light bulbs, and quite a few ground and solder connections. You could count the components by hand, but you would probably miscount them.

It's faster and more accurate to let AutoCAD help you do the counting. This process is called *attribute extraction*. Extracting attributes, unfortunately, is one of the toughest areas to learn in AutoCAD LT.

While other Windows programs boast "coaches" and "wizards" to step you through complicated procedures, AutoCAD LT provides no assistance in this important area of CAD. The attribute extraction command reckons back to AutoCAD v2.0 for DOS and is unchanged since 1985, except for the addition of a dialog box front end.

In this chapter, you learn how to extract attributes from a drawing. The process proceeds in three steps:

Step 1. **Template:** Create an attribute-extract template file.

Step 2. **Extract:** Set up the **AttExt** command to extract attributes.

Step 3. **Import:** Bring the attribute data into a spreadsheet.

Step 1: Create the Template File

Before AutoCAD can extract attribute data, it needs to know two things: (1) which data you want extracted; and (2) the format you want the data in. Unfortunately, AutoCAD has no default values; you must create a template file, which the AttExt command refers to for guidance in these matters. Here's how you create the template file:

1. If necessary, start Windows.
2. Start the Notepad application, the handy text editor that ships with every copy of Windows. From the taskbar, select **Start | Programs | Accessories | Notepad**.
3. Type the text, shown below, into Notepad. Make sure you type the number 0 (zero) and not the letter O in the right column:

BL:NAME C008000
BL:NUMBER N003000
PRODUCT C016000
STOCKNO C016000

There must be no blank lines.

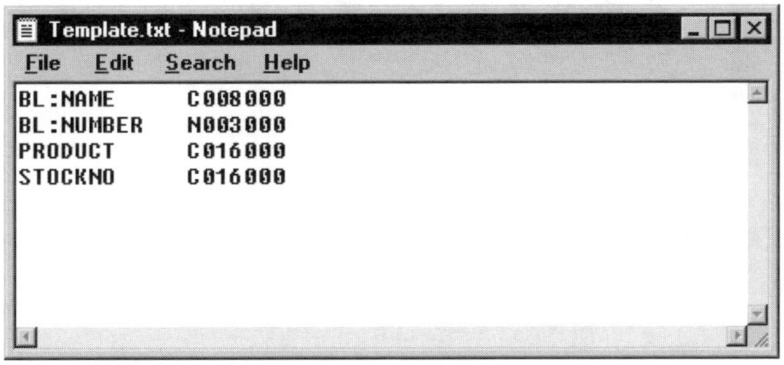

The attribute format codes in Notepad.

4. Save the file as TEMPLATE.TXT.

Note: The text in the template file has the following meaning:

Template Text	Meaning
BL:NAME	Extracts the name of all blocks found in the drawing.
BL:NUMBER	Counts the number of occurrences of each block.
PRODUCT, STOCKNO	Extracts the values of attributes with PRODUCT and STOCKNO tags.
C008000	Formats the data as characters (C), eight characters wide.
N003000	Formats the data as numbers (N), three places wide (003) and with zero decimal places (000).

Step 2: Extract Attribute Data

- Tools | Attribute Extraction

1. Launch AutoCAD LT, and open the Electric.Dwg drawing. If you did not complete the drawing in the last chapter, open the Electric.Dwg file from the companion CD-ROM.

2. To see the attribute text, use the AttDisp command. From the menu bar, select **View | Display | Attribute | On**. This allows you to see all the attributes. (This step is not necessary to extract attributes; AutoCAD extracts both visible and invisible attributes.)

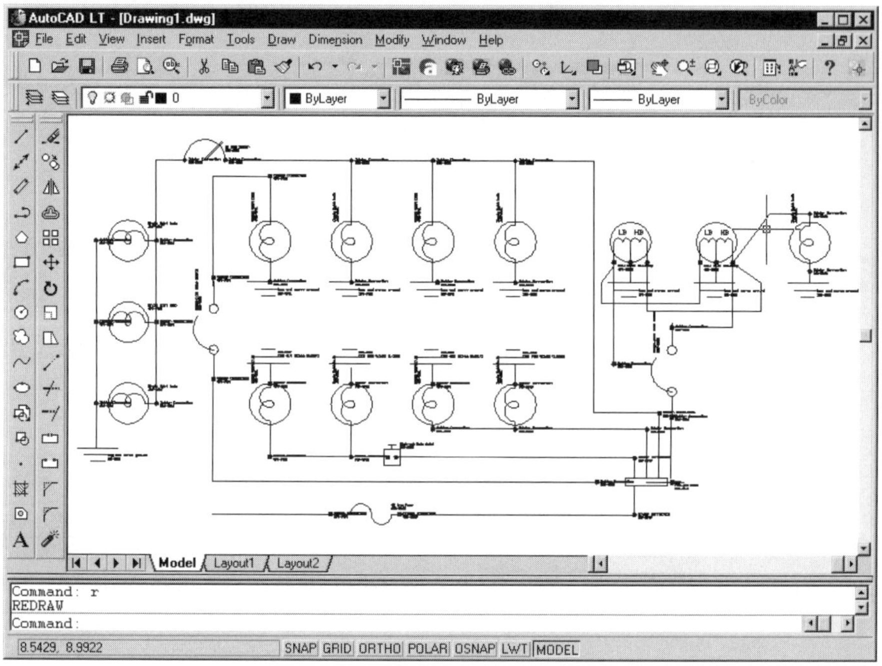

Note: The AttDisp command changes the way AutoCAD displays attribute text. The AttDisp command has the following options:

AttDisp	Meaning
On	Displays all attribute text.
Off	Displays no attribute text.
Normal	Displays attribute text, except when they are set to Invisible mode.

3. Let's start the attribute extraction. From the menu bar, select **Tools | Attribute Extraction**. AutoCAD displays the Attribute Extraction dialog box.

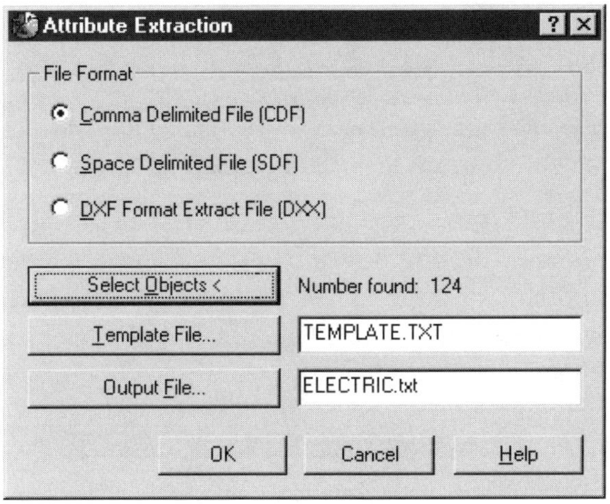

The Attribute Extraction dialog box.

4. Under File Format, you have the choice of three output formats:
   - **CDF** (short for "comma-delimited file") separates values by commas and is best suited for importing attribute data in spreadsheets and word processors.
   - **SDF** (short for "space-delimited file") separates values by spaces or tabs, and is best suited for importing data into database programs.
   - **DXF** (short for "DXF Format Extract File") exports data in AutoCAD's own DXF format and is best suited for programmers.

   Make sure that the CDF radio button is selected.
5. Click **Select Objects**. When AutoCAD prompts you, select all objects, as follows:

   Select objects: all
   124 found
   Select objects: [Enter]

   When you specify All, AutoCAD selects all objects in the entire drawing. This is faster than using Window or Crossing mode.
6. Click **Template File**.
7. When the Template File dialog box appears, select **Template.Txt**. You created this attribute extract template file at the beginning of this chapter.

8. Click **OK**. AutoCAD searches through the drawing, counting each instance of every component.
9. After a second or two, the results are deposited in the Electric.Txt file. You can examine it with Notepad, as shown in the figure below:

   For each block you placed in the Electric drawing, AutoCAD lists the block's name, the number of its occurrence, followed by the product name, and stock number. You've now created a rudimentary bill of material, which can be printed out on your printer or imported to a spreadsheet.

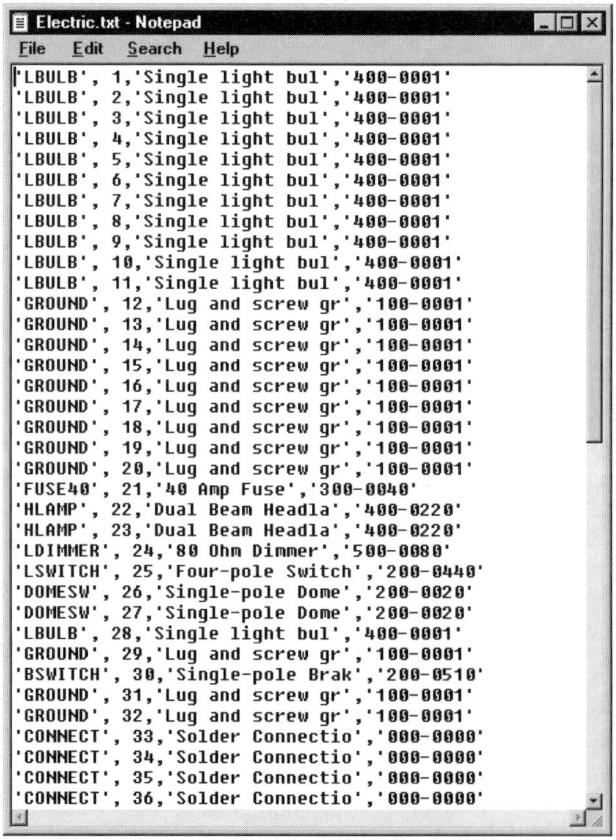


```
'LBULB', 1,'Single light bul','400-0001'
'LBULB', 2,'Single light bul','400-0001'
'LBULB', 3,'Single light bul','400-0001'
'LBULB', 4,'Single light bul','400-0001'
'LBULB', 5,'Single light bul','400-0001'
'LBULB', 6,'Single light bul','400-0001'
'LBULB', 7,'Single light bul','400-0001'
'LBULB', 8,'Single light bul','400-0001'
'LBULB', 9,'Single light bul','400-0001'
'LBULB', 10,'Single light bul','400-0001'
'LBULB', 11,'Single light bul','400-0001'
'GROUND', 12,'Lug and screw gr','100-0001'
'GROUND', 13,'Lug and screw gr','100-0001'
'GROUND', 14,'Lug and screw gr','100-0001'
'GROUND', 15,'Lug and screw gr','100-0001'
'GROUND', 16,'Lug and screw gr','100-0001'
'GROUND', 17,'Lug and screw gr','100-0001'
'GROUND', 18,'Lug and screw gr','100-0001'
'GROUND', 19,'Lug and screw gr','100-0001'
'GROUND', 20,'Lug and screw gr','100-0001'
'FUSE40', 21,'40 Amp Fuse','300-0040'
'HLAMP', 22,'Dual Beam Headla','400-0220'
'HLAMP', 23,'Dual Beam Headla','400-0220'
'LDIMMER', 24,'80 Ohm Dimmer','500-0080'
'LSWITCH', 25,'Four-pole Switch','200-0440'
'DOMESW', 26,'Single-pole Dome','200-0020'
'DOMESW', 27,'Single-pole Dome','200-0020'
'LBULB', 28,'Single light bul','400-0001'
'GROUND', 29,'Lug and screw gr','100-0001'
'BSWITCH', 30,'Single-pole Brak','200-0510'
'GROUND', 31,'Lug and screw gr','100-0001'
'GROUND', 32,'Lug and screw gr','100-0001'
'CONNECT', 33,'Solder Connectio','000-0000'
'CONNECT', 34,'Solder Connectio','000-0000'
'CONNECT', 35,'Solder Connectio','000-0000'
'CONNECT', 36,'Solder Connectio','000-0000'
```

The extracted attributes in CDF format.

Step 3: Importing the BOM into Excel

The steps shown here for importing the extracted attribute data (bill of material) into a spreadsheet is shown here for Microsoft Excel. Other spreadsheet programs, such as Lotus 1-2-3 and Star Office, have analogous steps.

1. Launch the spreadsheet program by double-clicking its icon.
2. From the menu bar, select **File | Open** to display the Open dialog box.

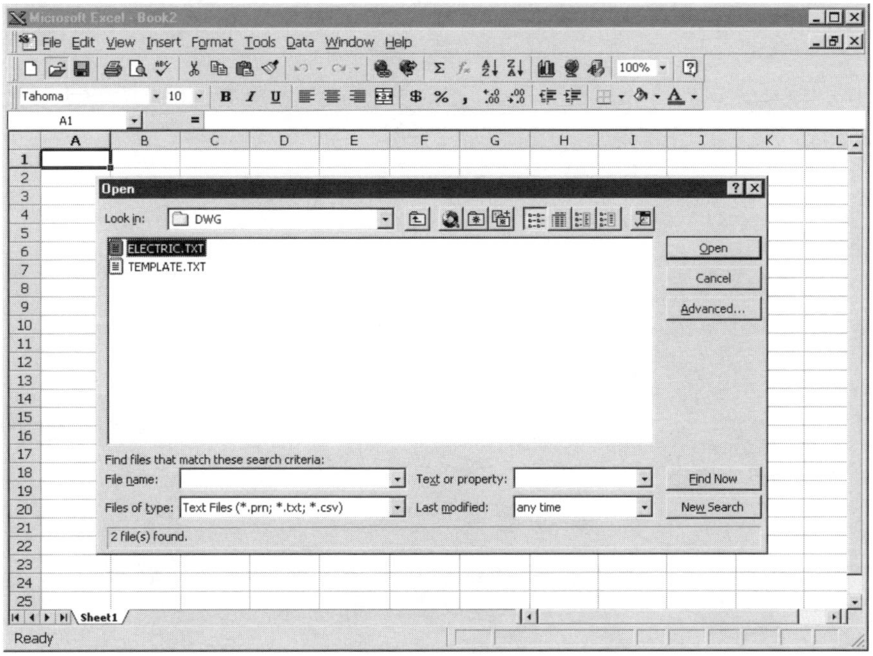

Excel's Open dialog box.

3. In the Files of type field, select **Text Files**.
4. Navigate to the folder holding the Electric.Txt file. (A copy of the file is available on the companion CD-ROM.)
5. Select the **Electric.Txt** file, and click **Open**. Notice that Excel displays the Text Import Wizard. These dialog boxes allow you to specify the format of the file. In the Step 1 dialog box:
   - Original data type: **Delimited**
   - Start import row: **1**
   - Click **Next**.

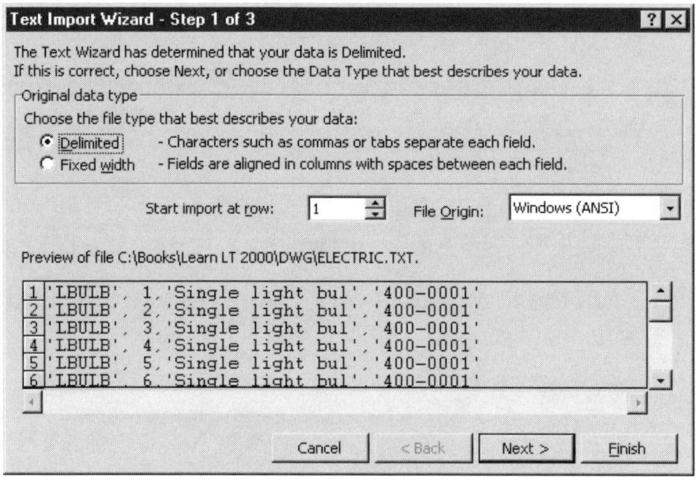

Text Import Wizard – Step 1 of 3.

6. In the Step 2 dialog box:
   - Delimiters: **Comma** (turn off **Tab**).
   - Treat consecutive delimiters as one: **Off**
   - Text Qualifier: **'** (single quote).
   - Click **Next.**

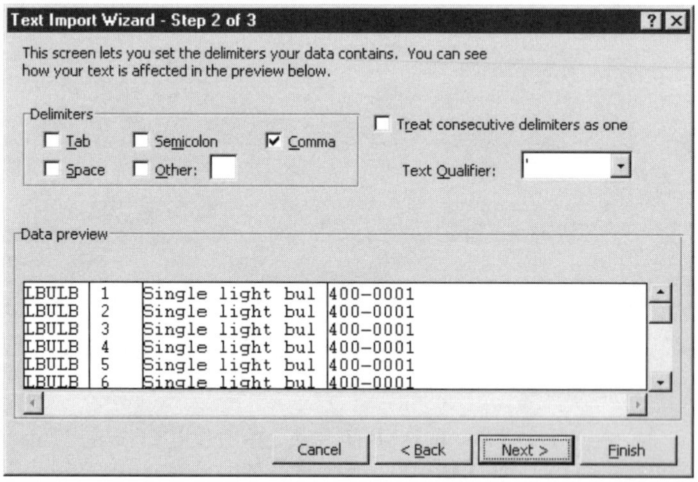

Text Import Wizard – Step 2 of 3.

7. In the Step 3 dialog box:
   - Select the first column, then click **Text.**
   - Select the third column, then click **Text** (leave the second column as General).

- Select the last column, then click **Text**.
- Click **Finish**.

Text Import Wizard – Step 3 of 3.

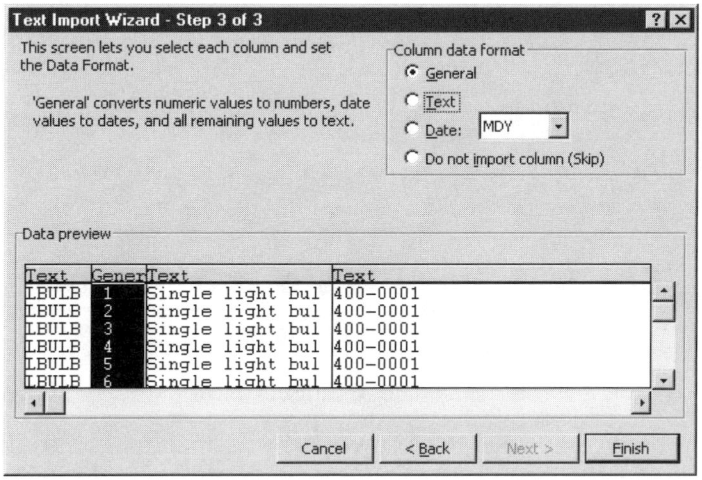

Notice that Excel loads the Electric.Txt file, displaying each field in its own column.

8. Select all four columns, then select **Format | Column | AutoFit Selection** to adjust the column widths.

Electric.Txt file loaded into Excel.

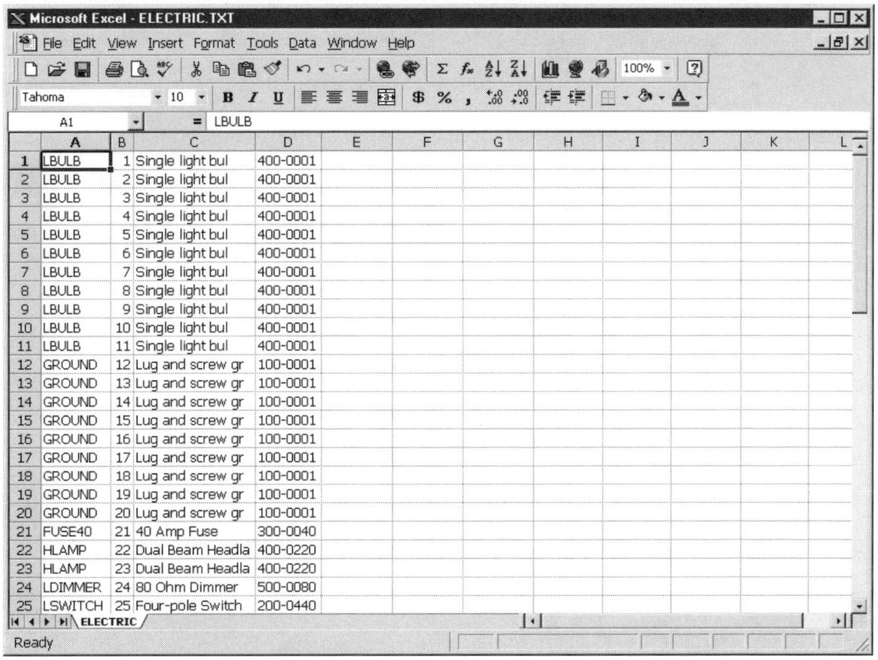

9. If you wish, add price and extension fields and a totals row, and format the text for lovely output.

> **Tip:** The spreadsheet formula to count the number of items in column B is:
>
> =count(B3:B72)

10. Remember to save the worksheet as an XLS file.

Importing Spreadsheet Data into Autocad LT

To bring the bill of material from Excel back into the AutoCAD drawing as a table, use the Windows Clipboard.

1. In Excel, select the rows you want to import into the drawing. You might want to select all the rows with data, or perhaps just the summary rows.
2. From Excel's menu bar, select **Edit | Copy** to copy the rows to the Windows Clipboard.
3. Switch to AutoCAD LT.
4. Reduce the size of the drawing to make some room for the incoming data. Try the Zoom 0.5x command.
5. Select the **Edit | Paste Special** command. AutoCAD displays the Paste Special dialog box.

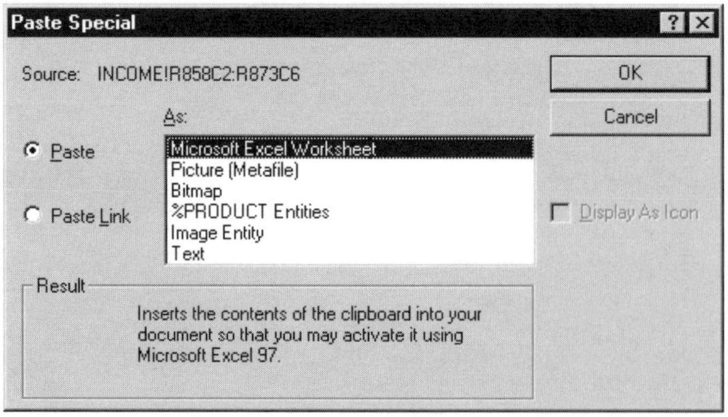

The Paste Special dialog box.

Note: Why not just use the Edit | Paste command? Because Excel places its data in 19 different file formats in the Clipboard, and because the Paste command gives you no choice over the best format for your need. In contrast, the Paste Special dialog box gives you the following six choices:

- Microsoft Excel Worksheet (linked object). By double-clicking on the image of the worksheet, the file is loaded back into Excel for further processing. This is the format placed by the simple Paste command.

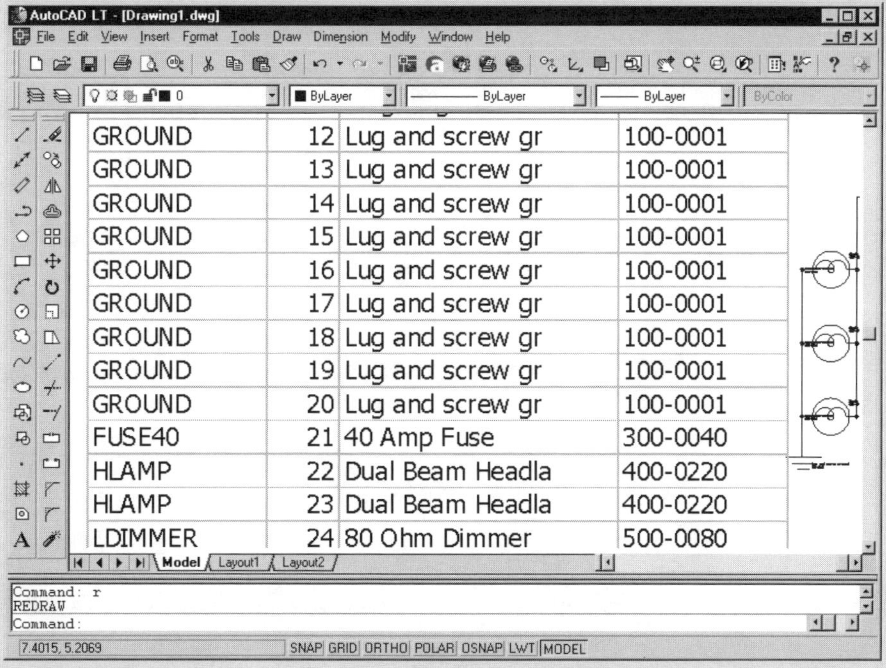

A close-up of the spreadsheet pasted as an Excel object.

- Picture (WMF metafile). The spreadsheet looks exactly like the linked object version but is not linked back to Excel, which saves memory.
- Bitmap (BMP raster image). A graphical representation of the spreadsheet, which looks like it does in Excel.
- AutoCAD LT entities. A bug in some early releases of LT 2000i may show this option as "%PRODUCT Entities." The spreadsheet's text is converted into AutoCAD text. While the formulae are lost, the text retains its formatting to a certain extent. (In a change from previous versions of LT, the object is no longer inserted as a block,

so you no longer need to use the Explode command before editing the text.) You can edit the text with the DdEdit command, and edit the style of the text with the Style command (the text styles all begin with the WMF prefix.)

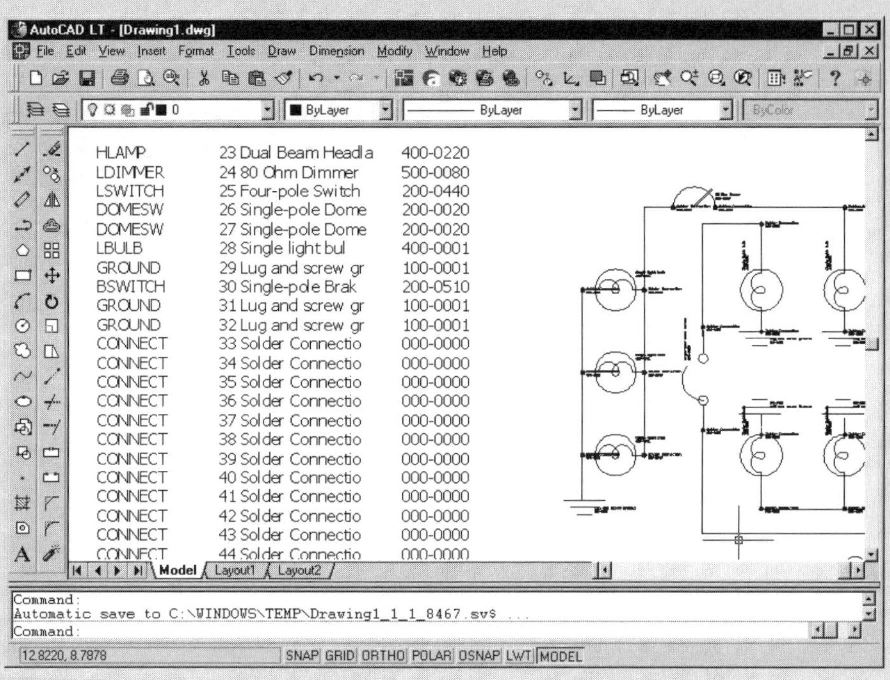

A close-up of the spreadsheet pasted as an AutoCAD LT object.

- Text. The spreadsheet is placed as an Mtext object, with greater loss of formatting than when placed as AutoCAD entities. The text uses the Standard style. You can edit the text with AutoCAD's DdEdit command, or format the text with the Properties command, such as its color and text style.
- Icon. The Excel icon is placed in the drawing, which redisplays faster than the actual spreadsheet.

6. Decide on the format that best suits you. Select the format name.
7. Click **OK**. AutoCAD copies the spreadsheet from the Clipboard and places it into the drawing.

Warning: The paste operation may fail when the amount of data in the Clipboard is too large for AutoCAD to handle. The failure ranges from nothing happening to AutoCAD crashing.

8. Notice the eight small black squares surrounding the object: drag the handles to resize and reposition the object in the drawing. AutoCAD LT has several commands specific to the pasted object. Right-click the pasted object; AutoCAD displays a cursor menu:

Editing an inserted Excel object.

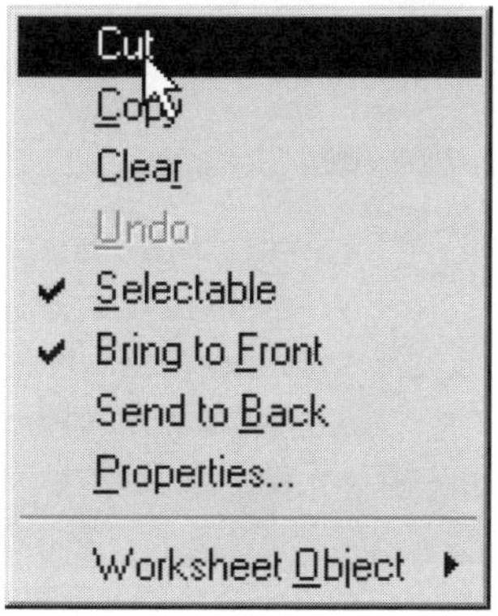

Cut: Cuts the object out of the drawing and sends it to the Windows Clipboard.

Copy: Copies the object from the drawing and sends it to the Windows Clipboard.

Clear: Erases the object from the drawing.

Undo: Undoes the previous command, including Cut and Clear.

Selectable: Toggles whether the object can be selected.

Bring to Front: The object appears in front of all other overlapping objects in the drawing.

Send to Back: The object appears behind other overlapping objects.

Properties: Displays the OLE Properties dialog box, which lets you change the size, scale, text font, and plot quality of the object.

Worksheet Object: Edit and Open launch the originating application and load the object for editing.

Convert: Converts the object into another format, or into an icon.

Chapter 9

Review Questions

1. AutoCAD LT can create a list of all blocks in the drawing.

   True / False

2. Which command extracts attributes from a drawing?

   a. AttDef
   b. AttExt
   c. AttDisp
   d. PasteSpec

3. Match the abbreviation with the best meaning:

a. TXT	i. Drawing interchange format
b. CDF	ii. Excel spreadsheet file
c. SDF	iii. Extension for an ASCII text file
d. DXF	iv. Space-delimited file
e. XLS	v. Comma-delimited file

4. A template file is optional for attribute extraction.

   True / False

5. Match the template abbreviation with the best meaning:

a. BL	i. 12-character word
b. C	ii. Numbers
c. N	iii. 12-digit number and two decimal places
d. N012002	iv. Characters
e. C012000	v. Block

6. When the AttDisp command is set to Normal, AutoCAD LT displays:

   a. All attributes not set to invisible.
   b. All attributes.
   c. All attributes defined by the user.
   d. All attributes defined by AutoCAD.

7. The CDF format separates values with:

   a. Tabs
   b. Spaces
   c. Commas
   d. Vertical lines

8. An Excel spreadsheet can be imported into an AutoCAD LT drawing.

   True / False

9. When a document is pasted in the AutoCAD drawing in AutoCAD LT Entities format, the text formatting is retained.

   True / False

10. When a document is pasted in the AutoCAD drawing in Text format, the text formatting is retained.

    True / False

Chapter 10

Region Modeling

In this chapter, you learn how to increase your CAD efficiency by working with regions and Boolean operations. These two AutoCAD features let you construct complex 2D shapes and analyze them. AutoCAD LT does this far faster than you would performing the calculations by hand.

Key Concepts

- Changing a collection of objects into a region
- Using point filters to find a coordinate
- Applying Boolean operations to regions
- Finding the mass properties of a region

Key Terms

Boolean—logical operations, such as AND, OR, and NOT

Mass properties—the properties of a mass, such as area, centroid, and radius of gyration

Point filter—returns a single coordinate

Region—a closed 2D area

Abbreviation

.x	X point filter

Commands	Shortcut	Menu Selection
DelObj	...	...
Intersect	in	Modify | Region | Intersect
MassProp Properties	Alt+TYR	Tools | Inquiry | Region/Mass
Region	reg	Draw | Region

Commands	Shortcut	Menu Selection
Subtract	su	Modify | Region | Subtract
Union	uni	Modify | Region | Union

Regions and Booleans

AutoCAD LT has the ability to work with regions. A *region* is a 2D closed area; the region cannot have intersecting curves. To create a region, you first draw its shape using drawing commands, such as Line, Arc, and Polyline. You can also use the Circle, Ellipse, and Spline commands to create its shape.

To convert the shape to a region, start the Region command and select the object. AutoCAD LT does the rest. After conversion, the object looks no different, but it has special properties, which we discuss later. The Explode command can be used to change a region back to lines, arcs, etc.

Warning: By default, AutoCAD erases the objects after they are converted to the region. To prevent this from happening, turn off the DelObj system variable (short for "delete object") by setting it to zero, as follows:

```
Command: delobj
Enter new value for DELOBJ <1>: 0
```

In addition, hatch associativity is lost. The region must be rehatched.

Boolean Operations

When you have two or more regions, you can perform *Boolean operations* on them. Boolean is a mathematical term describing logical operations known as AND, OR, and NOT. AutoCAD LT has three Boolean commands:

Intersect—Removes all but overlapping portions of two regions. The Boolean operation returns everything that is in region #1 AND region #2.

Subtract—Subtracts one region from another. The Boolean operation returns everything in region #1 NOT in region #2.

Union—Joins two regions into a single region. The Boolean operation returns everything that is in region #1 OR region #2.

The trickiest Boolean operation is Subtract, so this command is demonstrated in the following tutorial.

In the figure below, I drew a polyline, then used the PEdit command's Spline option to turn it into a smooth curve. I added the ellipse with the Ellipse command. I used the Region command on each closed shape. Then, I applied each Boolean operation on the two regions.

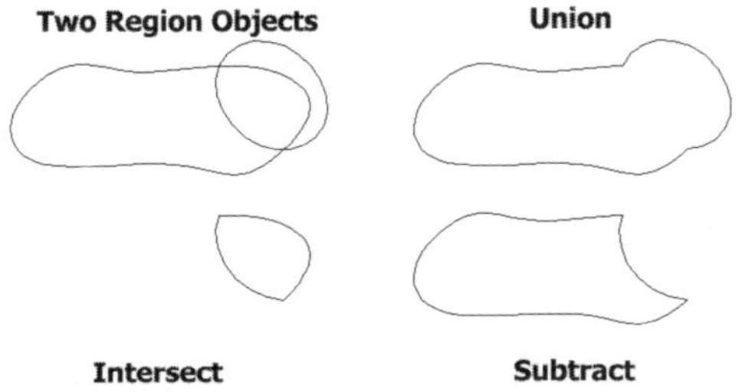

Applying Boolean operations to regions.

Creating a Waffle-Shaped Region

- Draw | Region
- Modify | Region | Subtract

The Region and the Boolean operation commands (Intersection, Subtract, and Union) are useful for working with areas of complex shape. In this tutorial, you create a waffle shape by applying the Subtract command to a group of regions, then finding its mass properties.

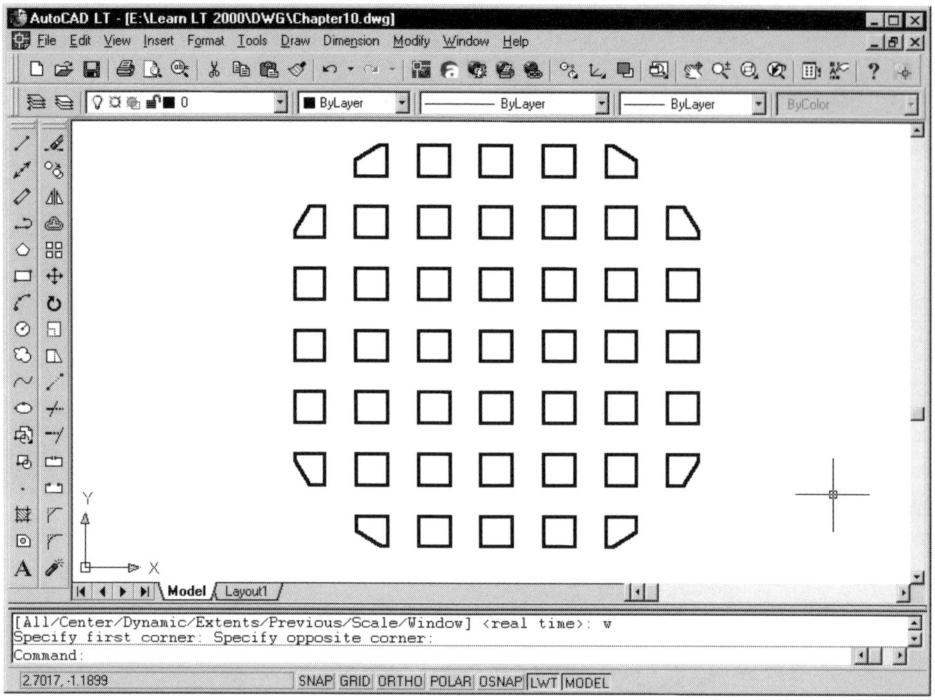

The object of the exercise: creating a waffle shape.

1. To start, draw a circle with a radius of 1.875 inches, as follows:

   Command: circle
   Specify center point for circle or [3P/2P/Ttr]: 0,0
   Specify radius of circle or [Diameter]: 1.875

2. Draw a rectangle, using object snap and point filters to draw it precisely, as follows:

   Command: rectang
   Specify first corner point or [Chamfer/Elevation/Fillet/Thickness/Width]: qua
   Of Specify other corner point or [Dimensions]: [pick circle's left edge]
   Specify other corner point or [Dimensions]: .x
   of qua
   of [pick circle's right edge]
   (need Y): .25

   Notice the use of Qua (quadrant) object snap, twice, to find precisely the circle's 180- and 0-degree points. The .x point filter allows you to pick the x-coordinate, then type the y-coordinate (0.25).

Creating a precise rectangle.

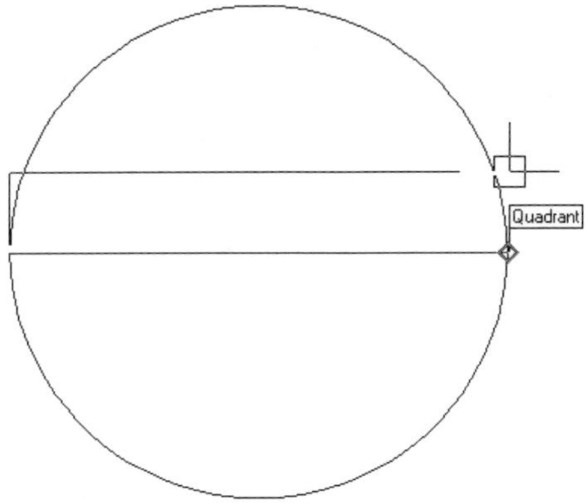

3. Move the rectangle to the bottom of the circle, again using object snap, as follows:

   Command: move
   Select objects: l
   1 found Select objects: [Enter]
   Specify base point or displacement: mid
   of [pick rectangle's lower leg]
   Specify second point of displacement or <use first point as displacement>: qua
   of [pick circle's lower edge]

   Notice the use of the L (last) selection mode to select the last drawn object visible in the viewport. The Mid (midpoint) and Qua object snaps are used to ensure precise vertical alignment.

Moving down the rectangle.

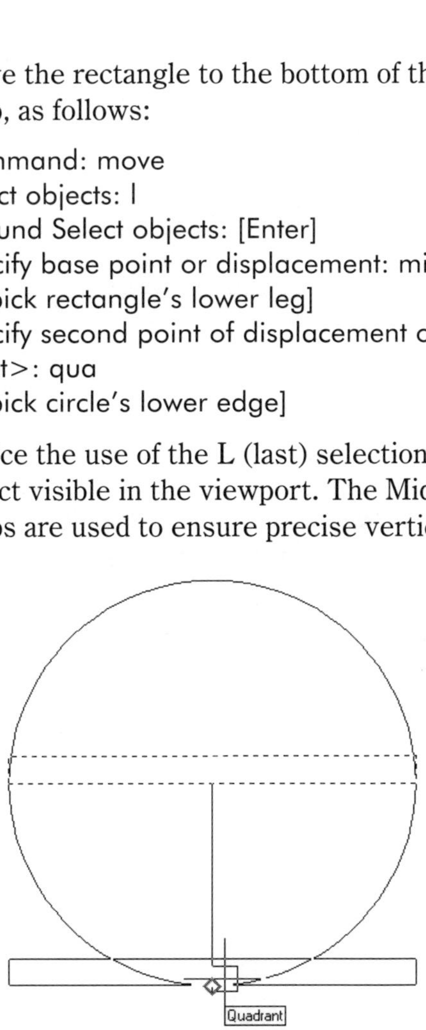

4. The next step is to create several copies of the rectangle. While we could use the Copy command, it is much faster to employ the Array command. We'll array rectangles twice in this tutorial: this time, we use the command-line version of the -Array command; later, we'll use the dialog box version.

   Array the rectangle vertically, as follows:

   Command: -array
   Select objects: l
   1 found Select objects: [Enter]
   Enter the type of array [Rectangular/Polar] <R>: [Enter]
   Enter the number of rows (---) <1>: 8
   Enter the number of columns (|||) <1> [Enter]
   Enter the distance between rows or specify unit cell (---): .5

   Notice that AutoCAD instantly creates seven copies of the rectangle. The hyphen prefix forces AutoCAD to display the prompts at the command line, rather than use the dialog box for this command.

Creating an array of rectangles.

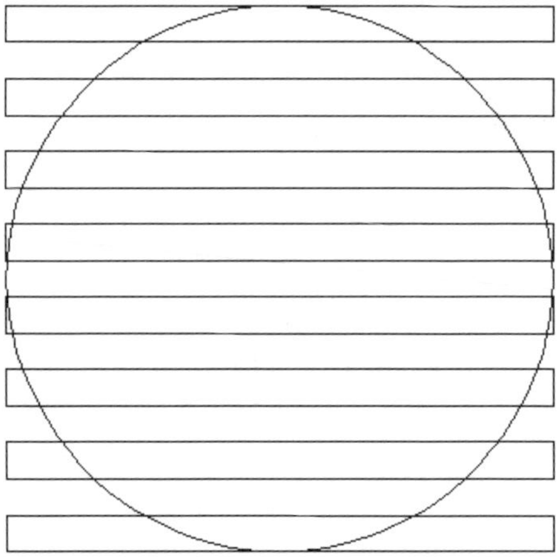

5. To create the other row of rectangles, mirror one horizontal rectangle, as follows:

   Command: mirror
   Select objects: p
   1 found Select objects: [Enter]
   Specify first point of mirror line: 0,0
   Specify second point of mirror line: @1<45
   Delete source objects? [Yes/No] <N>: [Enter]

Notice the use of the P (previous) selection set to select the same rectangle as selected for the Array command. If you had used L, AutoCAD would have selected the last drawn object (the final object drawn by the Array command).

Creating a vertical rectangle at a right angle.

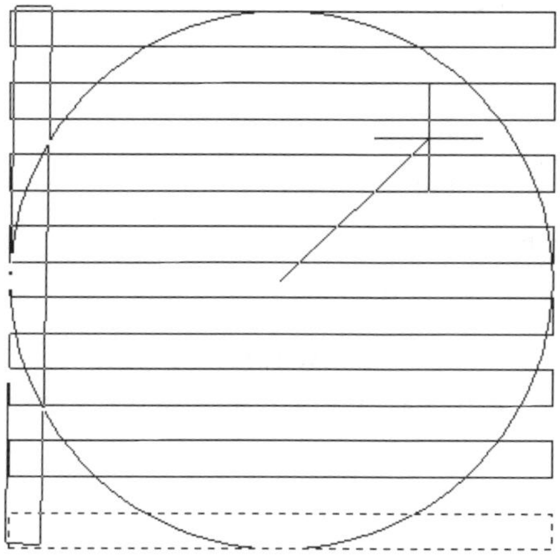

The placement of the mirror line is not crucial, as long it is at 45 degrees, hence the use of the relative polar coordinate: @1<45

6. Again, use the Array command. This time we use the dialog box, and this time we array the vertical rectangle. From the menu bar, select Modify | Array. Notice that AutoCAD displays the Array dialog box.

The Array dialog box.

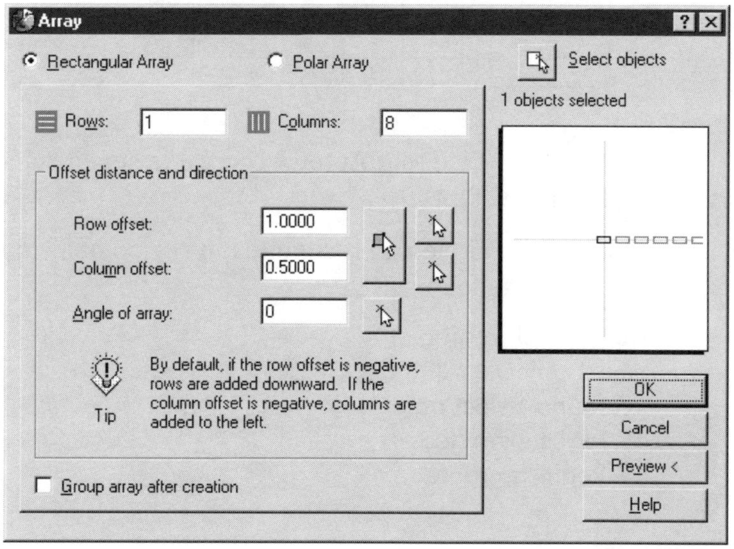

7. Make the following changes to the options in the **Array** dialog box:
   - **Array:** Select **Rectangular Array**
   - **Rows:** Change to **1**
   - **Columns:** Change to **8**
   - **Column offset:** Change to **0.5**
8. Click **Select Objects.** Notice that AutoCAD removes the dialog box, and prompts you at the command line:

   ```
   Select objects: l
   1 found Select objects: [Enter]
   ```

   Type **L** to select the last drawn object (the vertical rectangle), then press **Enter** to return to the Array dialog box.

Arraying the vertical rectangles.

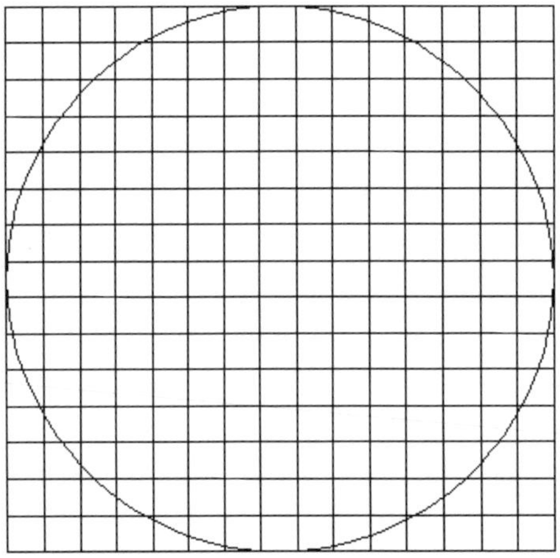

9. Click **OK.** Notice that AutoCAD creates seven copies of the vertical rectangle, instantly.
10. Convert the circle and rectangles to region objects with the Region command, as follows:

    ```
    Command: region
    Select objects: all
    17 found Select objects: [Enter]
    17 loops extracted.
    17 Regions created.
    ```

After conversion to regions, the objects look no different. The circle and rectangles are, however, circular and rectangular region objects. The Boolean operations would not work on regular circles and rectangles.

11. Subtract the rectangular regions from the circular region with the Subtract command, as follows:

    Command: subtract
    Select solids and regions to subtract from...
    Select objects: [pick circle]
    1 found Select objects: [Enter]
    Select solids and regions to subtract...
    Select objects: all
    Select objects: [Enter]

    To make it faster to pick all the rectangles, use All selection mode. The Region command ignores the fact that the circle has been picked twice.

    Notice the result: the Subtract command removes all overlapping regions, producing the waffle effect. That's one of the benefits to working with regions.

The Subtract command removes all overlapping regions.

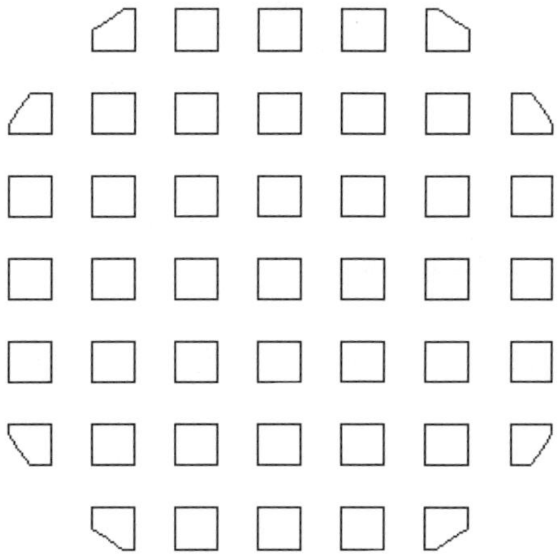

12. Another benefit to working with regions is being able to easily find their properties. If the waffle shape were made of regular 2D objects (such as rectangles and polylines), you would have to, for example, find the area of each individual square. In contrast, the waffle shape we created is a single region object.

To find the total area of the waffle pattern, use the MassProp command, as follows:

```
Command: massprop
Select objects: [pick waffle shape]
1 found Select objects: [Enter]
 ---------------- REGIONS ----------------

Area:                   2.7219
Perimeter:              44.3011
Bounding box:           X: -1.6250  --  1.6250
                        Y: -1.6250  --  1.6250
Centroid:               X: 0.0000
                        Y: 0.0000
Moments of inertia:     X: 2.3496
                        Y: 2.3496
Product of inertia:     XY: 0.0000
Radii of gyration:      X: 0.9291
                        Y: 0.9291
Principal moments and X-Y directions about centroid:
                        I: 2.3496 along [1.0000 0.0000]
                        J: 2.3496 along [0.0000 1.0000]

Write analysis to a file? [Yes/No] <N>: [Enter]
```

So now you know: the area of the waffle pattern is 2.7219 square units.

Note: AutoCAD assumes the density of the region is constant, with a value of 1.0. Some of the terms in the report generated by the MassProp command may be unfamiliar:

The bounding box and centroid of a region.

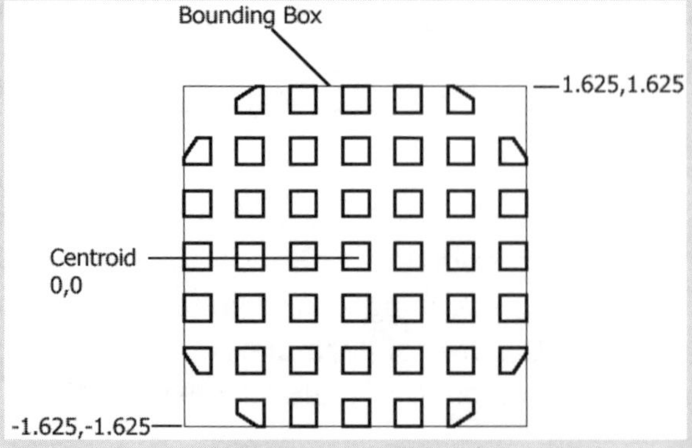

Bounding box is the rectangular limit of the region; the coordinates describe the lower-left and upper-right corners of an imaginary rectangle that tightly encloses the region.

Centroid is the center of mass of the region

Moments of inertia is a measure of a region's resistance to angular acceleration.

Two other commands that also report the area and perimeter information of regions are List and Area. The List command provides a more concise report of the region's area, perimeter, and bounding box. The Area command displays an even briefer report of just the area and perimeter. Use the Object option, and select the region:

```
Command: area
<First point>/Object/Add/Subtract: o
Select objects: [pick]
Area = 2.7219, Perimeter = 44.3011
```

Review Questions

1. The best definition for a region is:
   a. A 3D solid model created in AutoCAD 2000i.
   b. A view defined by the View command.
   c. A 2D closed area with no intersecting curves.
   d. A specific area of the drawing, such as the title block.
2. You create a region as follows:
   a. Draw the region's outline with the Region command.
   b. Draw the region's outline with drawing commands, then convert with the Region command.
   c. Set the DelObj variable to 0, then using the Region command.
   d. Draw the region's outline with the Region command, then convert with the PEedit command.
3. To convert a region back to a lines and arc, use the:
   a. Explode command.
   b. Convert option of the Region command.
   c. Explode option of the Region command.
   d. Regions cannot be converted back.

4. When turned on (set to 1), the purpose of the DelObj system variable is to:

   a. Prevent the user from deleting any objects from the drawing.
   b. Retain objects after they are converted to a region.
   c. Enables the Del key to erase objects in the drawing.
   d. Erase objects after they are converted to a region.

5. Hatch associativity is lost when a hatched area is converted to a region.

   True / False

6. Match the Boolean term with its best meaning:

a.	Intersect	i.	Join two regions into a single region.
b.	Subtract	ii.	Same as Intersect.
c.	Union	iii.	Same as Union.
d.	AND	iv.	Remove one region from another region.
e.	OR	v.	Remove all but overlapping areas of two regions.

7. In AutoCAD LT, Boolean operations apply only to region objects.

   True / False

8. Which command(s) is faster for making many copies of an object?

   a. Array
   b. Copy
   c. Union
   d. MRedo

9. The briefest report on the area and perimeter of a region is provided by the:

   a. Area command
   b. Region command
   c. List command
   d. MassProp command

10. The centroid of a region is its:

   a. Center
   b. Resistance to angular acceleration
   c. Rectangular limit of the region
   d. Center of mass of the region

Chapter 11

Accessing Autodesk's Portals

AutoCAD LT 2000i includes direct access to several Autodesk supported Web sites. You have already met one perhaps without realizing it: Point A forms the lower half of the AutoCAD LT Today window, which greets you when you first start AutoCAD LT.

Key Concepts

- Accessing Point A and other Web portals from inside AutoCAD
- Disabling and enabling the AutoCAD LT Today window
- Setting up an account with Buzzsaw.Com
- Learning about the RedSpark and AUGI Web sites

Key Terms

Object enabler—software plug-in that allows AutoCAD to edit proxy objects

Portal—a specialized Web site that provides information specific to your industry

Proxy object—an object defined by ObjectARX programming code

Abbreviations

AUGI	Autodesk User Group International
AEC	Architecture, engineering, construction
GIS	Geographic information system

You can, naturally, access any Web site you want with the Web browser you prefer. By making Web sites easily accessible from within AutoCAD LT, Autodesk hopes to draw you into using them more often than you might otherwise. The Web sites that are directly accessible from LT 2000i are:

Point A: A "portal" created by Autodesk to give you access to information specific to your type of CAD use, and provide updates and add-ons to AutoCAD LT as well as industry specific information.

Buzzsaw.com: A project management site. An Autodesk spinoff specific to AEC (architectural, engineering, construction) CAD users. Access from LT 2000i is via any file dialog box, such as File | Open.

RedSpark.com: An e-commerce site for construction material. Another Autodesk spinoff specific to mechanical CAD users. Access from LT 2000i is also via any file dialog box.

AutoCAD User Group International: The site for the Autodesk-sponsored user group for AutoCAD users. You can become a member, free, at www.augi.com. Access from LT 2000i is via **Help | Autodesk User Group International**, which launches your computer's default Web browser.

A fifth link is to Buy Online for making online purchases. At the time of writing this book, the BuyOnline command reports, "This command is not available for this version." This is a future link to Autodesk's eStore at www.autodesk.com/estore for purchasing LT 2000i and other products.

Curiously, there is no longer a direct access to the AutoCAD LT area of the Autodesk Web site (www.autodesk.com/autocadlt) as there was in LT 2000 and earlier.

To work with the tutorials in this chapter, your computer must: (1) be connected to the Internet; and (2) have Microsoft's Internet Explorer available as the Web browser.

Tip: You don't need to view the AutoCAD LT Today window each time you start LT 2000i. If you prefer the Startup dialog box found in LT 2000 and earlier, do the following:

1. From the AutoCAD LT menu bar, select Tools | Options.
2. In the Options dialog box, select the System tab. In the upper-right quadrant, notice the Startup option.
3. Select Show traditional startup dialog.
4. Click OK.

The next time you start AutoCAD LT, it will display the Startup dialog box. You will probably prefer this dialog box when your computer does not have a connection to the Internet. You can always open the AutoCAD LT Today window by selecting Tools | Today from the menu bar.

Point A

pointa.autodesk.com

Autodesk created Point A as a portal that helps you locate resources specific to your CAD discipline, including: general design and drafting; AEC (architecture, engineering, construction); land development and civil engineering; GIS (geographic information systems) and mapping; or manufacturing.

Point A can be accessed using your Web browser, or from within AutoCAD LT 2000i. Once there, Point A provides you with:

- Software updates and add-ons
- Industry news, links, and resources
- Searchable databases of parts, materials, and symbols
- Project-hosting services
- Hosted software applications
- Support documents, product tips, discussion groups, online training, and links to books

In addition, AutoCAD LT 2000i automatically goes to Point A when it opens a drawing containing proxy objects, looking for the appropriate plug-in that allows LT to correctly display the proxy object.

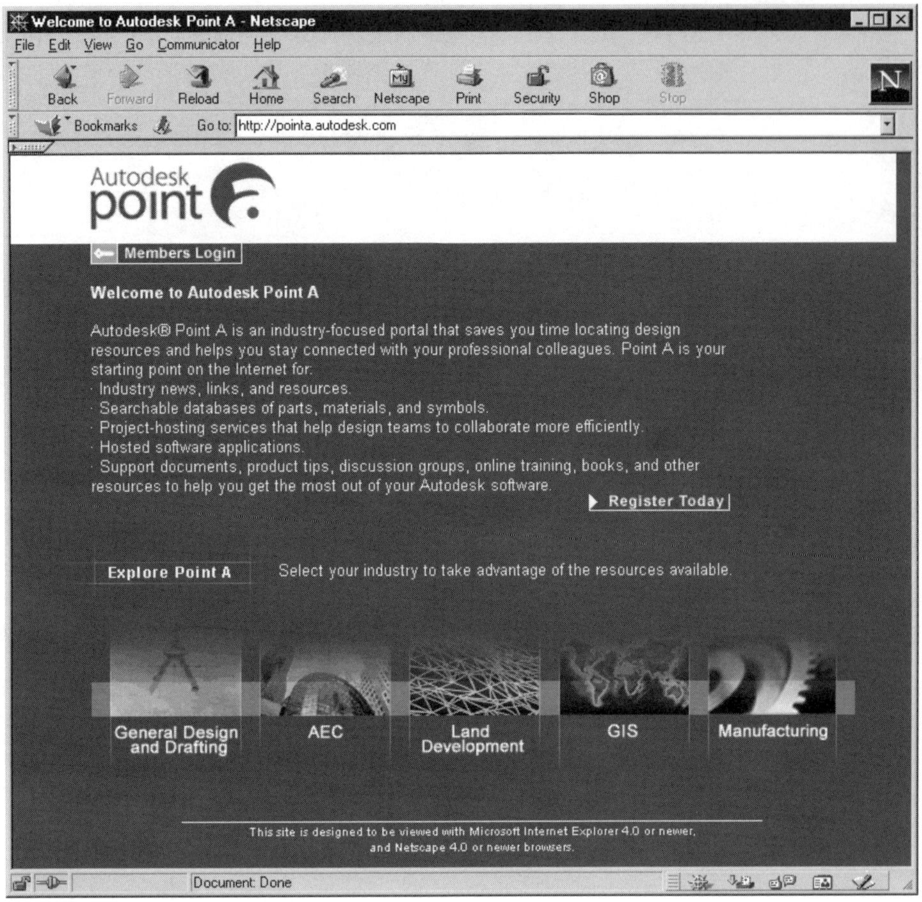

Accessing Point A

You can access Point A using your Web browser. Enter the following address: **pointa.autodesk.com** (notice that there is no "www" prefix as is normally found in Web addresses).

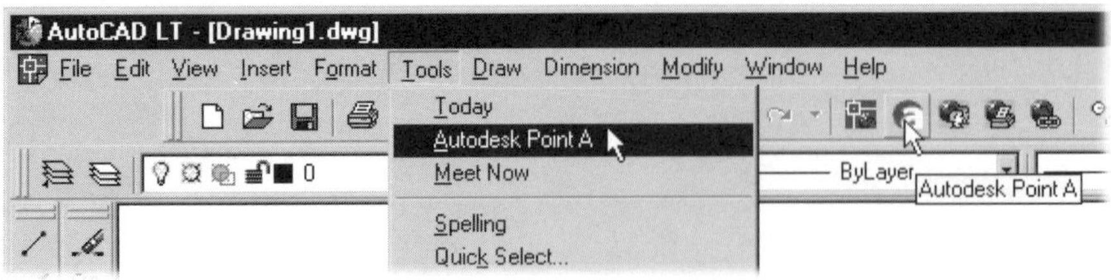

Within AutoCAD LT, there are several ways to access Point A:

- From the menu bar, select **Tools | Autodesk Point A.**
- From the Standard toolbar, click the **Autodesk Point A** button.
- In the AutoCAD LT Today window, click the Autodesk Point A logo.

Notice that the AutoCAD LT Today window becomes larger, and displays content from the Point A Web site.

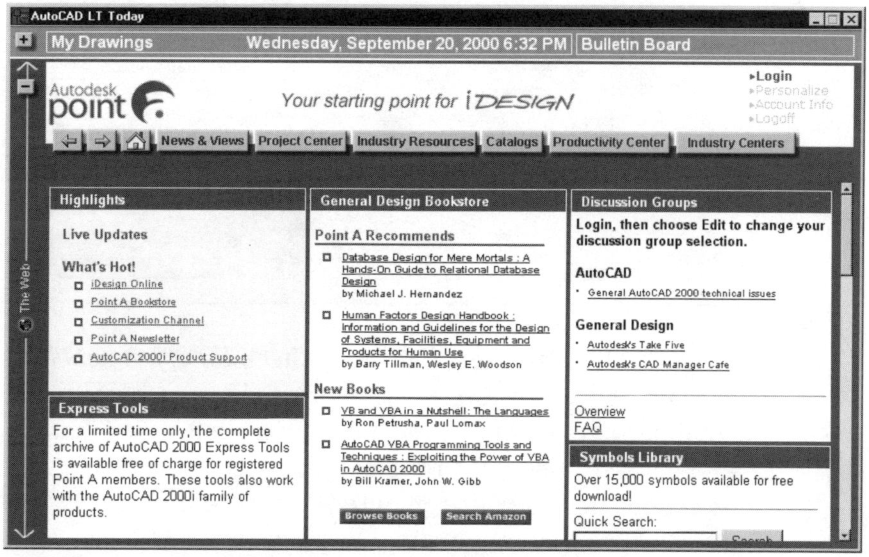

Tip: To maximize the AutoCAD LT Today window, click the Minimize My Workplace button next to My Drawings.

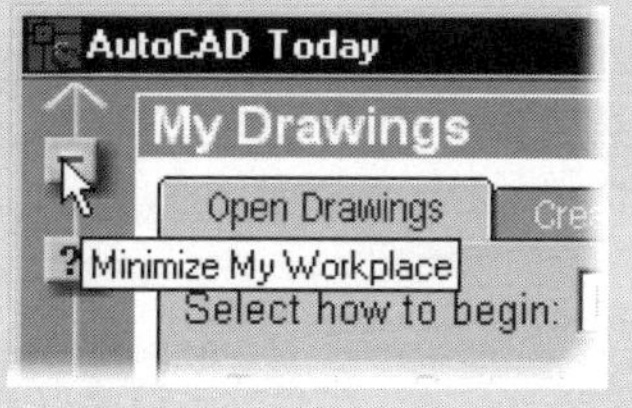

Automatic Updates and Plug-in Searches

From time to time, AutoCAD checks with Point A to see if updates are available. If so, you are asked permission to download code to your computer.

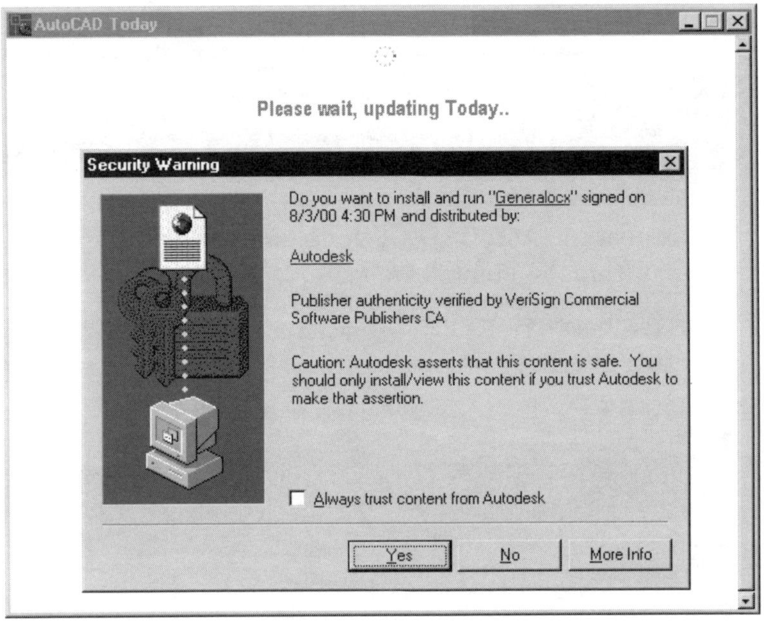

AutoCAD LT also uses Point A to check for *object enablers*. An object enabler allows AutoCAD LT 2000i to edit *proxy objects*. Proxies are programmer-defined objects that are controlled by programs written with Autodesk's ObjectARX programming interface. If LT cannot find the object enabler on your computer, it tries to find it at Point A (in accordance with your Live Enabler settings in the Options dialog). Enablers are available for proxy objects created by Achitectural Desktop and Land Development Desktop, although the search for enablers is not always successful, as indicated by the illustration.

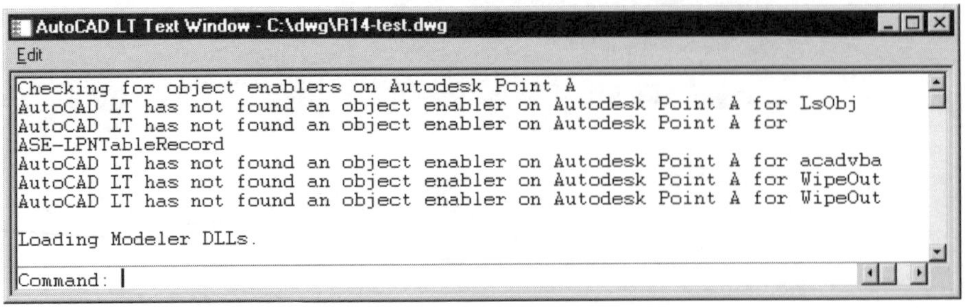

Buzzsaw.com

www.buzzsaw.com

In 2000, the new development in CAD was the *project Web site*. This is a specialized Web site that keeps track of projects. The Web site stores the project's CAD drawings and other documents, provides a meeting place for project participants, and allows you to keep a record of all project-related e-mail.

Buzzsaw provides a free project Web hosting service. Funded by Autodesk, buzzsaw.com was one of the very first of such sites. To use a project Web site, you: (1) choose a site name; (2) add the names and e-mail addresses of team members; (3) and upload your documents. As the project proceeds, team members e-mail each other (the messages are stored at the site), redline their drawings, post updated specs and drawing revisions. The Web site tracks all this. And since it is a Web site, the data can be accessed from anywhere you can get onto the Internet—you are no longer limited to tracking projects with your desktop computer.

Buzzsaw provides a "try me free" incentive. The first 100MB of disk space is free, as are toll free telephone and e-mail support. After you fill up those 100MB (about 100 drawings), the cost is US$199 per 250MB of data storage. Once the project is complete, you can have the project documents archived at US$150 per CD-ROM (each holds 650MB). If your free site is inactive for more than 90 days, it is shut down (after e-mail notification), and the data is erased.

Buzzsaw wants to provide a wide range of services to the architect, engineer, and contractor. For example, there is a Web Camera service (US$399 per month) that provides the hardware and the acquisition and storage of up to 1,000 images per month. And it doesn't need to be CAD-specific. The WebEx meeting software (free for an introductory period) allows your group to host online meetings, and redline documents concurrently.

Buzzsaw isn't limited to CAD. Its 100MB of free storage space could be a backup of your most important files: photos from your vacation that you share with family and friends; chapter files of your novel for access by the copy editor and publisher; and it can be accessed from any Web browser in the world.

> **Note:** It was during the A/E/C Systems '99 show that Autodesk announced its plans to seriously enter the Internet business. At the time, the only information they gave was a name: ProjectPoint. During the months following the announcement, www.projectpoint.com Web site lay dormant, except for the cryptic message, "Through this portal pass the finest fish in the world." CEO Carl Bass later told me its meaning: there was a fish market across from the restaurant where he and his staff went to plan ProjectPoint; the saying was posted over the market's entrance.
>
> When the Web site went live five months later, its name had changed to Buzzsaw.Com; ProjectPoint became the name of software for managing projects at Buzzsaw. In addition, Buzzsaw was spun off as an Internet start-up company with staff and million of dollars of investment money from Autodesk. That's why you'll often see the phrase, "Buzzsaw.Com, an Autodesk venture."

Trying Out ProjectPoint

Let's try out the ProjectPoint service provided by Buzzsaw. There are two ways you can approach Buzzsaw: (1) entering **www.buzzsaw.com** in your preferred Web browser; or (2) clicking the Buzzsaw button in one of LT 2000i's file dialog boxes.

To access Buzzsaw's ProjectPoint, you need two items:

- **Internet Explorer.** ProjectPoint works only with Microsoft's Web browser, version 4.01 or higher.
- **Internet.** Before you start this tutorial, ensure your computer is connected to the Internet.

The process involves these steps:

Step 1: Register yourself with Buzzsaw.com.

Step 2. Create a project workspace.

Step 3. Download the ProjectPoint software.

Step 4. Set up the project workspace.

Step 1: Register Yourself with Buzzsaw.com

1. From AutoCAD LT's menu bar, select **File | Open.**

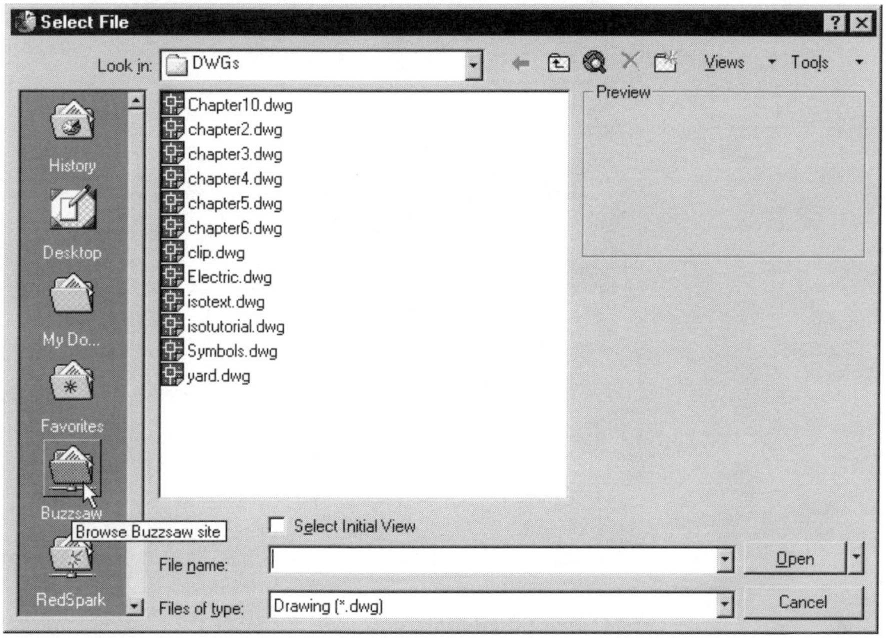

2. In the Select File dialog box, click the Buzzsaw icon (found on the left side). Notice that your computer's default Web browser is launched with www.buzzsaw.com/autocad/.
3. If this is the first time you have connected with Buzzsaw, LT displays a dialog box reading, "Once you have installed ProjectPoint click OK to continue." You might not notice the dialog box, since it may be obscured by the Web browser.

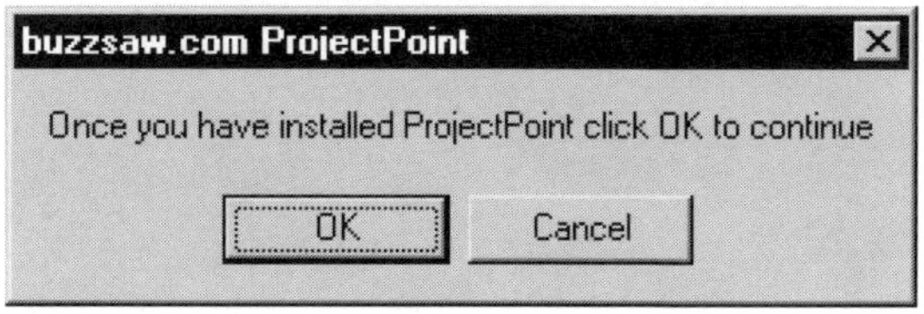

4. In the Web browser, click **Register now to start a project.**

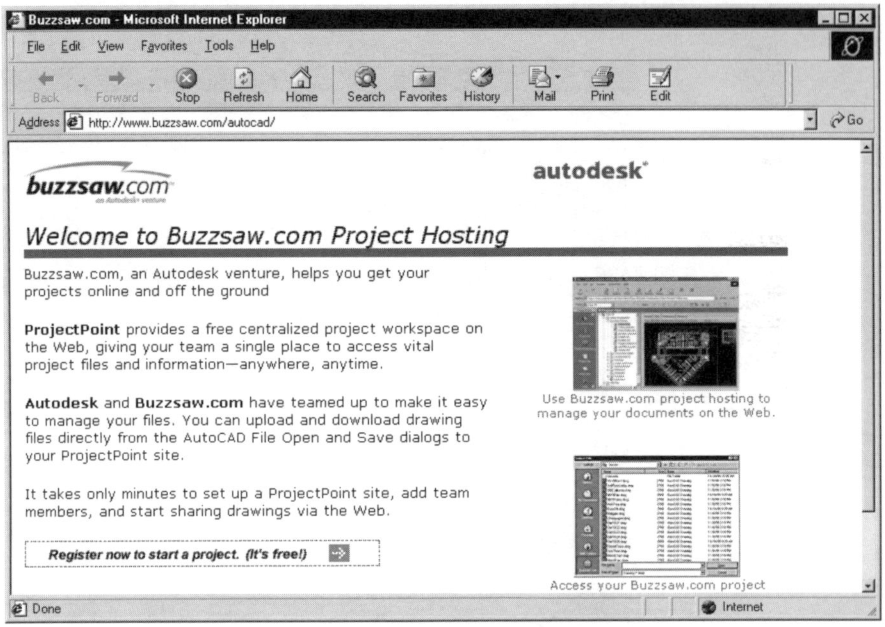

5. To register yourself with Buzzsaw, fill in the fields on the Web page. Those marked with an asterisk (*) must be filled in; the others are optional. For Password reminder, enter a phrase that reminds you of your password, such as "mother-in-law's maiden name." I specified the following:

   Username: learnautocadlt
   Password: wordware
   Password reminder: book publisher

   If you forget your password, you can get help at **www.buzzsaw.com/content/login/sendpass.asp.**

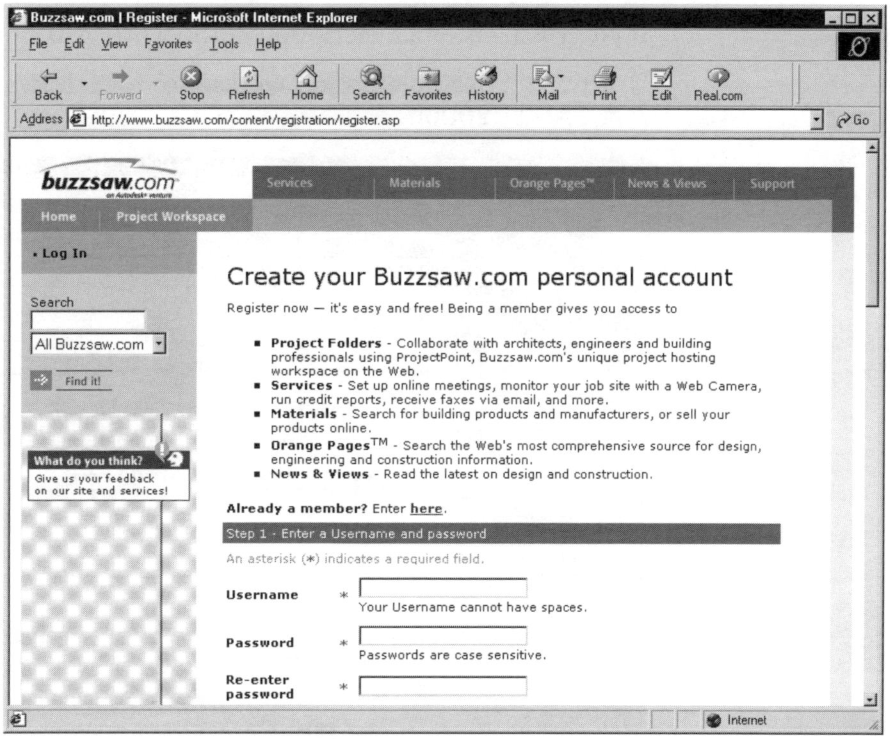

6. At the bottom of the page, read the Terms and Conditions, then click "I have read and ACCEPT the terms of the user agreement."

7. Click **Submit**. Shortly after you click this button, you will receive an e-mail from Buzzsaw welcoming you to their site.

Step 2. Create a Project Workspace

1. The next step is to create the *project workspace*. The workspace is where you manage your projects. You create a project workspace by specifying a *site name*. This can be any convenient name, such as your name or the name of your company. The name can be up to 32 numbers or letters long, but must start with a letter; special characters and spaces are not permitted. I am using the site name of "learnautocadlt."

   Site Name: learnautocadlt

   This means you will be able to go directly to this project workspace by entering the following URL: **http://folders.buzzsaw.com/client/learnautocadlt**.

Chapter 11

2. If you wish, enter a description of the workspace:

   Description: Example project workspace for readers of "Learn AutoCAD LT 2000i".

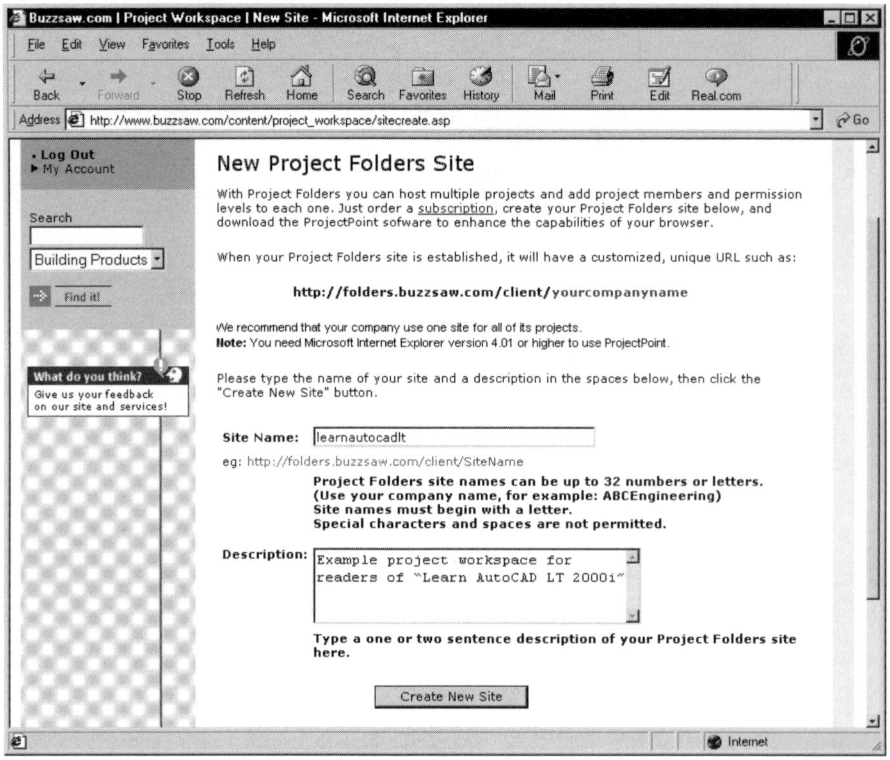

3. Click **Create New Site**. Wait for a few minutes while the site is created for you. The project site will remain available, as long as you use it. If you leave the site unused for 90 days, Buzzsaw will alert you with an e-mail before closing the site.

Step 3. Download the ProjectPoint Software

1. When the "Project Site Complete" page appears, click the link next to Site Name. This starts an automatic process that downloads a 2MB file containing the ProjectPoint software to your computer.

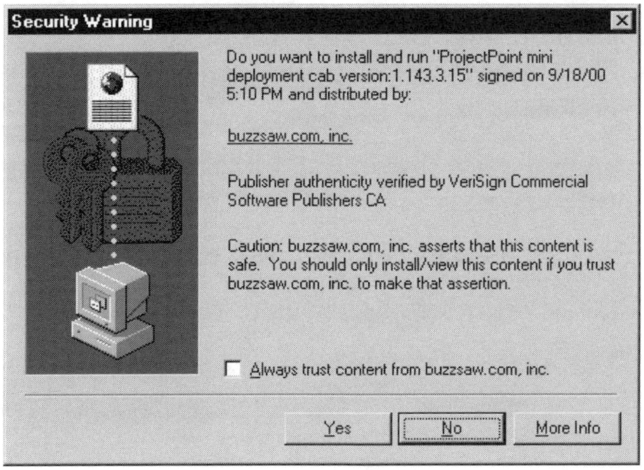

2. Your Web browser may display a warning dialog box. Click **Yes** to start the download process.
3. Once the software has been downloaded, a dialog box appears showing the ProjectPoint License Agreement. Click **I Agree.**
4. When the Buzzsaw.com ProjectPoint Setup dialog box appears, click **Install AutoCAD Mini-Client**. This allows you to access your project from within AutoCAD.

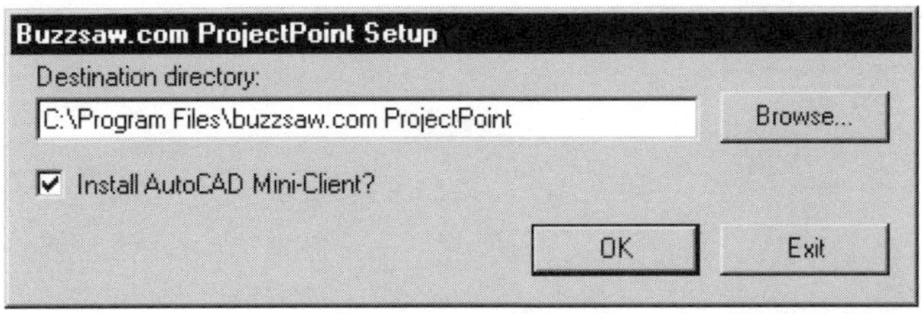

5. Click **OK** to accept the folder in which the software will be installed. If you wish, you may specify a different folder.
6. Once the software is installed, the ProjectPoint Login dialog box appears. Enter your username and password.

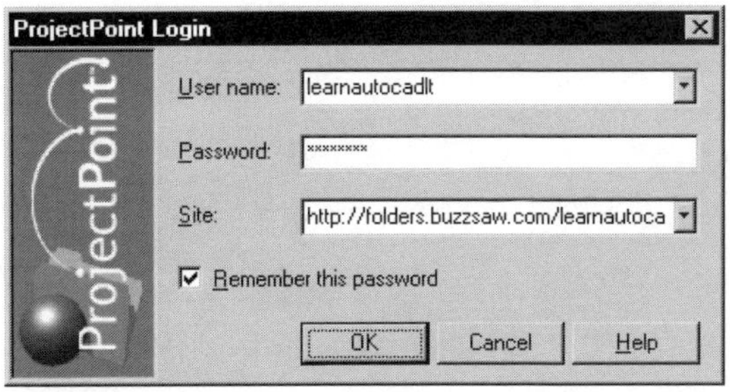

7. Click **Remember this password** to turn on the feature. This means you don't need to type the password each time you access ProjectPoint.
8. Click **OK**. Notice that Internet Explorer changes to display the ProjectPoint interface, which consists of:
   - Two new toolbars, **Project Selection** and **Standard.**
   - **Places list**, which takes you to various areas of the Buzzsaw Web site.

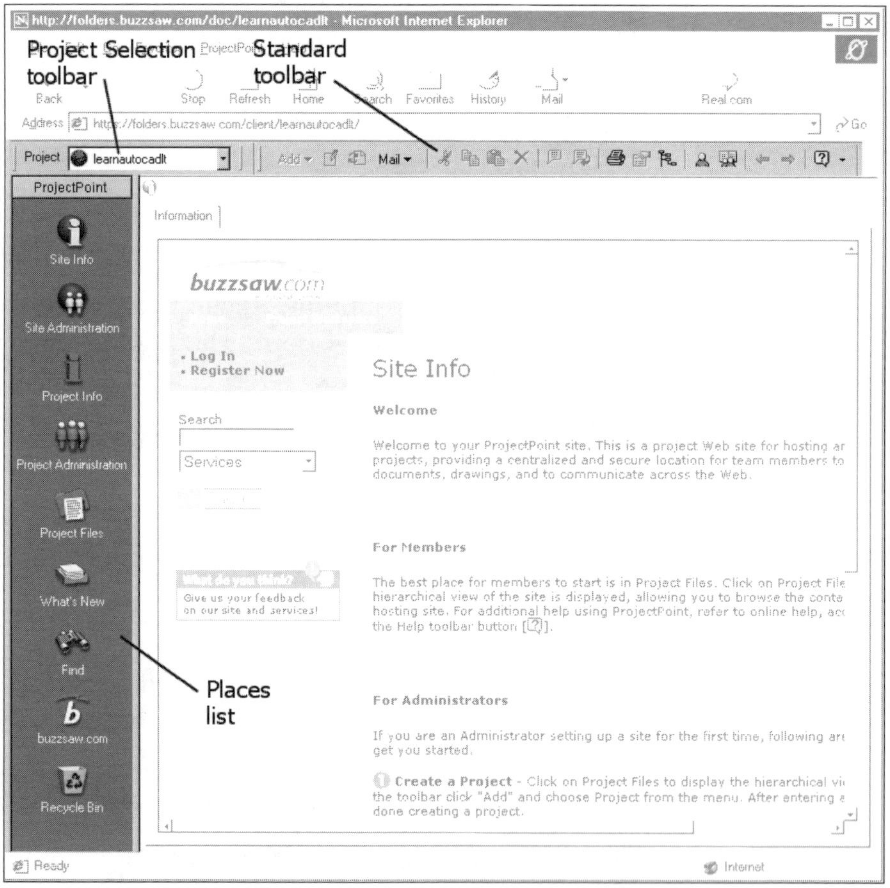

Step 4. Set Up the Project Workspace

Now that you are signed into your project workspace, you need to set it up. This consists of specifying the name of the project (each project workspace can contain many projects), providing access to other team members, and optionally changing the site info page.

1. To create a new project area, click **Project Files**, which you find about halfway down the places list. Notice that the Web browser now displays the Project Files view.
2. On the Standard toolbar, click **Add**.
3. From the pop-up menu, select **Project**.

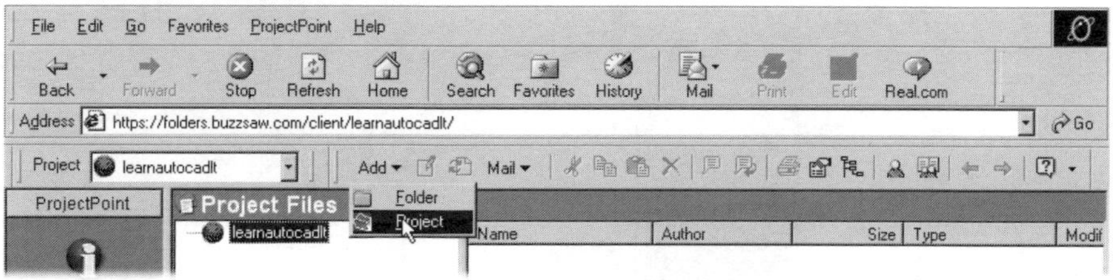

4. Enter the name for the project. The default is "New Project." The project is created instantly. The next step is to add the names of people permitted to access the project. If you are the only one, you can skip this step.
5. To assign personnel to the project, click **Site Administration** in the places list (second down from the top). Notice that the Web browser displays the Site Administration view.
6. Click the **Site Members** tab.
7. Click **New**.

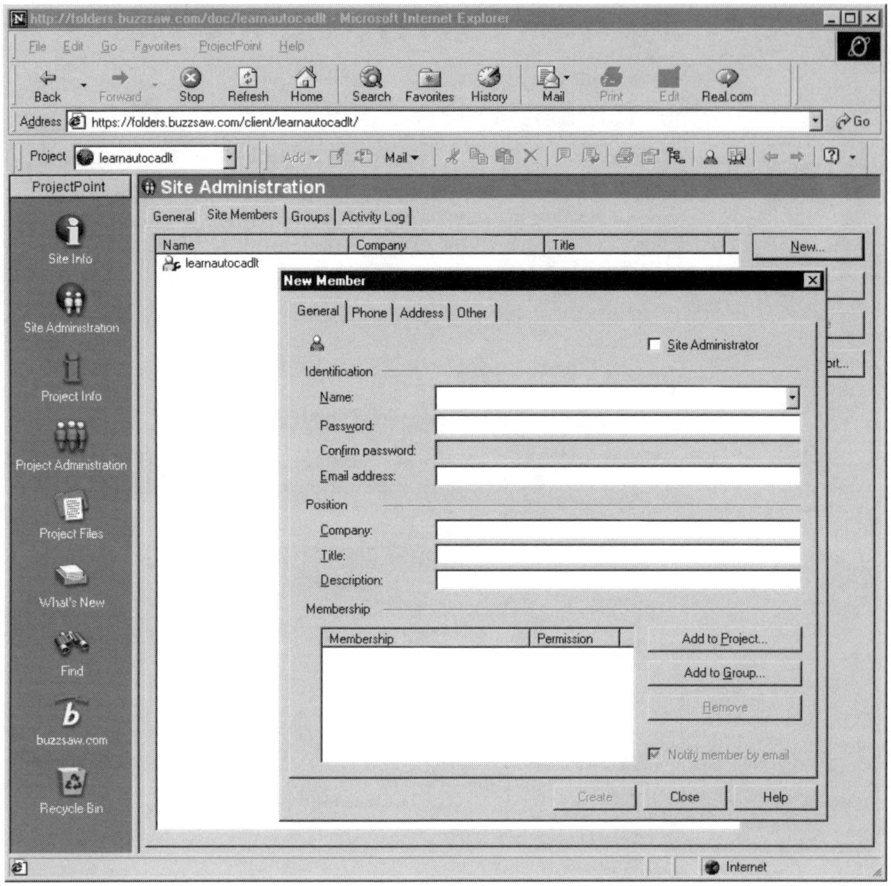

8. Fill out the dialog box for each person in the team. When done, click **Create**.

For additional help using ProjectPoint, use the online help accessed by clicking the **Help** button.

Tips: Read the privacy policy at **www.buzzsaw.com/content/common/privacy_policy.asp**.

At **www.buzzsaw.com/content/registration/profile_edit.asp**, you can edit your profile to no longer receive offers and mailings.

If you forget your password, go to the Buzzsaw.com login page, and ask to have your password e-mailed to you.

If you have problems signing in to Buzzsaw.com, e-mail **support@buzzsaw.com**.

To cancel your account, e-mail **support@buzzsaw.com** with your username. You will receive instructions on deleting your account.

Accessing Buzzsaw from AutoCAD LT

You can access your projects on Buzzsaw from within AutoCAD LT 2000i. Here's how:

1. Make sure your computer is connected to the Internet.
2. Start AutoCAD LT 2000i, and open a sample drawing.

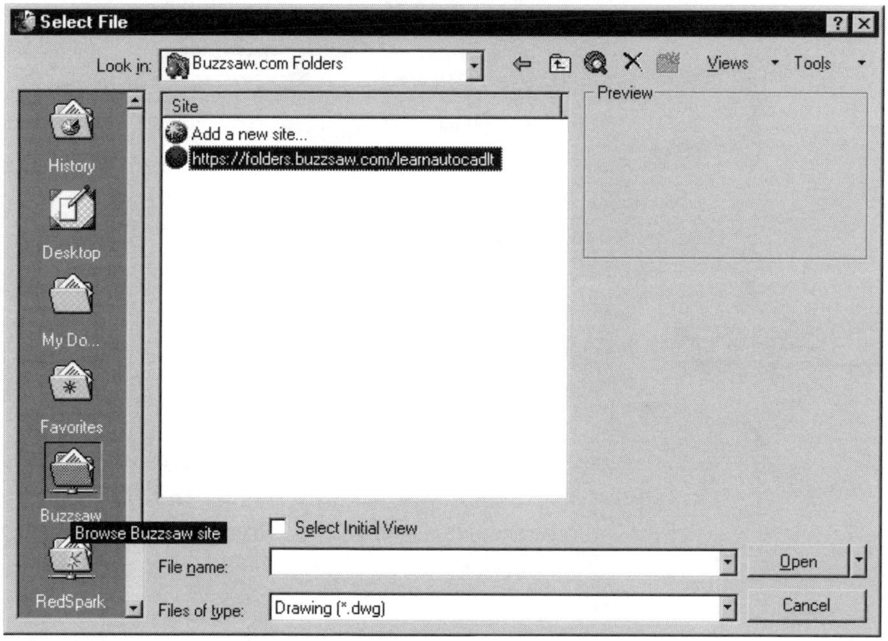

3. From AutoCAD's menu bar, select **File | Save As.**
4. In the Save Drawing As dialog box, click **Buzzsaw.** Notice that the file list changes to display the project workspace you created earlier.
5. Double-click the name of the project workspace. Notice that the ProjectPoint login dialog box appears. If necessary, fill in your username and password.
6. Click **OK.** Notice that the dialog box now displays the project folders within the workspace. Double-click the folder to open it.
7. Click **Save** to save the current drawing to your ProjectPoint folder.

Since the drawing has been saved to a Web site, it is now accessible from anywhere in the world with an Internet connection. You can access the drawing—as well as other documents, such as Word files, while on the road. International team members can also access the files.

> **Warning:** How safe is your data?
>
> When you enter a password or other personal information to the Buzzsaw Web site, it is encrypted using public key encryption. As data is transmitted between your computer and Buzzsaw's Web site, the ProjectPoint software uses Secure Socket Layer (SSL) technology, which supports up to 128-bit encryption (depending on the Web browser's capabilities). For your projects, you can specify security and user access options at the site, group, folder, and file levels.
>
> The physical hardware and software components of the Buzzsaw Web site is hardened against attack and natural disasters. The data center is guarded and under video surveillance on a 24/7 basis to ensure maximum access control and data protection.

RedSpark

www.redspark.com

The purpose of the RedSpark Web site is to provide a source for manufacturers—design and purchasing (internal to corporation) and fabricators and suppliers (external to the corporation). RedSpark plans to serve as a "vertical service provider" for manufacturing companies who buy and supply products and services.

At the time of this writing, RedSpark was not yet up and running, other than the introductory text at its Web site. RedSpark, however, plans to provide the following services:

- Custom and semi-custom engineered goods sourcing
- Catalog components sourcing
- Excess inventory exchange
- Used capital equipment exchange
- Request for proposals

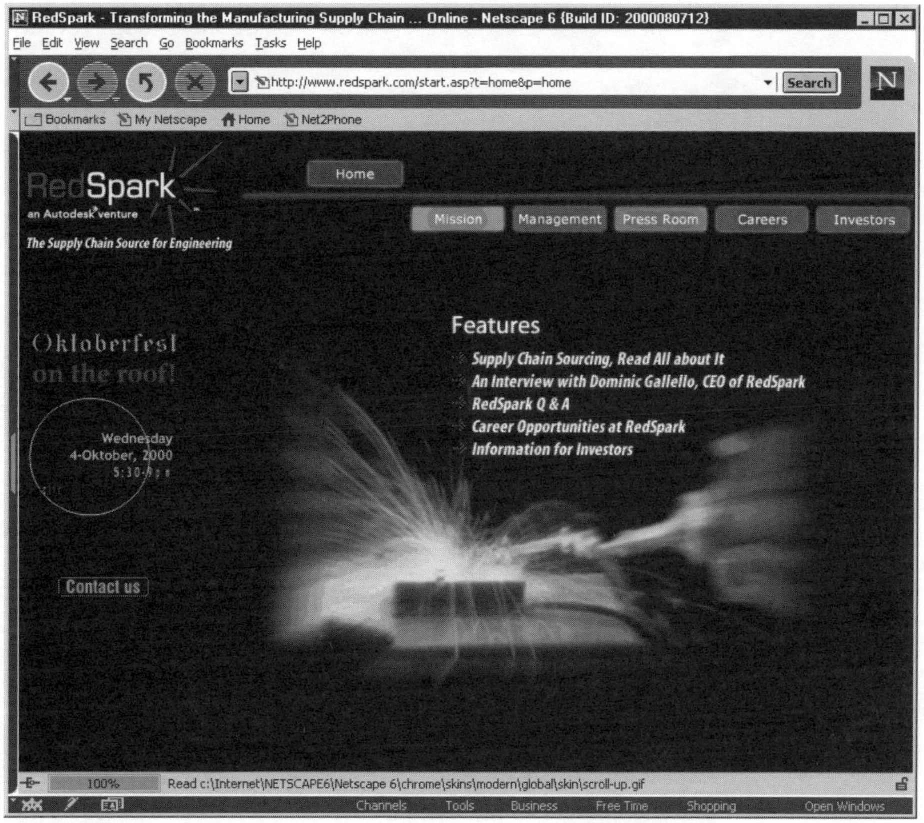

AUGI

www.augi.com

The Autodesk User Group International describes itself as "Autodesk's officially recognized global organization chartered with providing a two-way channel through which members can communicate their wishes and interests in Autodesk products and related issues directly to Autodesk and the industry. Our mission is to enhance members' productivity and knowledge with Autodesk software by creating and maintaining networking opportunities, educational programs, global communication channels and opportunities for user involvement."

Full access to the AUGI Web site provides resources aimed at educating Autodesk product users, including the AUGI Training Program, job resources and opportunities, and Local User Group resources. In addition, you receive:

- *WorldView*, the quarterly newsletter filled with technical information, resource channels, Autodesk news, and discounts on Autodesk-related products and services
- A CD-ROM with programs and other software utilities
- Participation in the official AUGI Wish List process for Autodesk software
- Industry Groups that bring together users with similar professional interests
- An annual conference and regional events
- A T-shirt

If you have questions about AUGI membership, contact **membership@augi.com** by e-mail.

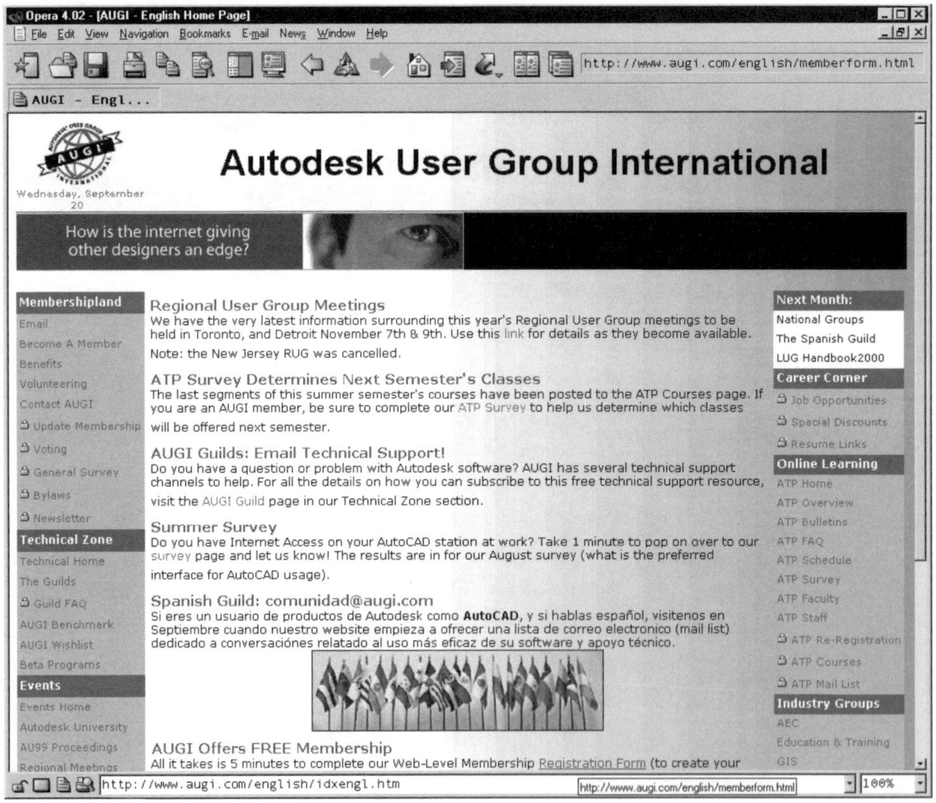

Review Questions

1. Select the best definition for a portal.
   a. A fish market.
   b. A specialized Web site that provides information specific to your industry.
   c. An object defined by ObjectARX programming code.
   d. A software plug-in that allows AutoCAD to edit proxy objects.
2. Select the best definition for an object enabler.
   a. An object defined by ObjectARX programming code.
   b. A password that allows you to access the object.
   c. A username that allows you to edit an object.
   d. A software plug-in that allows AutoCAD to edit proxy objects.
3. The Point A Web site can be accessed only from within AutoCAD.

   True / False
4. Point A does not cater to one of the following fields:
   a. General design and drafting.
   b. Architecture, engineering, construction.
   c. Accounting.
   d. Manufacturing.
5. A project Web site is a specialized Web site that keeps track of projects.

   True / False
6. The Buzzsaw.com Web site can be used only to store AutoCAD drawings.

   True / False
7. A password reminder is a:
   a. Phrase that reminds you of a forgotten password.
   b. Phrase that can be used in place of the username.
   c. Password that lets friends and relatives abuse your account privileges.
   d. A reminder sent to you to change your password.

8. You can lose all your project data if you leave your ProjectPoint account unused for 90 days.

   True / False

9. Other people can access the files you store in ProjectPoint, if you give them access.

   True / False

10. The purpose of AUGI is to:

    a. Help you work better with AutoCAD.
    b. Allow you to suggest new features to Autodesk.
    c. All of the above.
    d. None of the above.

Chapter 12

Connecting with the Internet

In this chapter, you learn how AutoCAD LT 2000i interacts with the Internet. You find out how to open and save drawings to Internet sites, how to attach hyperlinks to objects, and how to launch a Web browser from within AutoCAD LT.

Key Concepts

- Understanding the format of hyperlinks
- Starting a Web browser from within AutoCAD LT
- Accessing drawings from the Internet
- Adding and removing hyperlinks to objects in the drawing
- Learn about the undocumented SelectURL command

Key Terms

Download—transfer a copy of a file from an Internet site to your computer

Hyperlink—a reference to another file; when clicked, the file is loaded and displayed

Upload—transfer a copy of a file from your computer to an Internet site, usually by FTP

Web browser—software specifically designed to display Web pages

Abbreviations

URL	Uniform Resource Locator
HTML	Hypertext Markup Language

HTTP	Hypertext Transfer Protocol
FTP	File Transfer Protocol

Command	Shortcut	Menu Selection
Browser	Alt+HD	...
Hyperlink	Ctrl+K	Insert | Hyperlink
-Hyperlink	...	...
SelectUrl	...	...

Understanding Hyperlinks

AutoCAD LT lets you use hyperlinks in drawings for two purposes: accessing drawings over the Internet; and attaching hyperlinks to objects in drawings. A *hyperlink* lets you locate any file anywhere, whether stored on your computer or anywhere on the Internet. A hyperlink can point to a Web page, a news group, a compressed ZIP file, an audio clip, or a movie clip.

A hyperlink is more formally known as a URL, short for "Uniform Resource Locator." The *URL* is the universal file naming system of the Internet. You are probably most familiar with URLs that look like these:

http://www.autodesk.com is the Autodesk Web site.

http://www.wordware.com is Wordware Publishing's Web site.

http://upfrontezine.com is author Ralph Grabowski's Web site.

You usually don't need to type the "http://" routing prefix since Web browsers automatically add it. Autodesk, however, recommends that you type URLs in the following formats:

Resource	URL Format
Web Site	http://servername/pathname/filename
FTP Site	ftp://servername/pathname/filename
Local File	file:///drive:/pathname/filename
	file:///drive|/pathname/filename
	file://\\localPC\pathname\filename

Resource	URL Format
	file:////localPC/pathname/filename
Network File	file://localhost/drive:/pathname/filename
	file://localhost/drive|/pathname/filename

Looks complicated, doesn't it? It can make more sense when you pick apart the URL:

http Short for "Hypertext Transfer Protocol."

:// Designates a network connection.

servername Name of the server, such as www.autodesk.com.

pathname Name of a subdirectory or folder.

filename Name of the file.

The forward slash (/) is used by Unix systems, while the backward (\ ) slash is used by DOS-based systems, such as Windows 98. Other parts of the URL have these meanings:

ftp Short for File Transfer Protocol.

file The file is located on your computer.

drive The name of the disk drive, such as C:.

localPC The network name of your computer.

localhost The name of a network host computer.

Starting a Web Browser

■ Browser

To start a Web browser from within AutoCAD LT, use the Browser command. (As an alternative, you can switch to the Windows desktop and double-click your Web browser's icon.) The Browser command launches the default Web browser program installed on your computer—whether AOL's Netscape Navigator, Microsoft's Internet Explorer, Opera Software's Opera, or other.

1. Type the Browser command:

   Command: browser
   Location <http://www.autodesk.com/acltuser>:

2. AutoCAD prompts you for the Web location. The default is the AutoCAD LT 98 home page. Type a URL, and press **Enter**. Notice that your computer connects to the Internet. (If you have dial-up access, Windows displays the Dial-up Connection dialog box; click **Connect**.)

3. AutoCAD LT launches your computer's Web browser, and contacts the Web site.

When you select **Help | Autodesk User Group International**, the Browser command immediately launches the browser and displays information about AutoCAD user groups worldwide.

Opening, Inserting, and Saving Drawings from the Internet

- File | Open
- Insert | Block
- File | SaveAs

You access a drawing stored on the Internet with the Open command. (Older versions of AutoCAD LT used the OpenUrl command.) The difference is that you type a URL, instead of selecting a filename.

AutoCAD LT 2000i has several ways to open a drawing file located on the Internet. One method is to type a URL in the Select File dialog box; another method navigates an FTP site; and the third method uses LT's built-in Web browser to first access a Web site, then specify the file.

> **Note:** To work through these tutorials, your computer must have an Internet connection.

Method 1: Type URL

1. To open a drawing using a URL, type the Open command or select **File | Open** from the menu bar. AutoCAD LT displays the Select File dialog box.
2. In the File name field, enter a URL. For the purpose of this tutorial, you can enter the following:

   ftp://ftp.autodesk.com/pub/voloview/english/vve.dwg

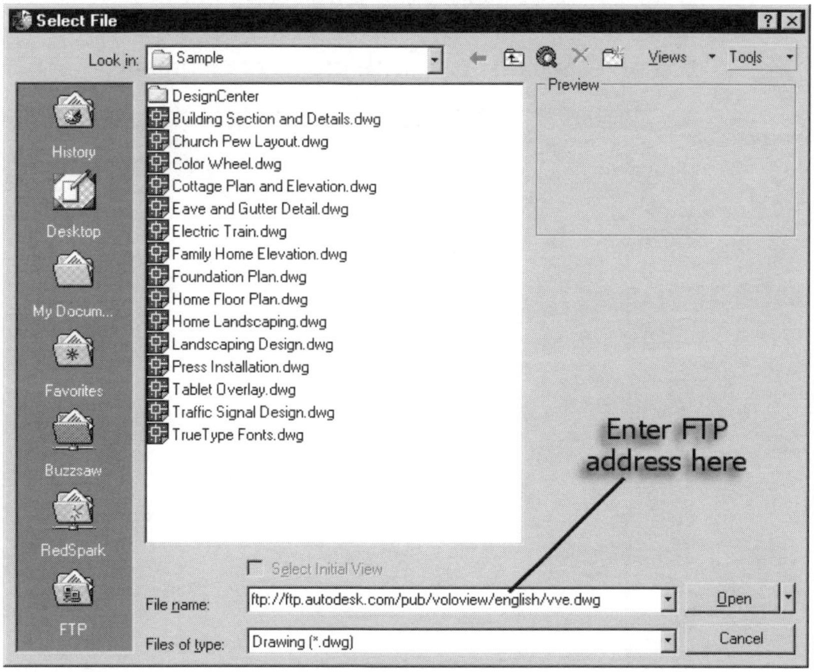

   This URL accesses a sample drawing file called Vve.Dwg at Autodesk's FTP (File Transfer Protocol) site.

3. Click **Open**. Notice that your computer connects to the Internet. (If you have dial-up access, Windows displays the Dial-up Connection dialog box; click **Connect**.)
4. AutoCAD displays the File Download dialog box, which provides you with a progress report on the file being downloaded to your computer. If you find it necessary to stop the download, click **Cancel**.

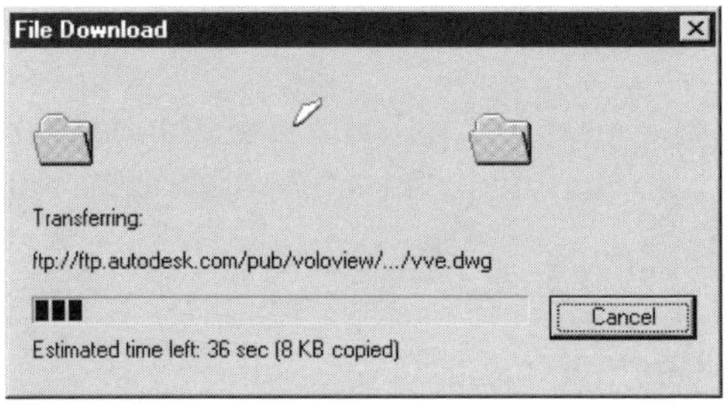

The File Download dialog box shows the progress of the file transfer.

AutoCAD LT displays the drawing in a new window. Notice the FTP filename on LT's title bar. If you wish, you can save the drawing to your computer's hard drive with the Save command.

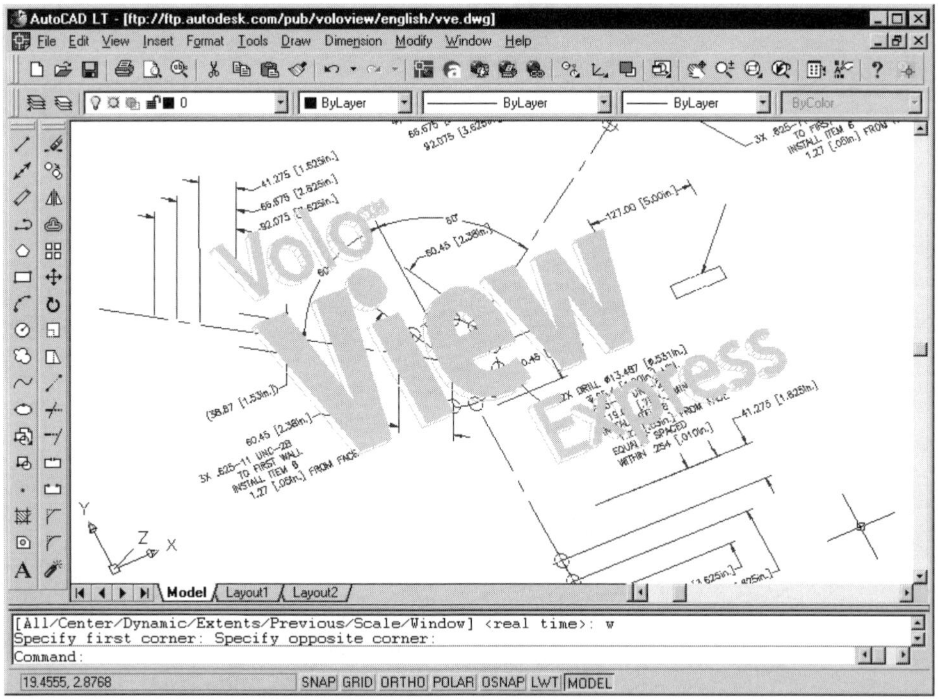

The drawing opened in AutoCAD LT.

Note: The Open command does not directly open the drawing file from the Internet. Rather, AutoCAD LT caches the drawing file by this two-step process: (1) LT copies the file from the Internet site to your computer's temporary Internet folder, such as \Windows 98\Temp\Internet; and (2) LT loads the drawing file from the hard drive. The file in the temporary folder is given a system-generated name, such as AC$81E0.Dwg.

Saving FTP Locations

It can get tedious entering FTP addresses, so LT provides a way for you to save them in a history-like list. Here's how:

1. In the Select File dialog box (accessed with the Open command), click **Tools | Add/Modify FTP Locations.** Notice that LT displays the Add/Modify FTP Locations dialog box.

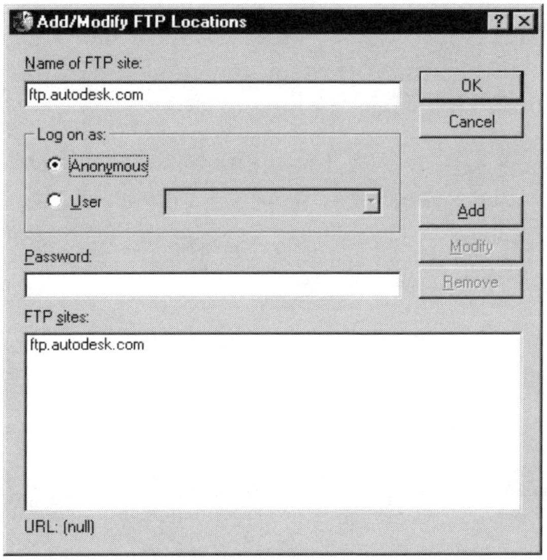

2. In the Name of FTP site field, type the address. Do not include the "ftp://" prefix, such as:

   Name of FTP site: ftp.autodesk.com

3. (Optional). There are two ways to access an FTP site. One method is as "anonymous," which means that no password is required. Other FTP sites require that you provide a password and username; if necessary, enter the password and username that the FTP site requires. (E-mail the Webmaster of that site for the info.)
4. Click **Add** to add the site to the list
5. Click **OK** to exit the dialog box.
6. Now when you want to access an FTP site, you can see the list of saved addresses by clicking the FTP button.

Warning: Note that LT only saves the root address (such as ftp.autodesk.com) and not a full path to a specific file, such as ftp.autodesk.com/pub/voloview/english/vve.dwg.

LT also does not save FTP addresses in the history list.

Method 2: Navigate the FTP Site

The first method is fine when you know the exact address of a drawing file located on an FTP site. If aren't exactly sure of the address, you can navigate the FTP site. Here's how:

1. Select **File | Open** from the menu bar. AutoCAD LT displays the Select File dialog box.
2. Click the **FTP** button, found under the RedSpark button (you may need to scroll down to see it).
3. Select an FTP address, such as the ftp.autodesk.com address that we added earlier.
4. Click **Open.** Notice the list of folder names, such as bin, dev, and etc.

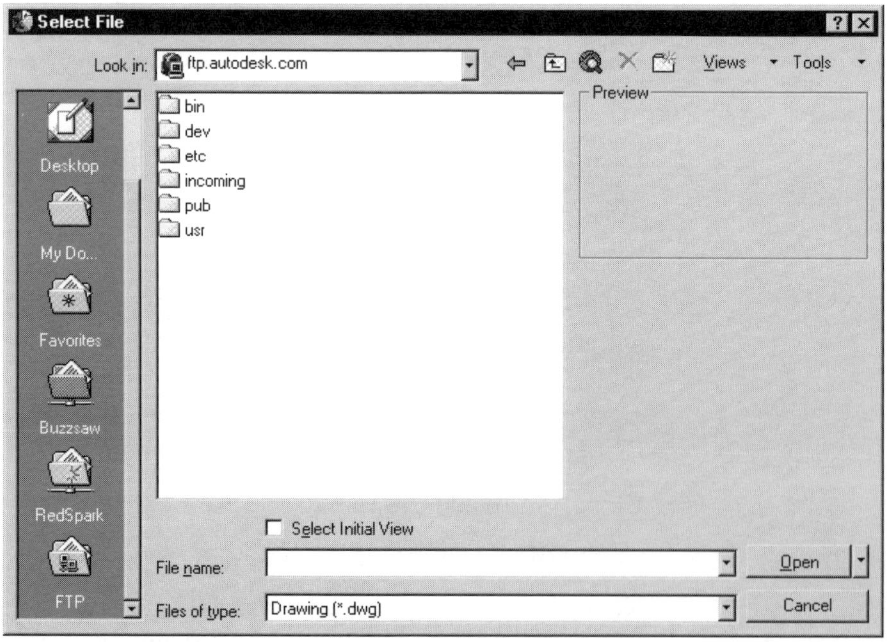

5. Navigating through an FTP site is exactly like navigating folders on your computer, except that it is a bit slower because of the data coming over the Internet:

   - Double-click to open a folder (go down one level).
   - Click the **Up One Level** button (or press Alt+2) to return.

   Pub (short for "public") is good folder to check out in any FTP site; this folder usually contains files available to the public. Sometimes the folder contains a .txt file, which is a text file containing explanations and warnings. To see details about files, select **View | Details.** This provides details such as the size of the file and the date the file was last updated.

6. When you have found the file you want, select it, and click **Open**.

Method 3: Search the Web

1. To open a drawing using the Web browser URL, select **File | Open** from the menu bar. AutoCAD LT displays the Select File dialog box.
2. Click the **Search the Web** button, third from the right. Notice that your computer connects to the Internet. (If you have dial-up access, Windows displays the Dial-up Connection dialog box; click **Connect**.) Notice that AutoCAD LT displays its own Web browser, which is a simplified version of Internet Explorer.
3. The Web browser automatically connects to Autodesk's home page for AutoCAD LT. To stop loading that page, click the **Stop** button (red button with x).
4. In the Look in field, type the address of the Web site you wish to access.
5. To access a specific file, type its URL in the Name or URL field.

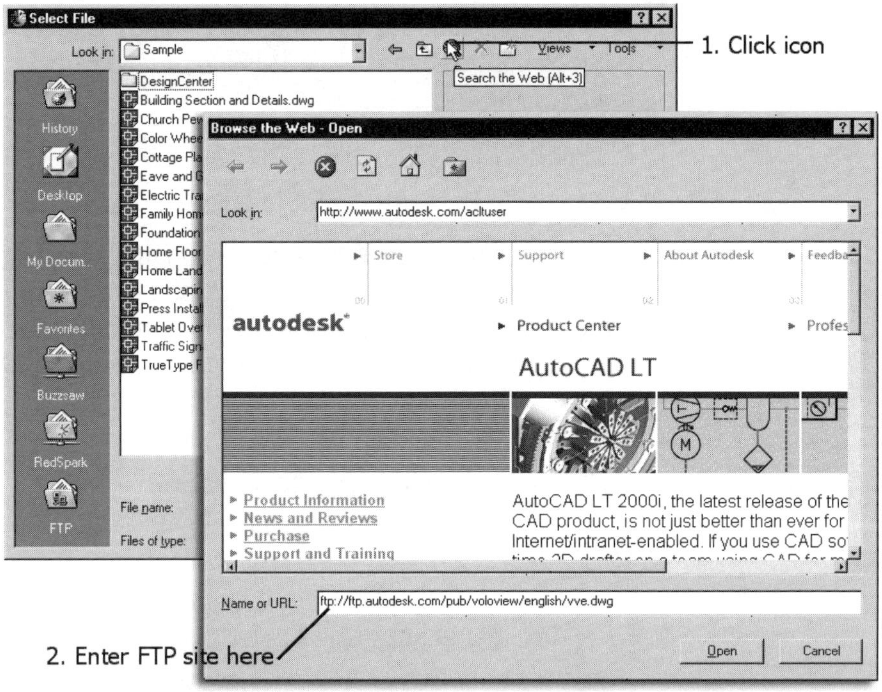

Note: The Web browser provided with AutoCAD LT has fewer user controls than you might be accustomed to Netscape Navigator or Internet Explorer. It has, for example, no menu bar, and just six buttons on the toolbar; there is no Options dialog box for customizing the browser to your liking. You cannot toggle cookies (for privacy reasons) nor the display of images (for faster loading).

From left to right, the buttons on the toolbar have the following meanings:

Back displays the previous Web page.

Forward displays the next Web page; available only if you previously clicked the Back button.

Stop stops the Web page from loading; useful when a page takes too long too load.

Refresh reloads the page from its Web site.

Home loads Autodesk's home page for AutoCAD LT; **www.autodesk.com/acltuser.**

Favorites displays the list of favorite Web sites; same as a "bookmark" in Netscape Navigator.

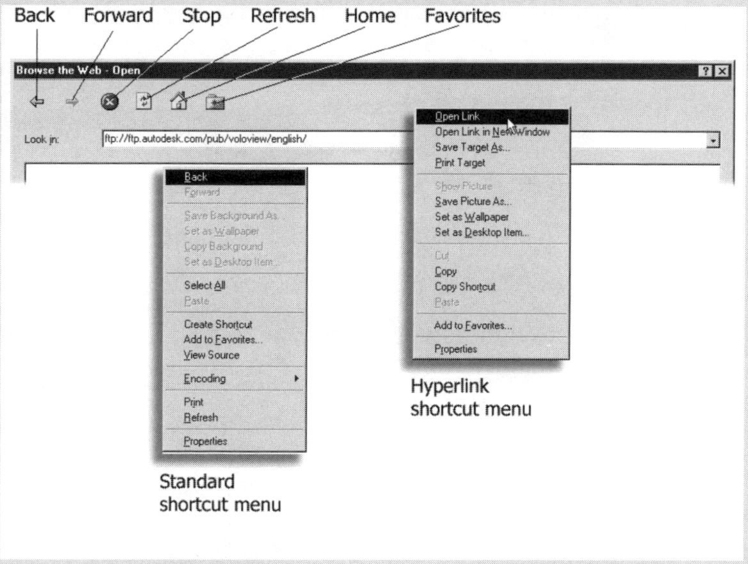

The user interface of the Web browser.

The Web browser has two different shortcut menus, depending on where you right-click: on the Web page, or on a hyperlink.

Right-click menu for Web page:

Back displays the previous Web page (same as clicking the Back button).

Forward displays the next Web page, if you previously clicked Back (same as clicking the **Forward** button).

Select All selects all of the Web page.

Create Shortcut creates a shortcut to this Web page.

Add to Favorites adds the Web page to the Favorites folder on your computer. You access favorite sites from the **Favorites** button on the browser's toolbar.

View Source displays the Web page's source HTML code in Notepad.

Encoding allows you to choose the language for displaying the text of the Web page.

Print prints the Web page on your printer.

Refresh reloads the page from its Web site (same as clicking the Refresh button).

Properties displays the Properties dialog box.

Right-click menu for hyperlinks:

Open Link opens the hyperlinked document in the Web browser (same as clicking the hyperlink).

Open Link in New Window opens the hyperlinked document in another copy of the Web browser.

Save Target As saves the document to a file on your computer.

Print Target prints the document.

Show Picture displays the image, if the display image feature is turned off.

Save Picture As saves the image to a file on your computer.

Set as Wallpaper places the image as wallpaper on your Windows desktop.

Set as Desktop Item places the Web page on your Windows desktop.

Copy copies the item to the Windows Clipboard

Copy Shortcut copies the item to the Clipboard as a shortcut.

Add to Favorites adds the Web page to the Favorites folder on your computer.

Properties displays the Properties dialog box.

Inserting a Drawing as a Block

- Insert | Block

The Insert command lets you access a block stored as a drawing file at an Internet site. The procedure is very similar to opening a drawing, except that you start with the Insert dialog box.

1. From the menu bar, select **Insert | Block**. Notice that AutoCAD displays the Insert dialog box.
2. Click **Browse**. Notice the Select Drawing File dialog box.

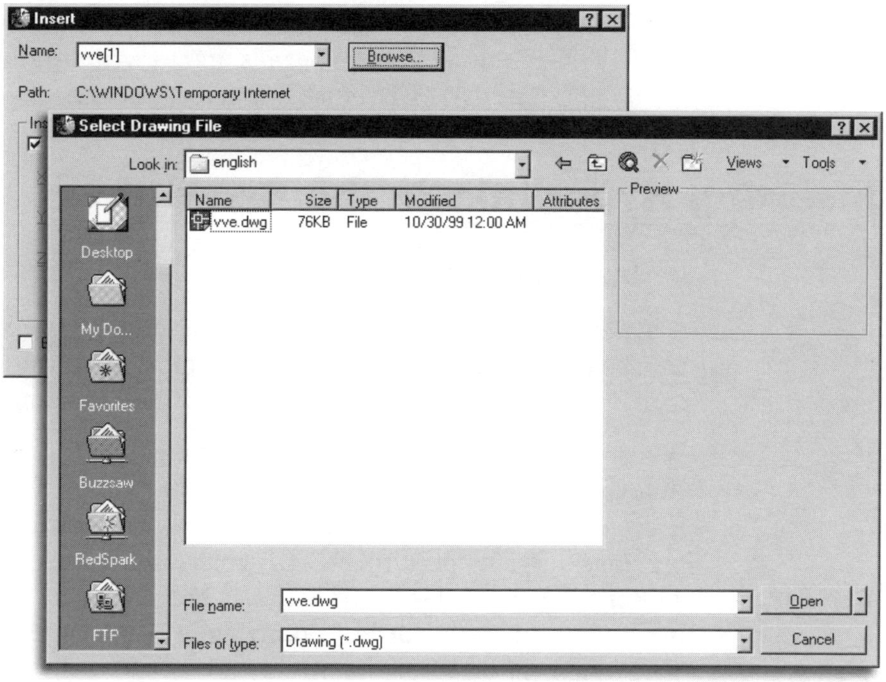

Click the Insert dialog box's Browse button to insert a block from the Internet.

3. As described above, either type a URL in the File name field, or click the **Search the Web** button. Notice the File Download dialog box, which shows the progress of the file transfer.
4. When the download is complete, notice AutoCAD gives the block its filename, such as VVE, in the Insert dialog box's Name field. You can then insert the block in the drawing using any traditional method.

Saving the Drawing to the Web

■ File | Save As

You save the drawing to a file server on the Internet with the SaveAs command.

1. Select **File | Save As** from the menu bar. AutoCAD displays the Save Drawing As dialog box.
2. Select a location from the FTP list, or enter a URL in the File name field. You must use the ftp:// prefix, rather than http://. That's because software cannot write a file to a Web site using the HTTP protocol; instead, software uses FTP (short for "File Transfer Protocol").

3. Click **Save**, and AutoCAD uploads the drawing file to the specified FTP site. You must have permission to write to the FTP site; if you do not, AutoCAD LT refuses to upload the file.

Inserting Hyperlinks in the Drawing

■ Insert | Hyperlink

A hyperlink allows you to jump from file to file. For example, clicking the title box in the drawing jumps to the company Web site. Clicking a standard detail jumps to the Web site of your local building code.

When you place a hyperlink in the drawing, you can use it directly within AutoCAD LT. In addition, the hyperlink remains active when the drawing is displayed by a Web browser, or has been exported in DWF format. (In previous versions of AutoCAD LT, you could not use a URL directly within a drawing. This restriction has been lifted in LT 2000). For more information about exporting a drawing in DWF format, see Chapter 13, "Drawings on the Web."

You can attach hyperlinks to a single objects, a selection set of objects, or to an area. You do this with the Hyperlinks command (if you prefer dialog boxes), or the -Hyperlinks command (the command-line version of the command gives you access to the Area option).

Note: A new drawing should be saved before you can attach any hyperlinks. This has to do with the use of relative URLs. Attempting to attach a hyperlink in a new drawing results in this dialog box:

The Hyperlink – Save Drawing dialog box.

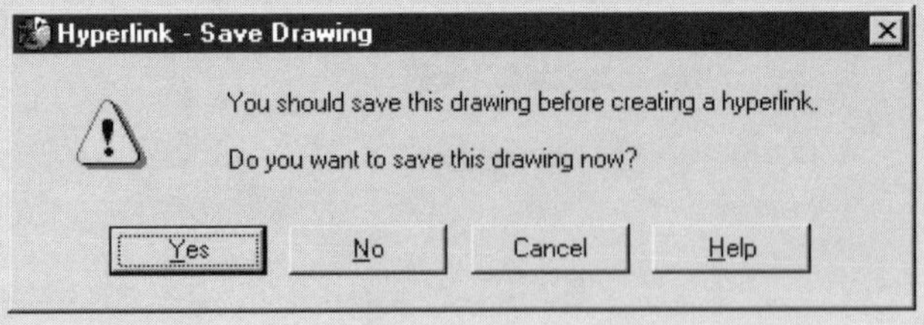

Attaching a Hyperlink via Dialog Box

The Hyperlink command displays a dialog box that allows you to attach one or more hyperlinks to one or more objects in the drawing.

1. From the menu bar, select **Insert | Hyperlink**. As an alternative, press **Ctrl+K**.
2. AutoCAD prompts you to select one or more objects to which the hyperlink should be attached:

   ```
   Select objects: [pick one or more objects]
   Select objects: [Enter]
   ```

   When you press **Enter** to end object selection, notice that AutoCAD displays the Insert Hyperlink dialog box. (This dialog box performs double-duty: when you select an object with a hyperlink, it becomes the nearly identical Edit Hyperlink dialog box.) The dialog box has one essential field, and several optional fields.

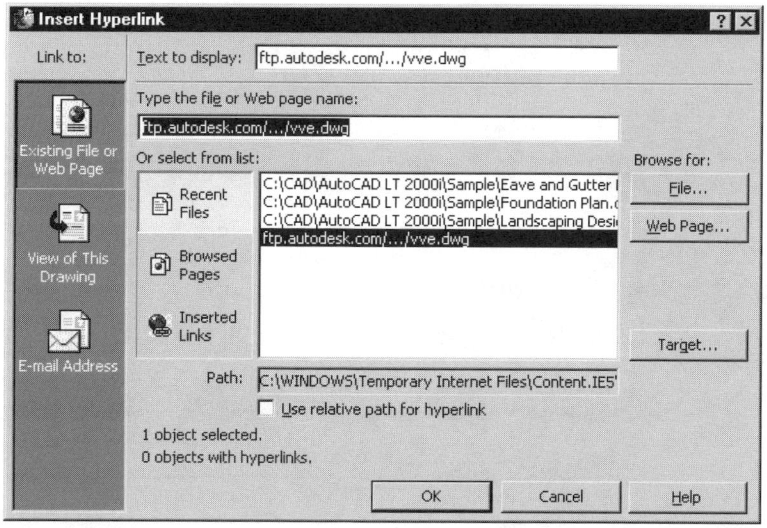

The one essential field is:

Type the file or Web page name is the field where you specify the filename or URL for a file stored on your computer, on another computer linked to yours via a network, or on the Internet. Click the Browse buttons—**File** or **Web page**—to select the file via dialog box or Web browser.

The optional fields are:

Target is an optional button. It allows you to specify a specific location within the file. This "specific location" varies, depending on the nature of the file. In AutoCAD, for example, it is the name of a layout or the name of a view; in a Web page, it is a target; in a word processing document, it is a bookmark; in a spreadsheet, it is a cell.

Text to display is an optional description of the hyperlink. You can enter any text that describes the hyperlink, such as "Project documents" or "Really kool Web site."

Use relative path for hyperlink toggles relative paths. When off, the full path to the file is stored with the hyperlink; when on, AutoCAD LT doesn't store the path and uses the path stored in the HyperlinkBase system variable. By default, the system variable has no value; in that case, AutoCAD uses the current drawing path. The Path field displays the path to the linked file. The Base field displays the default for relative hyperlinks.

3. In the **Type the file or Web page name** field, enter a URL. For the purpose of this tutorial:
   - Click **File.**
   - Navigate to the \AutoCAD LT 2000i\Sample folder.
   - Select the **Foundation Plan.dwg** file.
   - Click **Open.**
4. Click the **Target** button, and select **WINDOW.**

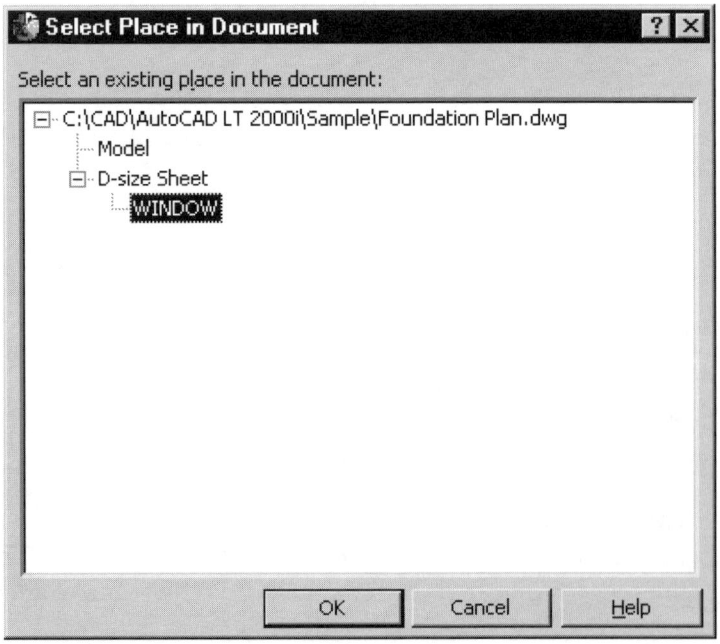

5. In the Text to display field, enter:

   Foundation Details

6. Click **OK** twice. The object has a hyperlink attached. To see this for yourself, pass the cursor over the object. Notice how the cursor changes to the chained-world icon, and a tooltip displays the hyperlink or its description.

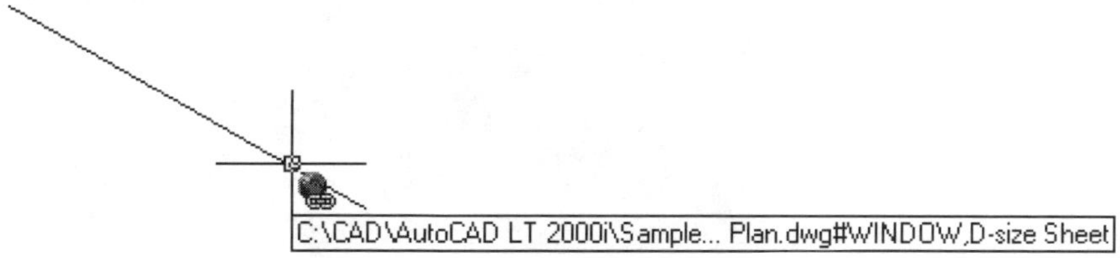

7. To use the hyperlink, select the object. Notice the highlighting. Right-click. From the shortcut menu, select **Hyperlink | Open "Foundation Details."**

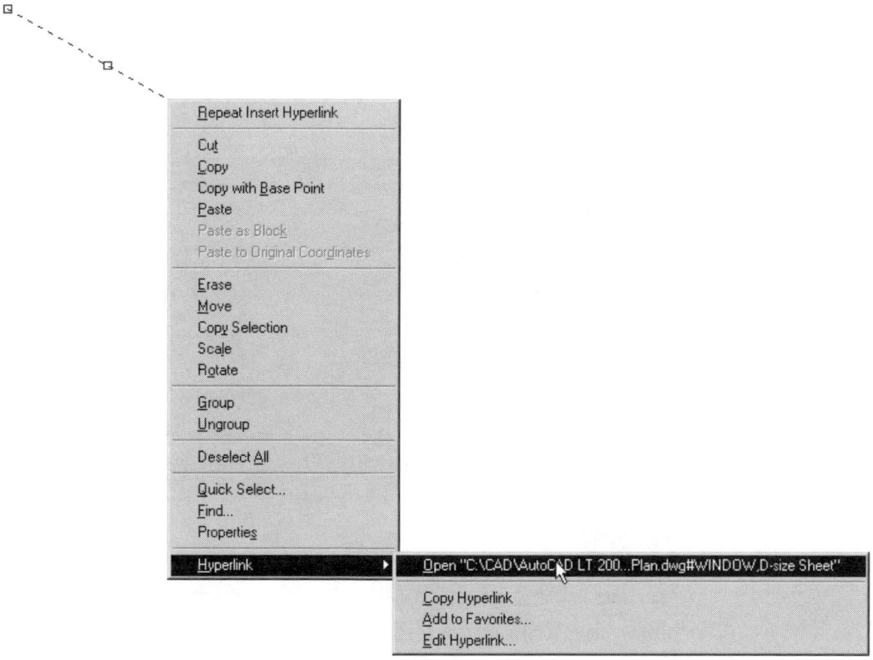

Notice that AutoCAD opens the Foundation Plan.dwg file in a separate window, then switches to layout mode, and displays the Window view.

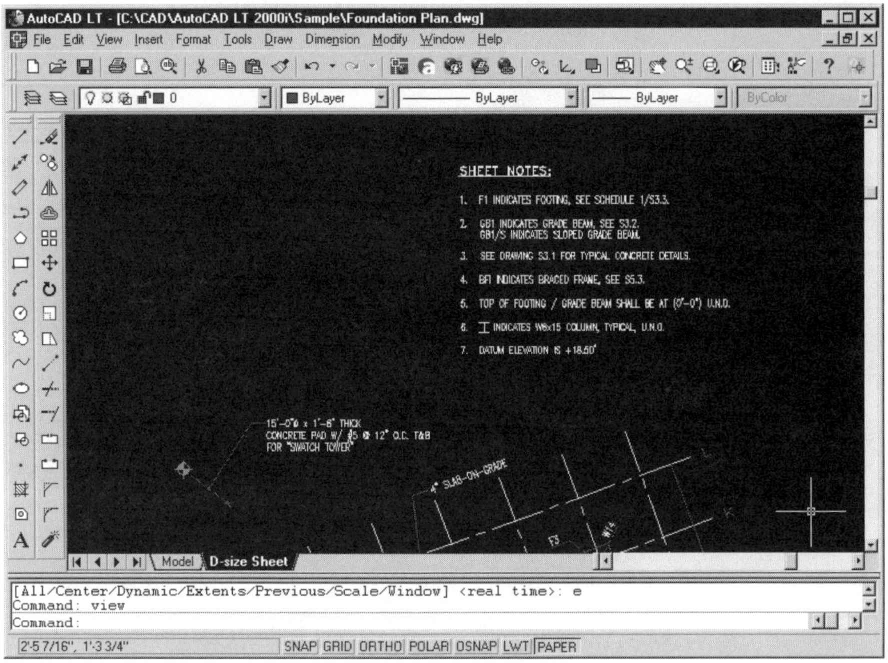

8. To help you navigate between linked drawings, open the Web toolbar:
   - Right-click the toolbar, and select Web from the menu.
   - Click the **Go Back** button to return to the previous drawing or document.

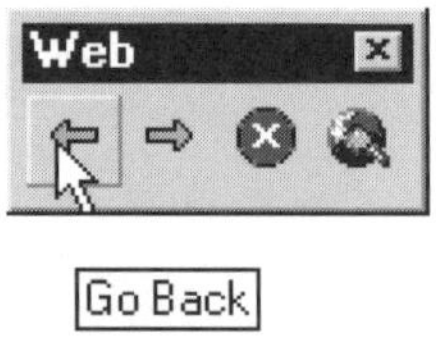

As an alternative, you can use the undocumented HyperlinkBack and HyperlinkFwd commands to traverse links.

Editing a Hyperlink

To edit a hyperlink, select the object and press **Ctrl+K**. AutoCAD displays the Edit Hyperlink dialog box. Notice that it is identical to the Insert Hyperlink dialog box, except for the Remove Link button. Click the button to remove the hyperlink from the selected object.

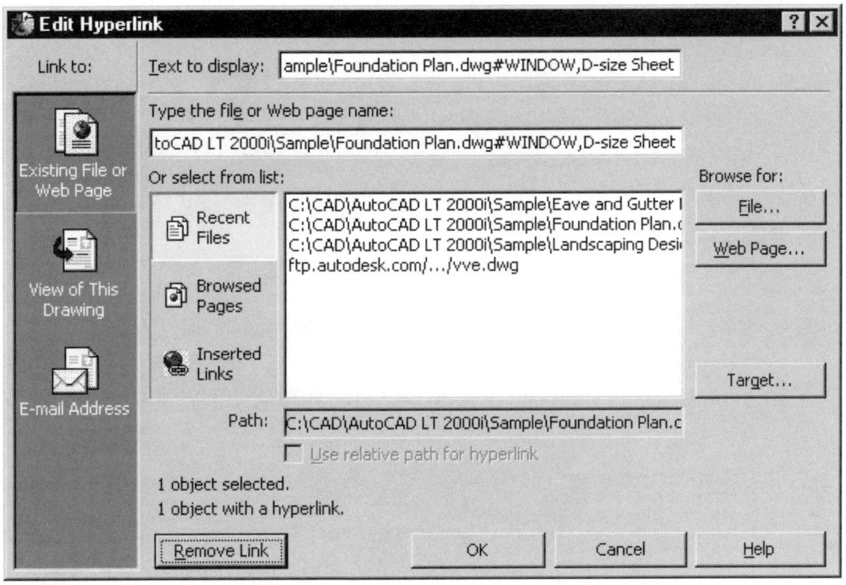

Attaching a Hyperlink to Areas

The -Hyperlink command is for attaching URLs to rectangular areas, as well as objects:

- An object URL is attached to one or more objects.
- An area URL is attached to a rectangular area, either around a group of objects or even where there are no objects.

1. Type **-Hyperlink** at the Command: prompt, as follows:

   ```
   Command: -hyperlink
   Enter an option [Remove/Insert] <Insert>: i
   ```

2. Type **A** to attach a URL to an area:

   ```
   Enter hyperlink insert option [Area/Object] <Object>: a
   ```

3. Pick two points that indicate the opposite corners of a rectangle or square shape:

   ```
   First corner: [pick]
   Other corner: [pick]
   ```

3. AutoCAD prompts you to type a URL:

```
Enter hyperlink <current drawing>: http://www.upfrontezine.com
Enter named location <none>: [Enter]
Enter descriptions <none>: Ralph Grabowski's Web site.
```

Notice that the hyperlink area shows as a red rectangle on layer URLLAYER.

If you find the red hyperlink rectangles distracting, freeze layer URLLAYER. If you erase or freeze the layer, it will not be exported in DWF format. Do not overlap URLs, either as objects or areas. This could result in the wrong hyperlink being activated later.

Warning: AutoCAD LT has these limitations in using URLs:

- AutoCAD LT does not check the hyperlink syntax you type in; if you make a typing error, the error is not caught.
- When a URL is attached to a block object, the hyperlink is erased when the block is scaled unevenly, stretched, or exploded.

Finding Objects with Hyperlinks

■ SelectURL

The bright red rectangle of areas with hyperlinks makes it easy to spot them, but objects with hyperlinks give no indication. To highlight objects and areas with hyperlinks, use the undocumented SelectUrl command.

1. Enter the command:

```
Command: selecturl
```

2. AutoCAD highlights all objects and areas with a hyperlink. The highlighting consists of dashed lines and cold handles.
3. Press **Ctrl+K** to display the Edit Hyperlink dialog box. Or, press **Esc** twice to remove the highlighting and handles.

Detaching URLs

To remove a hyperlink from an object:

1. Select the object. and press **Ctrl+K**. Notice that AutoCAD displays the Edit Hyperlink dialog box.

2. Click **Remove Link**. Notice that the dialog box disappears; the object has the hyperlink no more.

To return the hyperlink to the object, use the U command immediately.

The -Hyperlink command removes a URL from an area.

1. Type the command:

```
Command: -hyperlink
Enter an option [Remove/Insert] <Insert>: r
```

2. Pick the area with the hyperlinks, then select the hyperlink you want removed.

```
Select objects: [pick area]
1. http://www.upfrontezine.com (Ralph Grabowski's Web site.)
Enter number, hyperlink, or * for all: 1
```

Notice that AutoCAD reports the number of hyperlinks removed:

```
Remove, deleting the Area.
1 hyperlink deleted.
```

If the area contains no more huperlinks, AutoCAD erases the rectangle. When no more area URLs remain in the drawing, AutoCAD automatically purges URLLAYER.

Review Questions

1. A URL is also known as a:

   a. Unknown resort location
   b. Hyperlink
   c. File folder
   d. Universal read-line

2. To attach a URL to an area in the drawing, use the following command:

   a. -Hyperlink
   b. Hyperlink
   c. Rectang
   d. Area

3. FTP is short for "File Transfer Protocol."

   True / False

4. To open a drawing file located on the Internet, you use the following command:

   a. OpenFileLocatedOnTheInternet
   b. OpenWeb
   c. Open
   d. No command; LT cannot open files from the Internet.

5. Navigating through an FTP site is exactly like navigating folders on your computer.

   True / False

6. It is not possible to insert a block from the Internet.

   True / false

7. When you save a drawing to a location on the Internet, the following protocol must be used:

   a. SMTP
   b. POP
   c. HTTP
   d. FTP

8. Hyperlinks can be attached to:

   a. Objects in the drawing.
   b. Areas in the drawing.
   c. Both of the above.
   d. None of the above.

9. The shortcut keystroke for accessing the Hyperlinks dialog box is:

   a. Ctrl+H
   b. Ctrl+K
   c. Ctrl+I
   d. Ctrl+W

10. To find all objects and areas with hyperlinks, use the following command:

    a. SelectURL
    b. ShowURL
    c. HighlightURL
    d. Hyperlink

Chapter 13

Drawings on the Web

Several years ago, Autodesk developed a file format called DWF (short for "drawing Web format"). It is designed to display AutoCAD drawings on the Internet. In this chapter, you learn how to use DWF file with AutoCAD LT.

Key Concepts

- Using the Publish to Web wizard
- Plotting the AutoCAD drawing as a DWF or JPEG file
- Viewing DWF files in a Web browser
- Adding a DWF file to a Web page

Key Terms

Plot file—a plot of the drawing saved in a file on disk

Plug-in—a small software program that allows a Web browser to display unfamiliar file formats

Abbreviations

DWF	drawing Web format
JPEG	Joint Photographic Expert Group
WHIP	Windows High Performance

Web File Formats

Can you display AutoCAD's own DWG files on the Internet? Yes, you can. All your Web browser needs to view AutoCAD drawings is a *plug-in* that handles DWG files. Plug-ins are available from Autodesk (Volo) and third-party developers.

If DWG files can be displayed on the Internet, why is a different file format needed? Autodesk invented DWF to overcome some of the drawbacks to using DWG files. DWF is a smaller, more secure representation of your AutoCAD drawing. There are several advantages to using DWF, and some disadvantages.

Pros and Cons to DWF

A DWF file is as much as eight times smaller than the equivalent DWG file. Smaller means it transmits over the Internet faster. This is an important consideration because many Internet users have relatively slow modem access.

A DWF file is secure because it cannot be edited in the same way as your original DWG file. While it is possible to reverse-translate the DWF file back to DWG format, much information is removed, such as attribute data, some entity information, and so on. Since the people viewing the DWF file are not seeing the original drawing, they cannot tamper with the data. You can think of DWF as a *plot file*, when you plot the drawing to a file on disk, instead of to the plotter. When you hand out copies of the drawing plotted on paper, the original cannot be tampered with.

You should be aware, however, that the DWF format has drawbacks:

- There is the extra step of translating from DWG to DWF. When you deal with two versions of the same drawing, you get into "version control"—is the DWF you're viewing up to date with the original DWG?
- A DWF file does not always represent the drawing as you might expect, because it is strictly a 2D-wireframe file format. In particular, DWF cannot display 3D data (although DWF displays flattened 3D viewpoints), rendered, or shaded drawings.
- If someone browsing the Web does not have the DWG plug-in installed, they cannot see the DWF file. For this reason, LT 2000i adds the option to export the drawing in JPEG format.

Pros and Cons to JPEG

In addition to DWF, AutoCAD LT 2000i can also publish drawings in JPEG format. JPEG is short for "Joint Photographic Experts Group," which isn't terribly meaningful to anyone. JPEG is one of the two most common formats used to display images on the Web; the other is known as GIF, but Autodesk has chosen not to include it because of licensing issues.

Like DWF, JPEG has its pros and cons:

- JPEG images are read by all Web browsers.
- JPEG files are compressed, although not as efficiently as DWF files. The method of compression results in *artifacts*, which are things you see in the image that don't exist in the original drawing (see illustration).
- When you zoom into a JPEG image, the lines and text appear to become small squares, called *pixels*. This is because JPEG is a raster format.

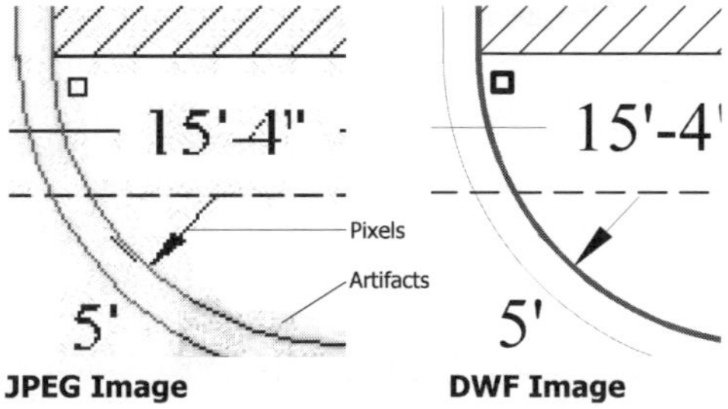

Publish to Web Wizard

To make it easier to output your drawing to the Web, AutoCAD LT 2000i has added a Publish to Web feature. It helps you through the steps of creating a complete Web page with DWF or JPG images of AutoCAD drawings. In addition, the wizard modifies Web pages created previously by the wizard. Perhaps a better name would have been "Publish as HTML" because the wizard generates the images and HTML code necessary for a complete Web page.

You publish one or more drawings as a Web page, as follows:

1. From the menu bar, select **File | Publish to Web**. (Alternatively, you can enter the PublishToWeb command or click the **Publish to Web** button on the toolbar.) Notice the Publish to Web dialog box.

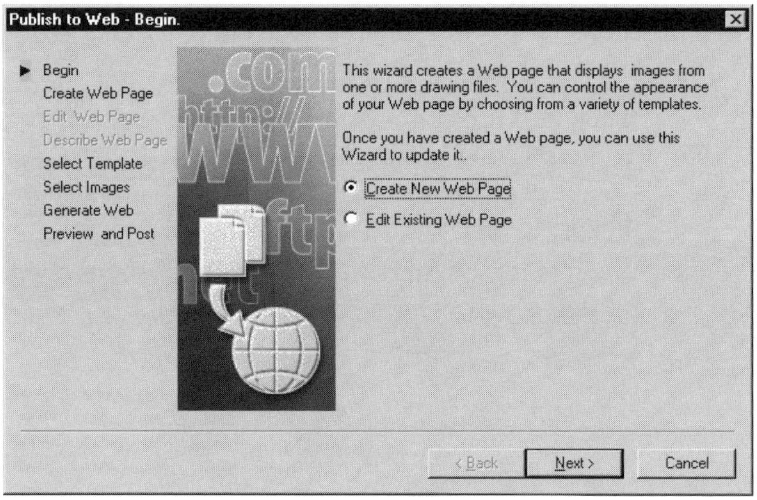

2. Ensure **Create New Web Page** is selected, and click **Next.**

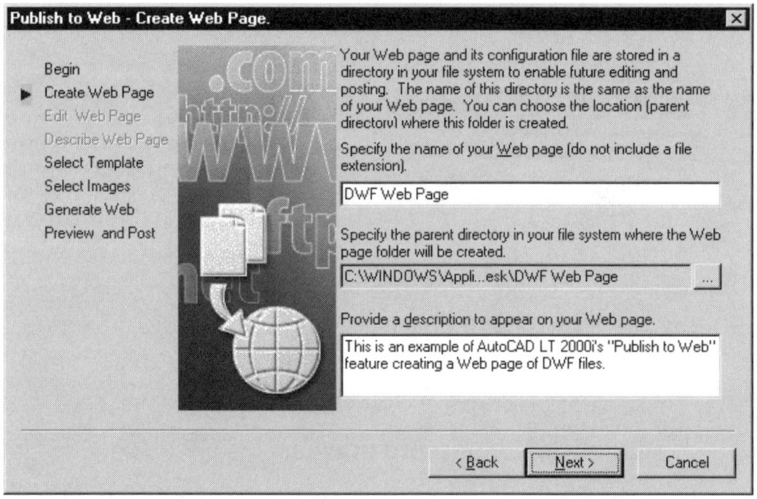

3. Enter a name for the Web page, which AutoCAD uses to create a folder on your computer's hard drive. This folder contains all the HTML, JPEG, and DWG files needed to represent a single Web site. This name also appears on the Web page as its title.

   You can use the default parent folder (C:\Windows\Applications Data\Autodesk), or click the ellipsis (...) button to select a different parent folder. The parent folder holds all the folders created by this wizard.

Type a description of the Web page. This text will appear on the Web page under the title.

Click **Next**.

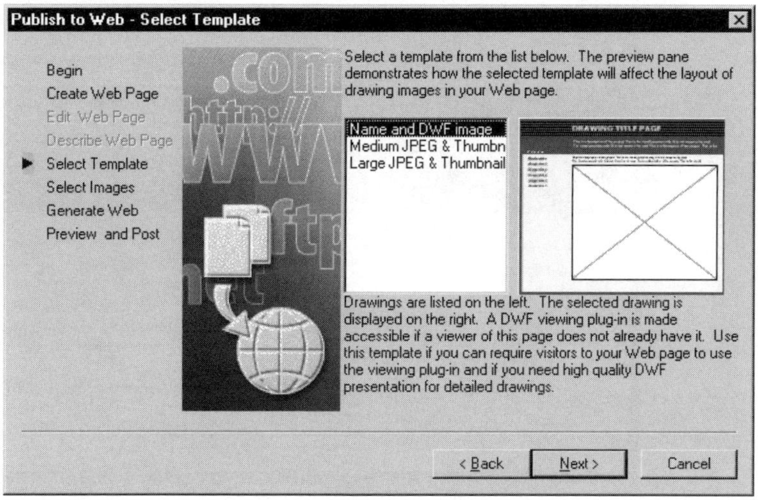

4. Here you have three choices of how you want the Web page to look:

   Name and DWF Image generates a Web page with:

   - Drawing names listed on the left side.
   - The selected drawing is displayed in DWF format on the right.

   Medium JPEG and Thumbnails generates a Web page with:

   - A home Web page displays an array of JPEG *thumbnails* (small images, 123x95 pixels in size, and 3KB to 6KB each), four columns wide.
   - Each thumbnail is linked to another Web page, which displays the drawing as a medium-size JPEG (size is 789 x 610 pixels, and 30KB to 100KB in filesize, depending on complexity).

   Large JPEG and Thumbnails generates a Web page with:

   - A home Web page displays an array of JPEG thumbnails, four columns wide.
   - Each thumbnail is linked to another Web page, which displays the drawing as a medium-size JPEG (size is 1024 x 768 pixels, and 50KB to 200KB in filesize).

The illustration below shows the DWF version. The JPEG version is shown on the following page.

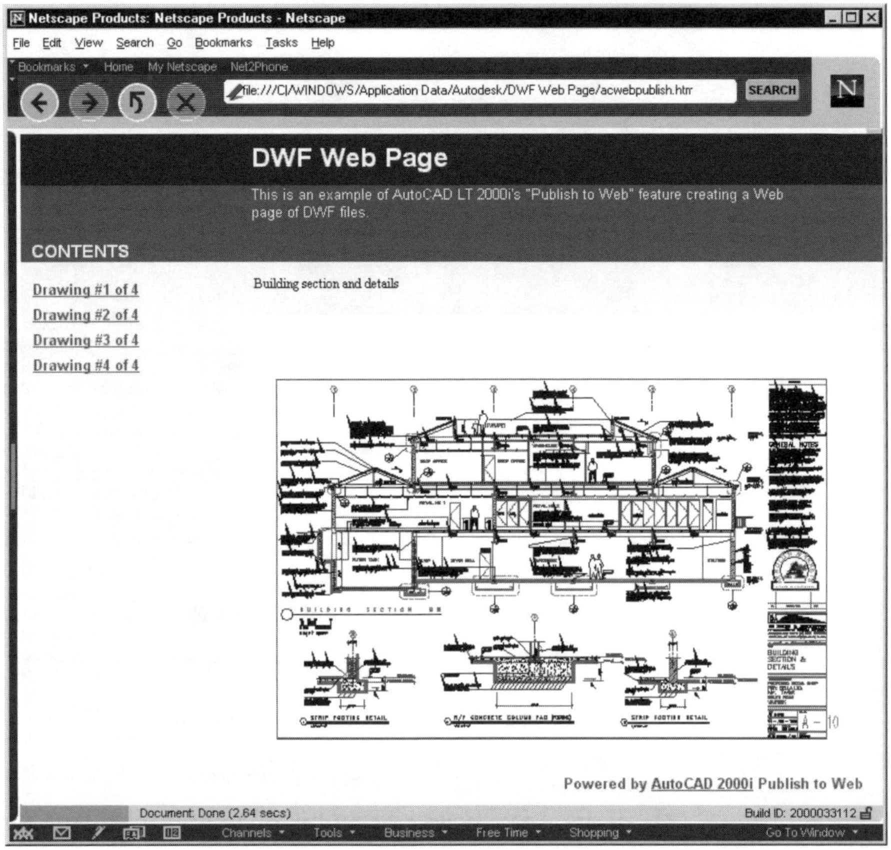

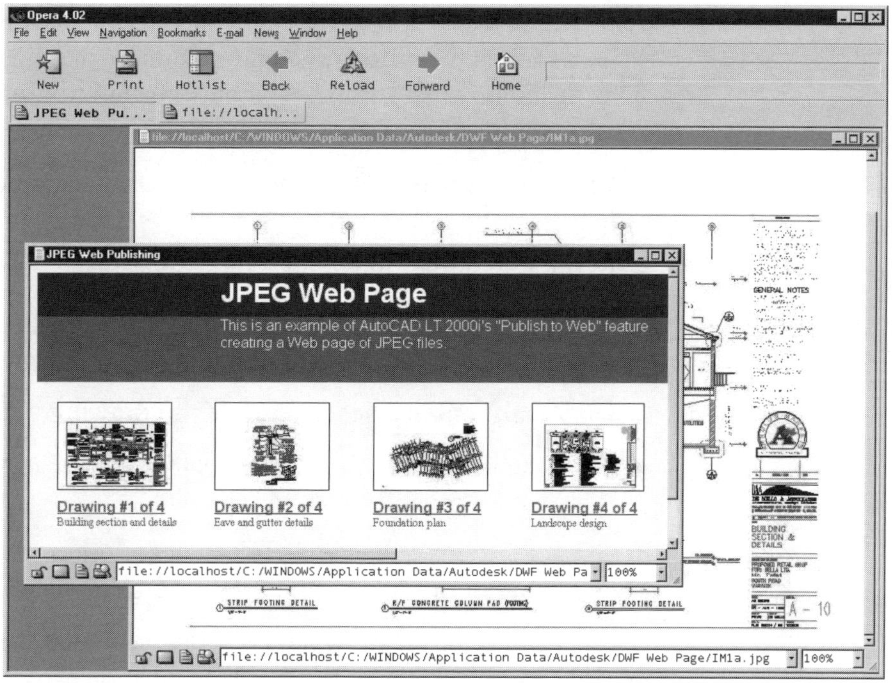

For this tutorial, select **Medium JPEG and Thumbnails**, and click **Next.**

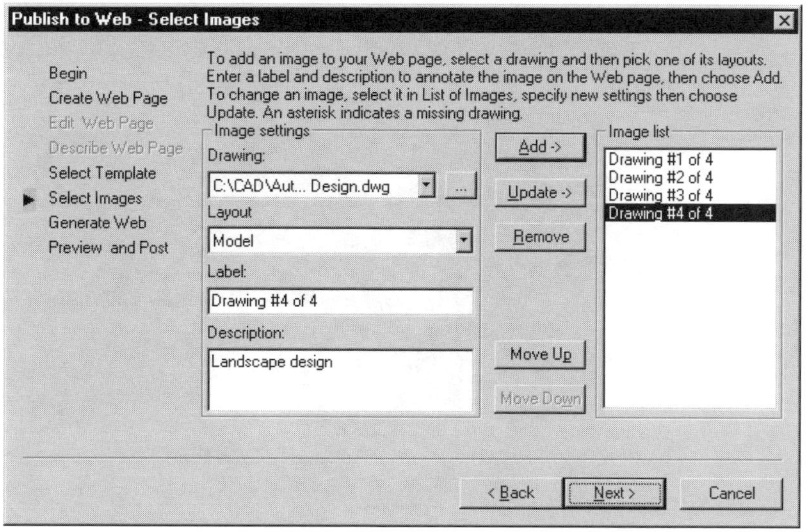

5. Now you select all the drawings you want displayed by the Web page. You can select a single drawing, or as many drawings as you wish. This dialog box has a number of buttons and text entry fields that must be employed in the correct order:

   - Click the ellipsis (...) button next to Drawing, and select a DWG file.
   - In the Layout list box, select Model or a layout mode.
   - Enter a title for the drawing next to Label; each drawing must have a label, otherwise you cannot add it to the Image list.
   - Optionally, enter a description of the drawing under Description.
   - Now you can click the **Add** button to add the drawing to the Image List.

   If you need to make a change to the description or anything, select the drawing in the Image list, make the changes, then click **Update**.

   To change the order in which the drawings appear, select the drawing in the Image List, and click the **Move Up** or **Move Down** buttons.

   LT does not, unfortunately, allow you to select an entire folder of drawings; you have to pick each drawing one by one. Once you have added all the drawings you want, click **Next**.

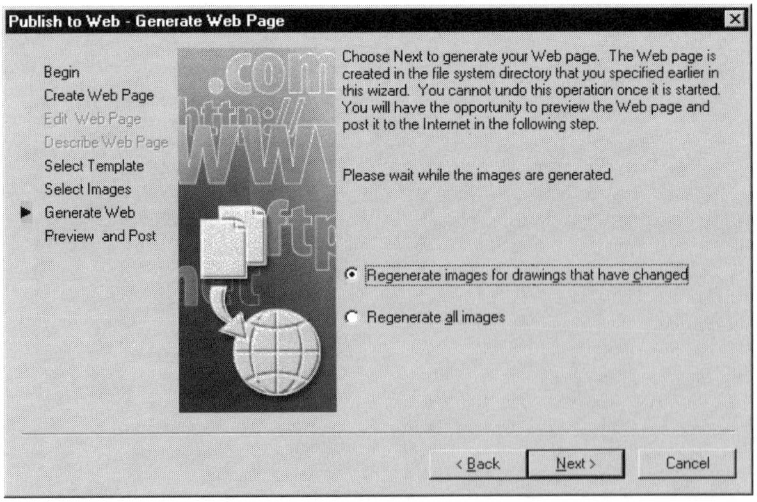

6. In this step, AutoCAD generates the JPEG or DWF images for the Web page. You have two options. Regenerating just the changed images is faster; regenerating all images is more certain to catch any drawing that has changed. Click **Next**, and wait while AutoCAD opens and "plots" the drawings as JPEG or DWF files, and generates the HTML code. This process can take some time, depending on the number of drawings and the speed of your computer. Using a 400MHz computer, for example, takes roughly a half-minute per drawing.

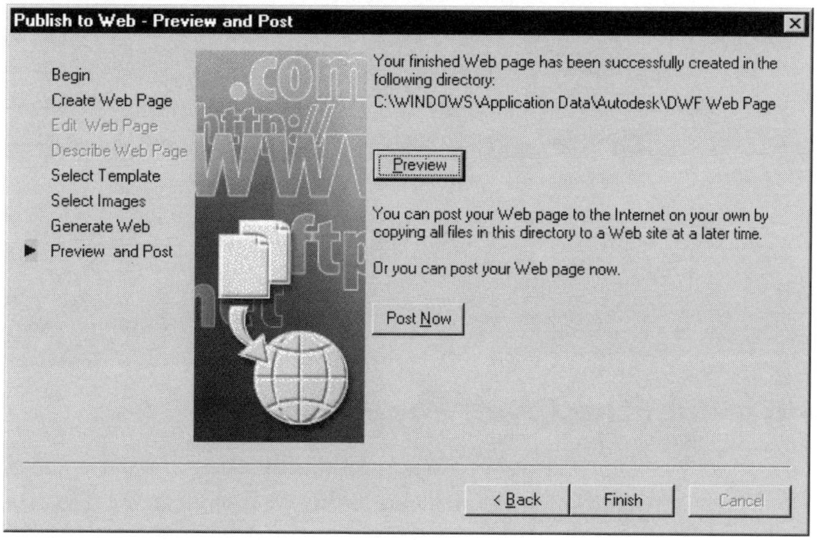

7. When AutoCAD is done, click the **Preview** button to launch your computer's default Web browser.

Click **Post** to have AutoCAD upload the files to an Internet site. Click **Finish** to end the wizard.

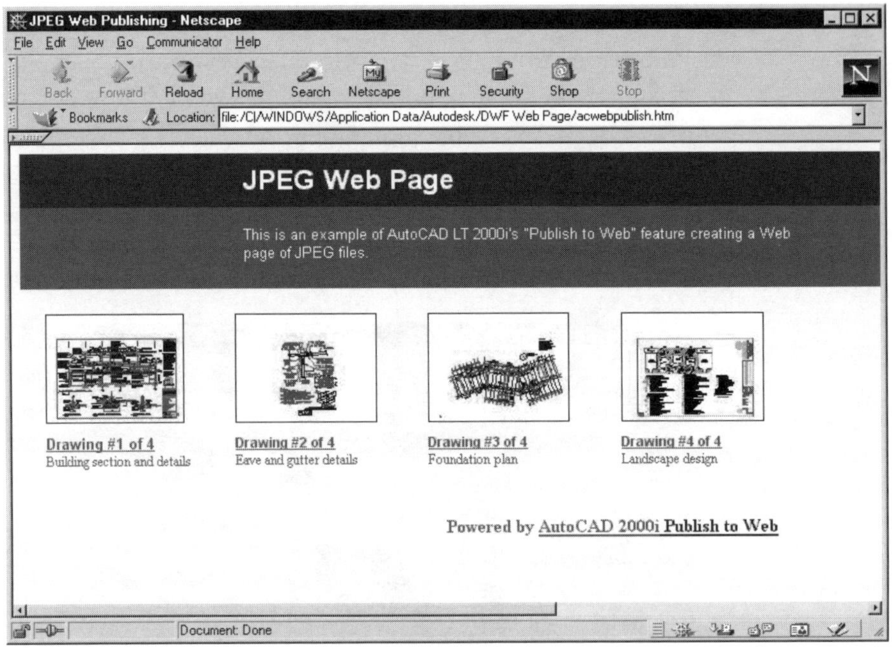

Editing a Published Web Page

The Publish to Web wizard can edit the pages it created. Technically, AutoCAD stores data about the Web page in a PTW file. When you select the wizard's Edit option, you are asked to select the PTW file to open. Make your changes, such as editing the description of drawings or changing the list of drawings. AutoCAD then replots the files that have changed.

> **Note:** AutoCAD generates generic filenames for the HTML, DWF, and JPEG files. The "home" HTML file is named AcWebPublish.Htm.
>
> The thumbnail JPEG files are named IM1.jpg, IM2.jpg, etc., while the corresponding full-size JPEG adds an "a" to the filename, such as IM1a.jpg. IM is short for "image."
>
> When outputting as DWF files, the home HTML file contains the first drawing. The other drawings are stored in HTML pages named PTWTemplateA1.htm, etc. The DWF files use the same naming convention as JPEG images: IM1.dwf, etc.
>
> AutoCAD uses the PTW file to correlate the generic filenames with the actual filenames, as well as store the descriptive data displayed by the HTML page.

Plotting Drawings as DWF and JPEG

The Plot command also exports drawings in DWF and JPEG formats for use by Web sites. (Older versions of AutoCAD LT used the DwfOut command.) You can think of this as the manual version of the Publish to Web wizard, which only generates a single DWF or JPEG—and not any of the extras, such as the HTML code or thumbnails.

1. Type the Plot command or select **File | Plot** from the menu bar. AutoCAD displays the Plot dialog box.
2. In the Plot Device tab, click the list box next to **Name.**

Select DWF ePlot.pc3 as the plotter configuration name.

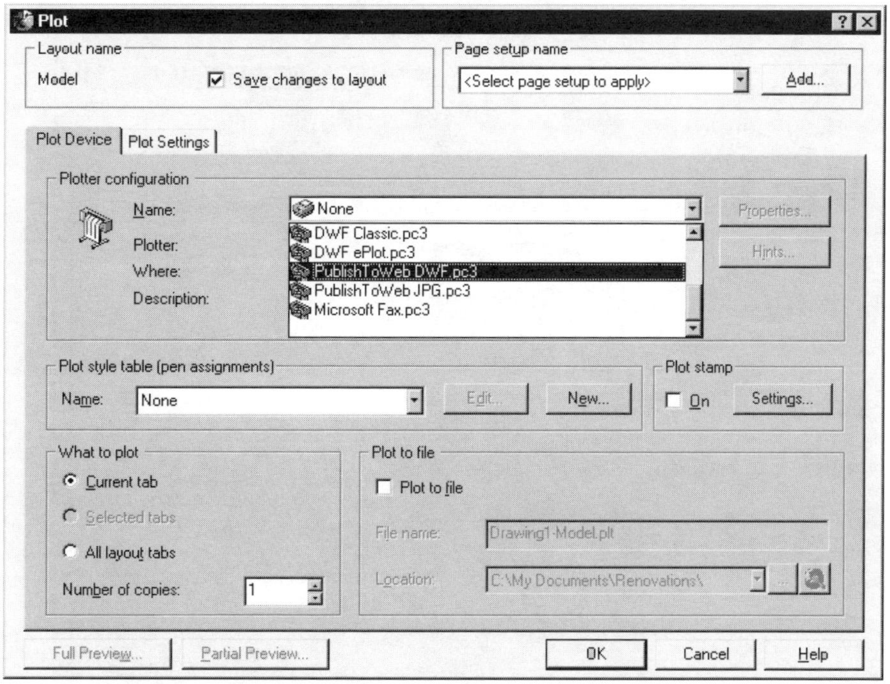

3. In the list box, select “PublishToWebDWF.pc3” to create a DWF image, or “PublishToWebJPG.pc3” to create a JPEG image.

Notice that AutoCAD LT includes three plotter configuration files for creating DWF files: “DWF Classic.pc3” uses a black drawing background, while “DWF ePlot.pc3” uses a white drawing background and shows a paper boundary. As far as I can tell, “DWF ePlot” is identical to “PublishToWebDWF.”

Tip: When you don't specify a lineweight, AutoCAD LT 2000i applies a lineweight of 1.5mm (0.06 in) when plotting, which may affect the look of the DWF files when zoomed in. If you don't want this, turn off the Plot with Lineweights option on the Plot Settings tab in the Plot dialog box.

In some cases, you may not want to use a lineweight of zero, such as for text, which would otherwise be too bold when zoomed in a DWF. Since text should be on its own layer, set that layer's lineweight to a thickness of 0.0.

4. To change the settings, click **Properties**. Notice the Plotter Configuration Editor dialog box.

The Plotter Configuration Editor dialog box.

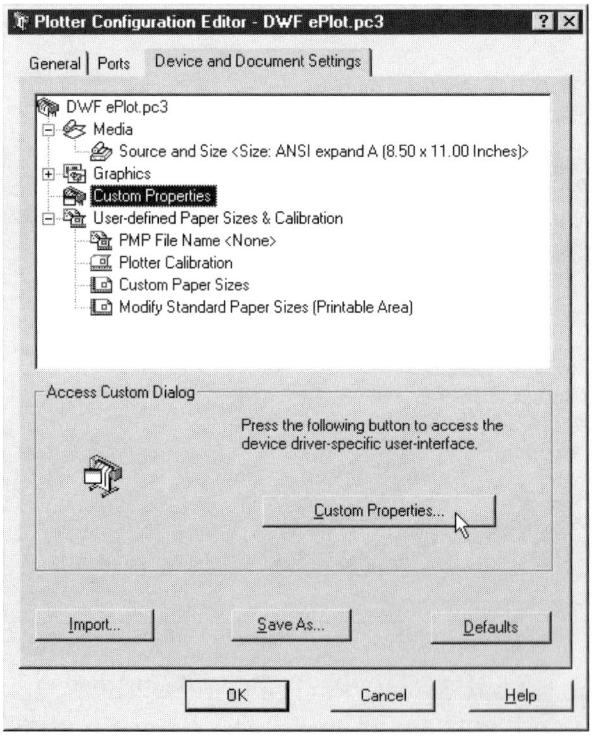

5. Ensure the **Device and Document Settings** tab is selected. In the tree view, select **Custom Properties**, then click the **Custom Properties** button when it appears in the lower half of the dialog box.

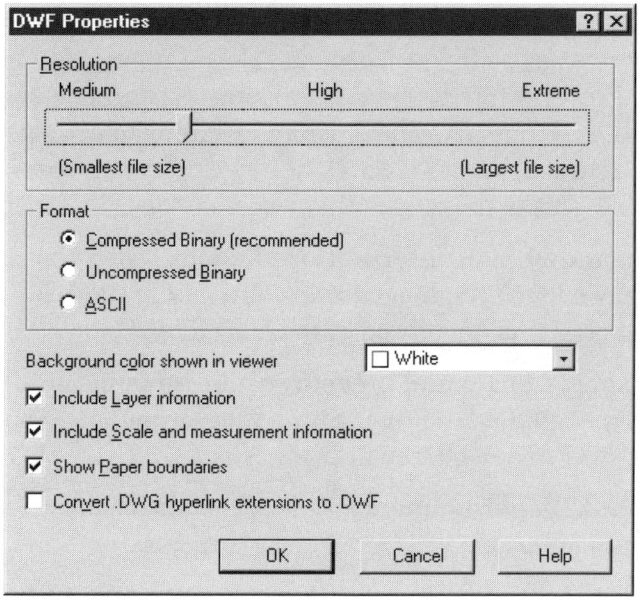

The DWF Properties dialog box.

6. In the DWF Properties dialog box, you can select the following options:

   Resolution: **Medium** precision is good enough for all except for extremely detailed drawings. Medium precision is 40% smaller than Extreme precision, so that it takes 40% less time to transmit over the Internet.

   - **Medium** precision creates a DWF file with 16-bit integers. This creates a smaller file size, but may result in a less accurate image.
   - **High** precision uses 20-bit integers.
   - **Extreme** precision uses 32-bit integers.

   Format: The only time you would not select **Compressed Binary** is when a software application cannot otherwise read the DWF file.

   - **Compressed Binary** plots the DWF file in a compressed, binary format; compression does not cause data loss.
   - **Uncompressed Binary** plots the DWF file in binary format, without compression; creates a larger file size.
   - **ASCII** plots the DWF file in ASCII (plain text) format, without compression; creates the largest file size.

Background color shown in viewer: Some users prefer to work with black as the background color in AutoCAD. On the Internet, however, white works better as the background color. Choose a color from the list box. The default is white; you can also choose **Drawing Background**, or select **Other** to display the Windows Color dialog box. This is the only option available for JPEG images.

Include Layer information: Retains layers that are turned on and thawed; you can toggle layer visability in the DWF viewer. When off, layer data is not stored in the DWF file.

Include Scale and measurement information: Scale data is included in the DWF file; allows measurements to be taken in the DWF viewer. When off, the drawing is not scaled.

Show Paper boundaries: Displays the same sheet border as in layout mode.

Convert .DWG hyperlink extensions to .DWF: All .DWG filename extensions found in hyperlinks are changed to .DWF.

7. Click **OK**, **OK**, and **OK**. Notice that AutoCAD exports the drawing as a DWF or JPEG file. AutoCAD displays a dialog box to show the progress of the translation process.

> **Tip:** Normally, you would plot the drawing to a DWF or JPEG file, then use FTP or other software to copy the file to your Web site. AutoCAD is able to save the files to a location on your network or the Internet.
>
> To save to a network location, click the ellipsis (...) button next to Location, and select a location from the Network Neighborhood.
>
> To save to a location on the Internet, click the Browse the Web button (looks like a earth with a magnifying glass). This displays LT's internal Web browser, which allows you "to search for a Web location to save the current plot file," according to Autodesk documentation.

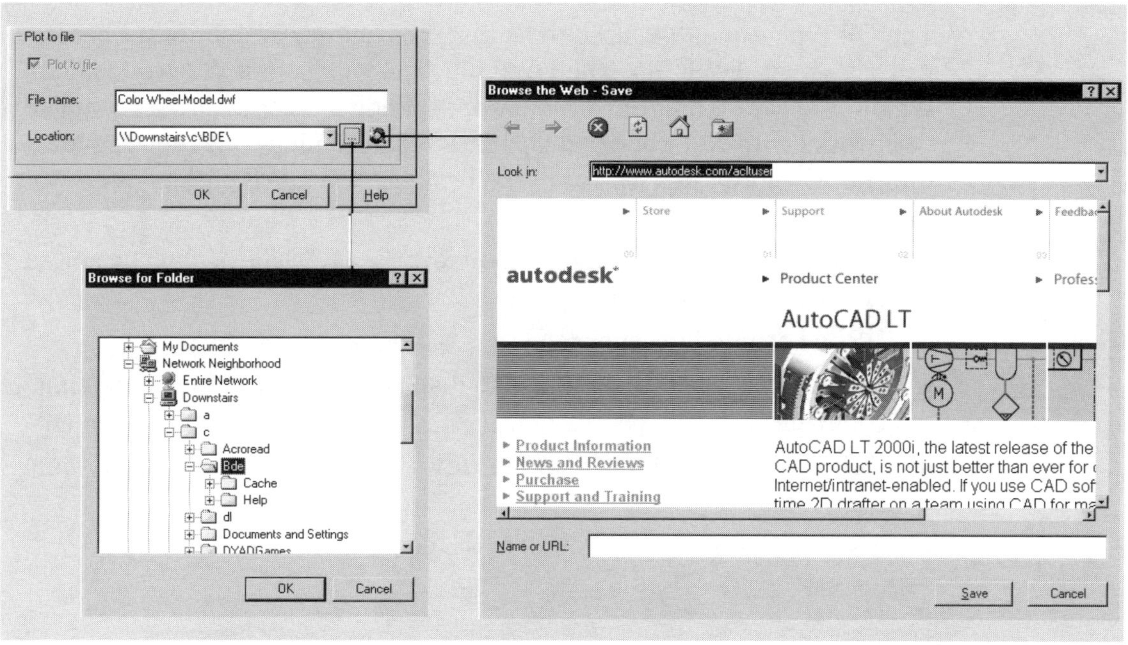

Warning: AutoCAD LT itself cannot display DWF files, nor can it import a DWF file. To view the DWF file, you must use a Web browser with the appropriate plug-in. Or, use file viewing software that supports DWF.

AutoCAD LT cannot export 3D objects with a z-coordinate to a DWF file, although it does flatten 3D views.

Shaded views and views with hidden lines removed are not exported to the DWF file.

As of AutoCAD LT 98, DWF files can be made in layout mode (formerly paper space).

Viewing DWF Files

As noted, AutoCAD LT itself cannot display nor import DWF files. Instead, you use a Web browser (with the appropriate plug-in) to see the DWF file. A *plug-in* allows a Web browser to correctly display any file. When the browser does not understand a file it has loaded, it checks its list of plug-ins. In most cases, you can download plug-ins free for just about any file format.

The DWF plug-in is required to be installed in your Web browser before you can view a DWF file. The DWF plug-in from Autodesk is called Whip!. A separate plug-in is required, depending on the browser you use. One works with Netscape and Opera; a separate plug-in is required for Internet Explorer. Regardless of the browser, the DWF plug-in offers these features:

- Real-time zoom and pan lets you view the DWF file as quickly as the drawing in AutoCAD LT.
- URLs become hyperlinks.
- Right-clicking the DWF image to displays a cursor menu of viewing commands.
- If the drawing contained named views, these are available from the cursor menu.
- The DWF file can be printed, or saved in DWF or BMP (bitmap) format.

Note: If you don't know whether the DWF plug-in is installed in your Web browser, select Help | About Plug-ins from the browser's menu bar. You may need to scroll through the list of plug-ins to find something like this:

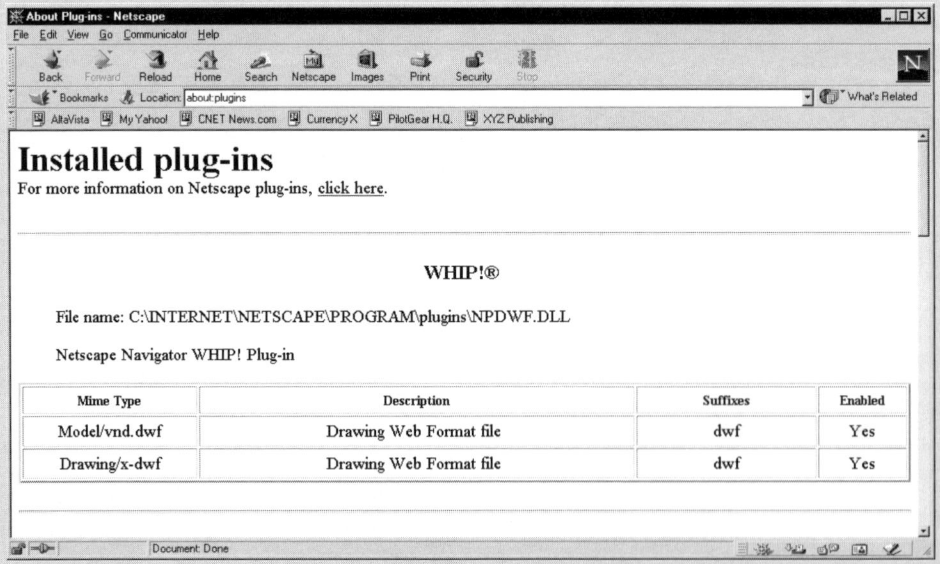

Checking for the DWF plug-in in Netscape Navigator.

Current versions of AutoCAD LT automatically install the DWF plug-in; versions prior to LT 98 do not. If the plug-in is not present, download it from the Autodesk Web site by following the steps below.

The DWF Plug-in

The DWF plug-in has several commands that let you view, print, and save the DWF file.

1. Move the cursor over the DWF image.
2. Right-click to display the cursor menu.
3. Move the cursor over a command and press the left mouse button.
4. To pan the drawing, select **Pan.** The cursor changes to an open hand. Panning in the Web browser is the same as real-time panning in AutoCAD LT.
5. To zoom the drawing, select **Zoom.** The cursor changes to a magnifying glass. Zooming in the Web browser is the same as real-time zooming in AutoCAD LT.

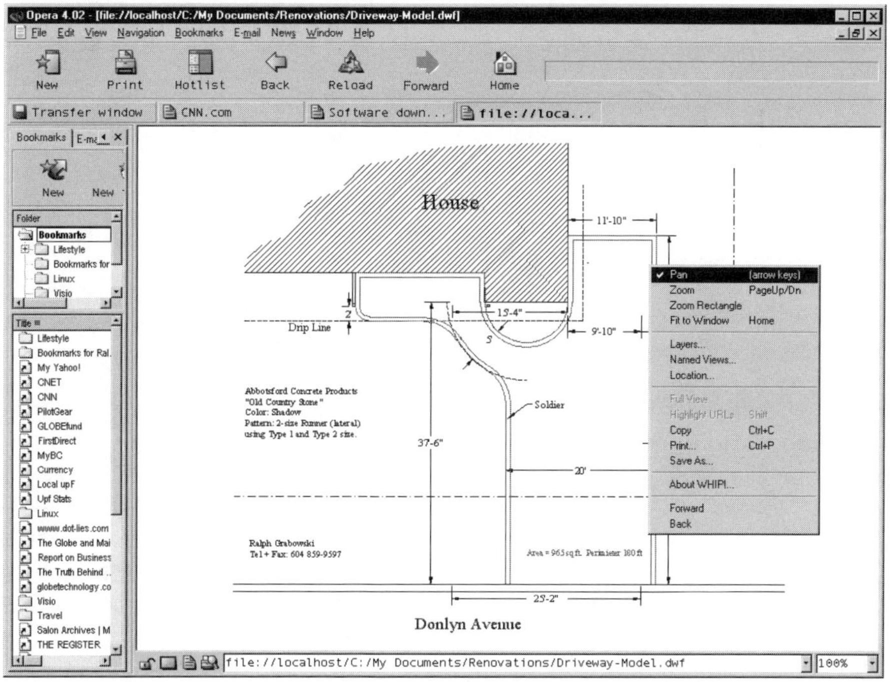

6. To zoom a windowed area, select **Zoom Rectangle.** The cursor changes to a plus sign (+). Zoom rectangle in the Web browser is the same as Zoom Window in AutoCAD LT.

7. To see the entire image, select **Fit to Window,** which is the same as Zoom Extents in AutoCAD LT.

8. To see the image all by itself in the browser, select **Full View.** This command lets you see the DWF image as large as possible. To return to the Web page, right-click and select **Back.**

9. To toggle layers, select **Layers.** The Whip! plug-in displays the Layers dialog box. To turn off a layer, click the light bulb icon next to the layer's name.

Toggling the display of layers.

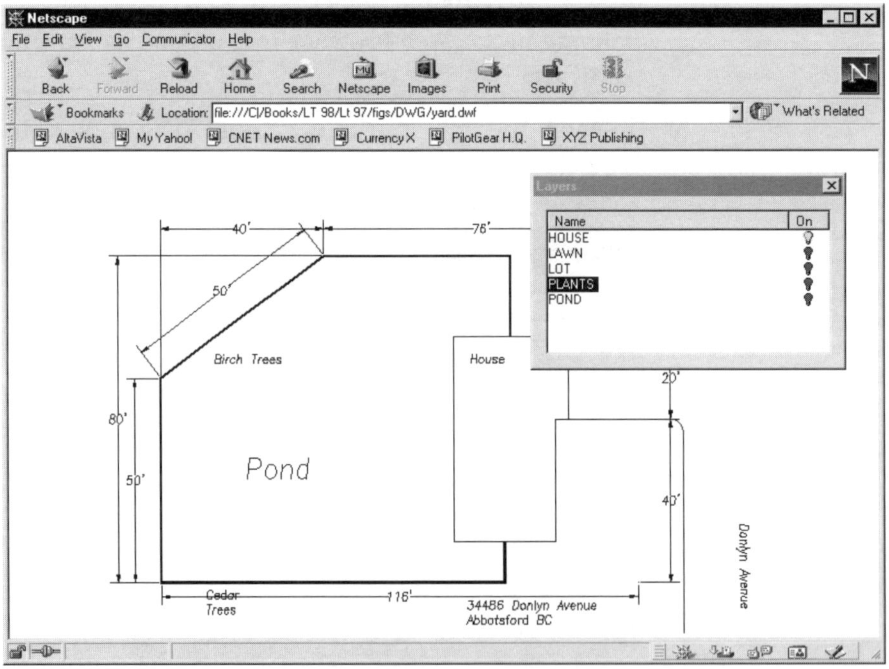

10. To see named views, select **Named Views,** which brings up the Named View dialog box. The dialog box lists named views, if you created them while the drawing was still in AutoCAD LT (with the View or DdView command). Even if there are no named views, the dialog box displays "INITIAL," which is the same view you see when the DWF image first loads. Double-click a named view to see it; click the small **x** to dismiss the dialog box.

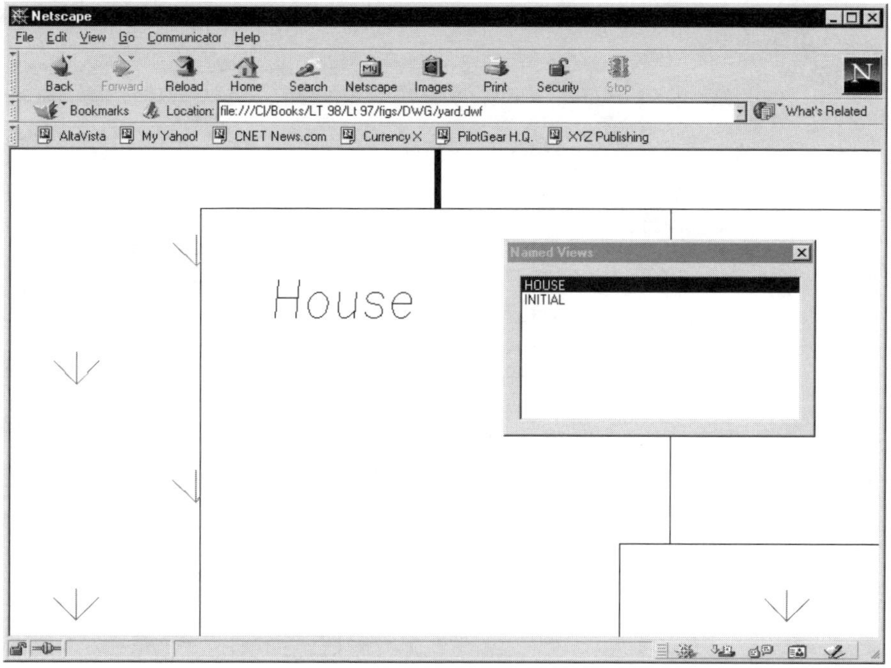

Selecting a named view.

11. The Location option on the cursor menu displays a coordinate dialog box. It shows the current x,y,z-coordinates under the cursor, just like the AutoCAD status bar.

The Location toolbar.

12. To see where hyperlinks are located in the DWF image, select **Highlight URLs**. This causes gray rectangles to flash over the objects with hyperlinks. To see the actual hyperlink name, move the cursor over a highlight area and look at the browser status line.

Highlighting hyperlinks in the DWF file.

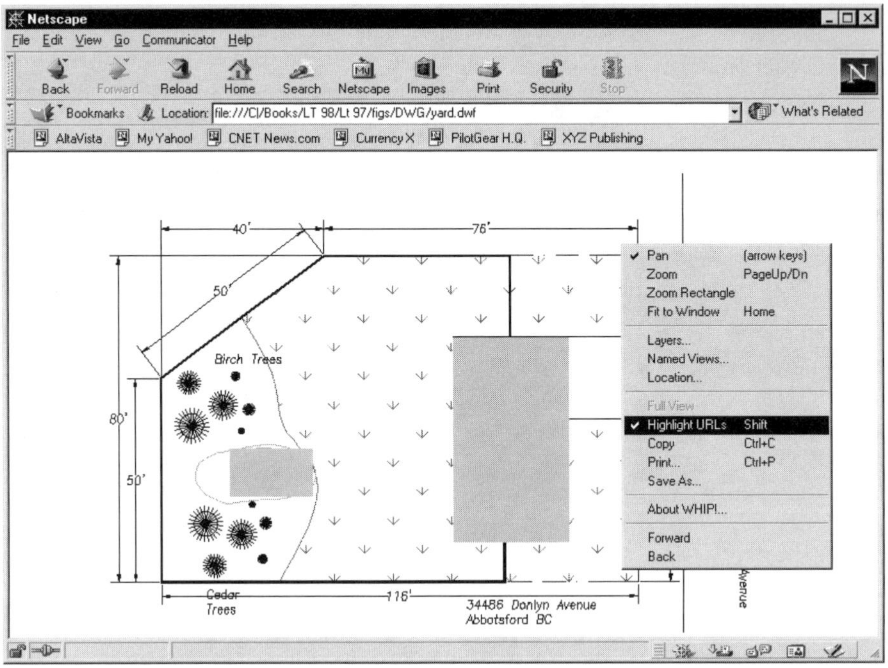

13. To copy the DWF image to the Windows clipboard, select the **Copy** command.

14. To print the DWF image, select the **Print** command. Notice that the Web browser displays the Windows Print dialog box.

15. To save the DWF image, select **SaveAs,** which saves the DWF file in three formats:

    - **DWF**: Drawing Web format
    - **BMP**: Windows bitmap, a raster format
    - **DWG**: AutoCAD drawing format

    The DWG option works only if the DWG file is located in the same folder as the DWF file. In this case, the DWF file is not translated to DWG format; instead, the original DWG file (if it exists) is copied.

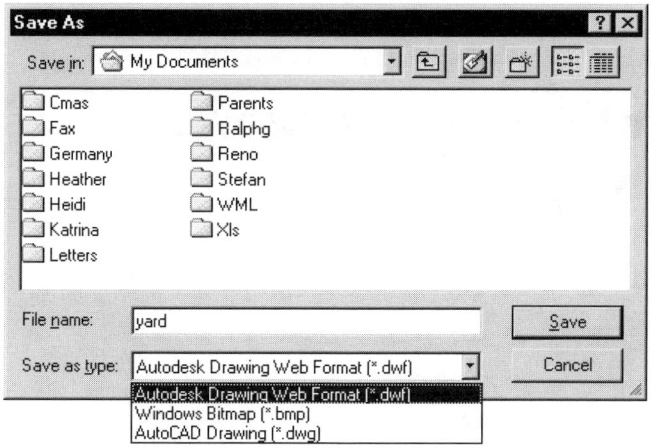

Saving the DWF file in several formats.

16. To see information about the DWF file and the DWF plug-in, select **About WHIP!**. The dialog box displays these elements, among others:
    - **DWF file revision number**: The version number of the DWF file.
    - **Description**: A description that can be generated by the DWF export software.
    - **Author**: An author name that can be generated by the DWF export software.
    - **Creator**: A creator name that can be generated by the DWF export software.
    - **Source filename**: The filename of the DWG file from which the DWF file was created.
    - **Creation time**: Date and time the DWF file was created.
    - **Modification time**: Date and time the DWF file was last modified.
    - **Source creation time**: Date and time the source DWG file was created.
    - **Source modification time**: Date and time the source DWG file was last modified.
    - **Current view left**: Left extents of the view in DWF units.
    - **Current view right**: Right extents of the view in DWF units.
    - **Current view bottom**: Bottom extents of the view in DWF units.
    - **Current view top**: Top extents of the view in DWF units.

Click **Whip! Home Page** to jump to the Autodesk Web site. Click **OK** to exit the dialog box.

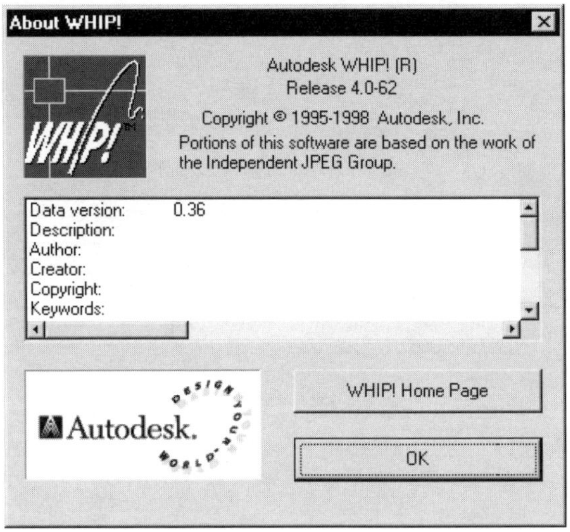

The Whip! plug-in's About dialog box.

17. To move back to the previous Web page, select **Back.**
18. To move forward (after moving back), select **Forward**.

Embedding a DWF File

Here is the HTML code you need to place a DWF file in your Web page.

1. The <embed> tag is an HTML tag for embedding a file in a Web page:

```
<embed src="yard.dwf">
```

   The quotation marks (") are required.

2. You can specify the size of the DWF image by adding Width and Height options:

```
<embed width=640 height=480 src="yard.dwf">
```

   The 640 and 480 specify the size in pixels.

3. If a user has turned off their Web browser's ability to load images, you can still include a textual description of the image when the browser does not load images:

```
<embed width=640 height=480 description="The Yard.Dwg from
Learn AutoCAD LT 2000i" src="yard.dwf">
```

Review Questions

1. A DWF file cannot be edited by AutoCAD.

   True / False

2. Match the abbreviation with its best meaning:

a.	DWF	i.	Joint Photographic Experts Group
b.	JPEG	ii.	Windows High Performance
c.	HTML	iii.	Drawing Web format
d.	WHIP	iv.	Hypertext Markup Language
e.	DWG	v.	Drawing

3. Which one of the following statements about DWF is not true:

   a. Compresses a drawing by as much as 8x smaller.
   b. Becomes pixelated when zoomed in.
   c. Can be displayed by a Web browser when the appropriate plug-in is installed.
   d. Protects the original DWG from being edited.

4. The Publish to Web wizard converts an AutoCAD drawing into the following file formats:

   a. DWF and JPEG
   b. DWG and GIF
   c. DXF and PNG
   d. DGN and PCX

5. One of the following statements is *not* true about JPEG:

   a. Uses compression to make the file smaller.
   b. Lines and text become pixels when zoomed in.
   c. Can be displayed by any Web browser.
   d. Can be edited by AutoCAD LT.

6. The Plot command can create a DWF or JPEG file from a drawing.

   True / False

7. When viewing a DWF file in a Web browser, you cannot zoom and pan the image.

   True / False

8. Named views and layers are retained in the DWF file.

   True / False

9. Named views and layers are retained in the JPEG file.

   True / False

10. Which is the best HTML tag for placing a DWF file in a Web page?

    a. Src
    b. Image
    c. Place
    d. Embed

Chapter 14

Isometric Drafting

AutoCAD LT 2000i has a limited capability for drawing in three dimensions ("3D" for short). When CAD software does not have full 3D functions, it is sometimes called "2-1/2D" drafting. LT's 3D capabilities allow you to:

- Input 3D coordinates to many drawing and editing commands
- Use the 3dPoly command to draw a 3D polyline
- Give all objects an elevation (height above the x,y-plane) and a thickness with the Elev and Thickness commands
- Set 3D viewpoints via the DView and View commands
- Set user-defined 3D coordinate systems with the UCS command
- View 3D solid models created in AutoCAD 2000i (but not edit them)
- Shade a 3D model with the Shade command

In the days of manual drafting, it wasn't possible to create a 3D drawing. To represent an object in 3D, manual drafters commonly used isometric representation. It shows a view of the object showing its front, right, and top sides. To show all three side at the same time, they are drawn at 120-degree angles to each other.

AutoCAD LT provides little help for isometric drafting. It displays an isometric grid and cursor, provides a command to switch between isometric planes, and draws isometric circles, but that's all. This chapter shows you how to set up a drawing for isometric drawing, which includes creating the appropriate text and dimension styles.

Key Concepts

- Simulating 3D design by isometric drafting
- Setting up LT for isometric drafting
- Understanding the three isoplanes
- Creating isometric text and dimension styles

Key Terms

Isometric—a form of drafting that allows you to see three sides of a 3D model

Isoplane—the three drawing planes, situated at 120 degrees to each other

Oblique—the slant of text

Abbreviations

2D	two-dimensional
3D	three-dimensional
Dimstyle	dimension style
Iso	isometric

Command	Shortcut	Menu Selection
Isoplane	Ctrl+E	...

What is Isometric Drafting?

In brief, isometric drafting is a way to simulate a 3D drawing. It shows three sides of an object, making it easier to visualize than standard 2D drawings. Isometric drawings are commonly used for piping layouts. *Isometric* means "equal projection." Isometric drafting is sometimes called "axonometric projection."

The front, top, and right sides of an object are drawn at 120 degrees to each other. To make isometric objects look "right," they have to be drawn skewed. A square is drawn as a rhombus. A circle is drawn as an ellipse. Arcs and fillets are drawn as elliptical arcs. Text is slanted by 30 degrees. And all are drawn at a different skew for each of the three isometric planes.

Note: In addition to isometric, there are two other forms of drafting that represent 3D:

Oblique projection draws the front view as normal, then draws the top and right sides at any angle, although 30 or 45 degrees are most common. You can create oblique drawings in LT using methods similar to that of isometric drafting.

Perspective projection attempts to simulate the effect of objects decreasing in size the further they are from your eye. One- and two-point perspectives are commonly used for artists conception drawings, but cannot be used for construction since distances cannot be measured. AutoCAD LT allows you to display a drawing in perspective mode using the DView command.

The Acltiso.Dwt file is a template drawing containing ISO standards; ISO is short for International Organization for Standardization. The template has nothing for creating isometric drawings.

AutoCAD LT provides half of the tools you need for creating isometric drawings. On the plus side, AutoCAD provides:

- An isometric grid and cursor for each of the three isometric planes. Use the Snap command's Style option to turn on Iso mode.
- A shortcut key for quickly switching between isometric planes. Press **Ctrl+E** to switch between the isometric planes: top, right, and left.
- Ellipses and elliptical arcs that draw correctly as isometric circles and arcs in the appropriate isoplane. Use the Ellipse command's Isocircle option to draw isometric circles; use the Arc option to draw isometric arcs.

On the minus side, AutoCAD does not provide:

- Isometric text styles
- Isometric dimension styles

In the next section, I'll show you how to turn on Iso mode, switch between isometric planes, and practice drawing an isocircle. In the following section, I show you how to create isometric text and dimension styles.

Setting Up LT for Isometric Drafting

Start AutoCAD LT with a new drawing. To turn on isometric mode, use the Drafting Settings dialog box, as follows:

1. From the menu bar, select **Tools | Drafting Settings**. Notice that AutoCAD displays the Drafting Settings dialog box.
2. Select the **Snap and Grid** tab.

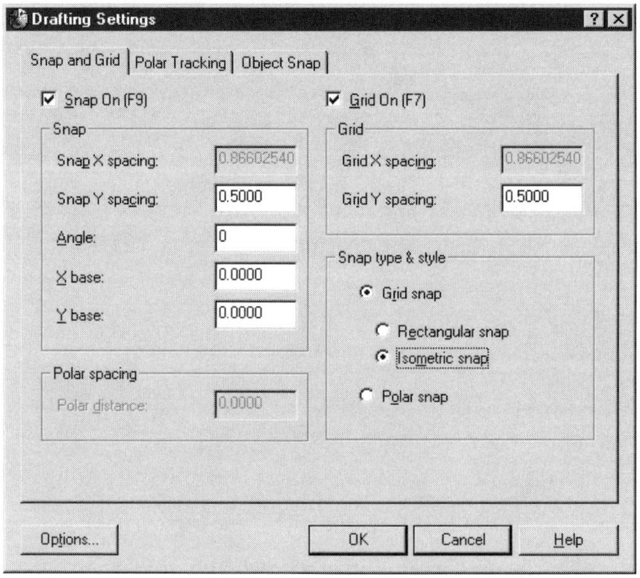

The Snap and Grid tab of the Drafting Settings dialog box.

3. In the **Snap type & style** area, select the **Isometric snap** radio button. Notice that **Snap X spacing** and **Grid X spacing** fields turn gray. You can rotate the isometric grid, but AutoCAD does not allow you to specify different spacing in the X and Y directions.
4. Click **OK**. AutoCAD is now in isometric mode. Notice that the grid and the cursor are no longer orthogonal (at 90 degrees). Instead, the grid dots and cursor crosshairs are at 120 degrees. AutoCAD starts off in the left isoplane.

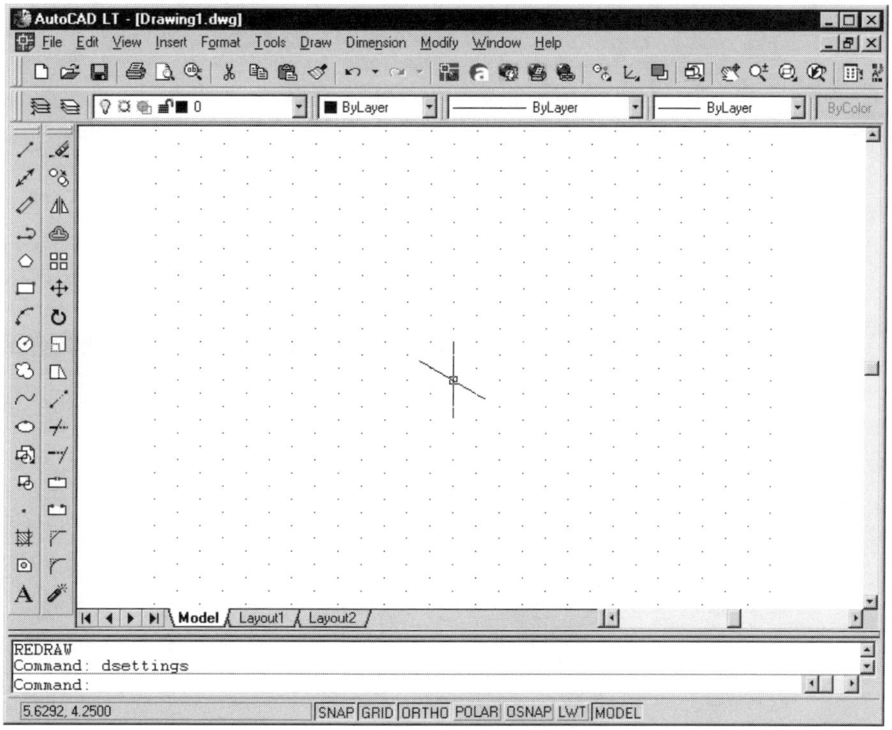

The isometric grid and cursor are at 120 degrees.

5. Press **Ctrl+E** (hold down the **Ctrl** key and press **E**). Notice that the grid and the cursor change their orientation. AutoCAD now displays the top isoplane, and in the command prompt area, AutoCAD notes:

   <Isoplane Top>

6. Function key F5 is an alternative to pressing Ctrl+E for changing the isoplane. Press **F5**. Notice that the grid and the cursor change their orientation again. AutoCAD is now displaying the right isoplane, and in the command prompt area, AutoCAD reports:

   <Isoplane Right>

7. Press **F5** (or **Ctrl+E**) again and again. As you do, AutoCAD switches the display between the three isoplanes: left, top, right, and back to left:

   <Isoplane Left>
   <Isoplane Top>
   <Isoplane Right>

8. It is very helpful to turn on ortho mode when creating isometric drawings. On the status bar, click **ORTHO**.

9. If you wish, save the drawing as an isometric template drawing, as follows:

   - From the menu bar, select **File | Save As.**
   - When the Save Drawing As dialog box appears, select **Drawing Template File (*.dwt)** from the Files of type list. Notice that AutoCAD automatically selects the Template folder.
   - For the File name, type **Isometric.**

Saving the drawing as a template file.

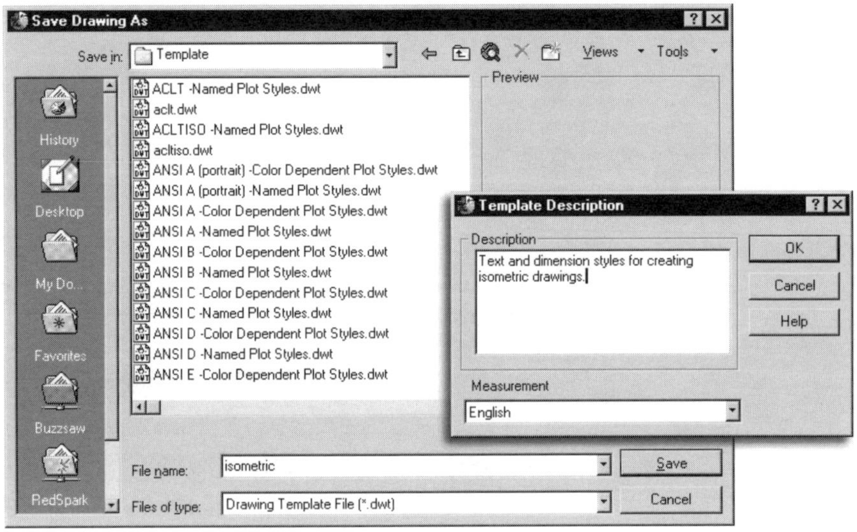

   - Click **Save.** Notice that AutoCAD displays the Template Description dialog box.
   - In the Description field, type something useful, such as **Text and dimension styles for creating isometric drawings.**
   - From the Measurement list box, select either **English** or **Metric**, whichever you prefer.
   - Click **OK.** Notice that the filename on AutoCAD LT's title bar changes to:

     AutoCAD LT – [isometric.dwt]

Drawing Isometric Circles and Arcs

Earlier I mentioned that all objects are drawn skewed in an isometric drawing. AutoCAD provides a command to draw two of those objects, the isometric circle and arc; there are no similar commands for drawing isometric rectangles or other basic shapes.

To draw an isometric circle or arc, you use a "hidden" option of the Ellipse command. Here's how:

1. At the Command: prompt, type the Ellipse command:

   ```
   Command: ellipse
   Specify axis endpoint of ellipse or [Arc/Center/Isocircle]:
   ```

   Notice that AutoCAD has added the Isocircle option, which you haven't seen before. The Isocircle option appears only when AutoCAD is in isometric mode. You cannot access the Isocircle option from the menu bar.

2. Type **I** to specify an isocircle. Click points on the screen when AutoCAD prompts you for the center and radius (or diameter) of the isocircle.

   ```
   Specify axis endpoint of ellipse or [Arc/Center/Isocircle]: i
   Specify center of isocircle: [pick]
   Specify radius of isocircle or [Diameter]: [pick]
   ```

   Notice that AutoCAD draws the isocircle.

An isocircle is drawn with the Ellipse command.

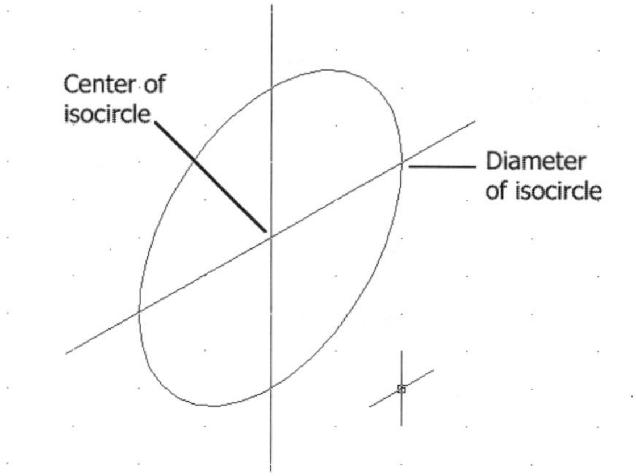

3. Press **F5** to change to another isoplane.
4. Repeat the **Ellipse** command. Notice that the isocircle is drawn in another orientation.
5. Press **F5** again, and repeat the **Ellipse** command. This time the isocircle is drawn in the third orientation.

AutoCAD's isocircles are sensitive to the current isoplane.

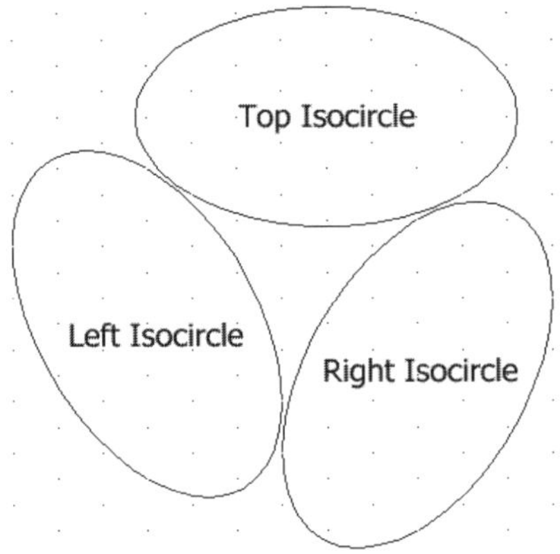

6. To draw an arc or a fillet in isometric mode, use the Arc option of the Ellipse command, as follows:

```
Command: ellipse
Specify axis endpoint of ellipse or [Arc/Center/Isocircle]: arc
Specify axis endpoint of elliptical arc or [Center/Isocircle]: isocircle
Specify center of isocircle: [pick]
Specify radius of isocircle or [Diameter]: [pick]
Specify start angle or [Parameter]: 0
Specify end angle or [Parameter/Included angle]: 180
```

Like isocircles, AutoCAD draws isometric arcs appropriate to the current isoplane.

Creating the Isometric Text Styles

As noted earlier, AutoCAD LT does not include text and dimension styles appropriate for isometric drafting. For this reason, we have to create them ourselves. We cannot simply place the text rotated by 30 degrees because it looks "wrong." Even when adjusted for the isometric angle (by +30 or –30 degrees), the text looks as if it is slanted. The solution, as we will see, is to create a style that *obliques* (slants) the text by 30 degrees.

Normal text looks "wrong" when placed in an isometric drawing.

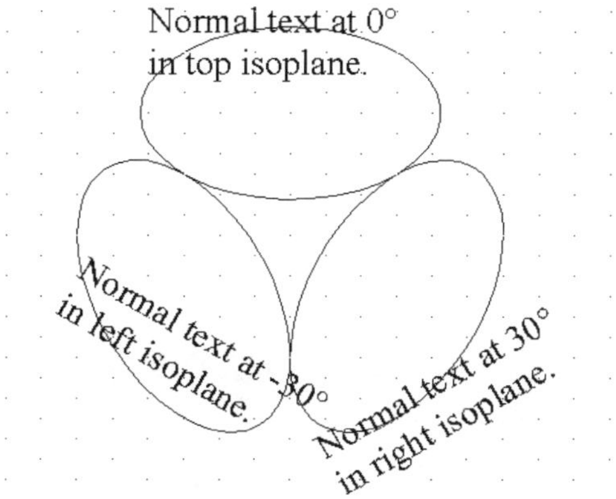

To place text that looks correct in an isometric, you need to create three new text styles: one for each isoplane. In each style, the obliquing angles must be set to +30 or -30 degrees. Later, when placing the text in the drawing, you need to specify a rotation angle of +30 or -30 degrees. The following table summarizes the six angles:

Isoplane	Style: Oblique Angle	Text: Rotation
Left	-30°	-30°
Top	-30°	30°
Right	30°	30°

You use the Style command to create a text style, as follows:

1. From the menu bar, select **Format | Text style**. Notice the Text Style dialog box.
2. Click **New.** Notice the New Text Style dialog box.
3. In the Style Name field, type **leftiso.**

The New Text Style dialog box.

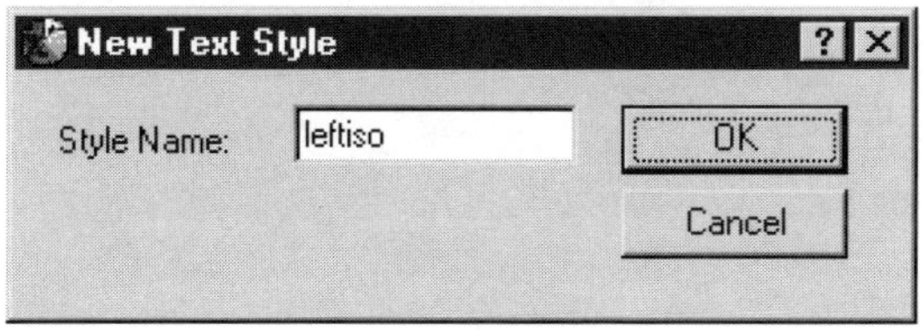

4. Click **OK.**

5. Back in the Text Style dialog box, set the following options:

   Font Name: **Times New Roman** (or any other font that you fancy)

   Oblique Angle: **–30**

   Click **Apply.**

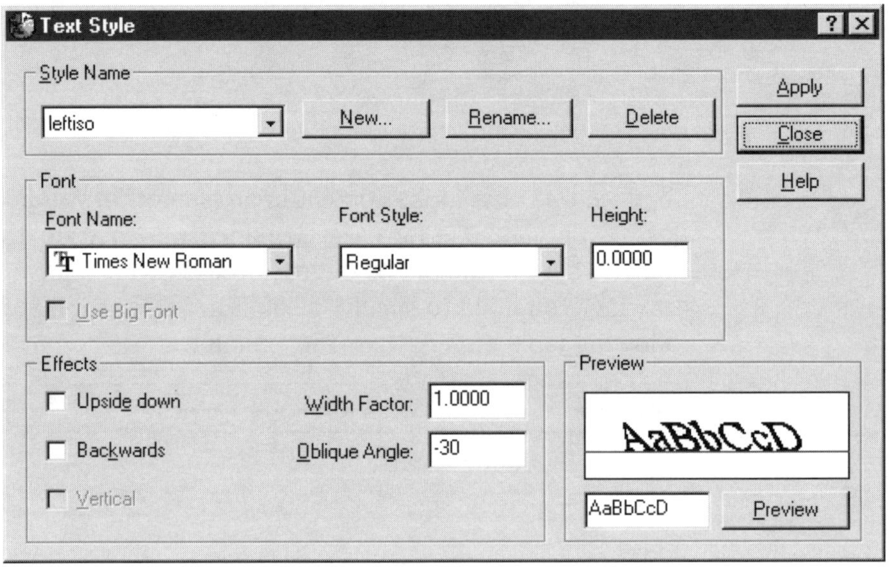

The Text Style dialog box.

6. Repeat the process two times:
   - Click **New.**
   - Style Name: **topiso**
   - Click **OK.**
   - Font Name: **Times New Roman** (or the same font as before).
   - Oblique Angle: **–30**
   - Click **Apply.**
   - Click **New.**
   - Style Name: **rightiso**
   - Click **OK.**

- Font Name: **Times New Roman** (or the same font as before).
- Oblique Angle: **30**
- Click **Apply.**

7. Click **Close.** You now have three text styles, one for each isoplane.
8. Use the Save command to save the Isometric.Dwt template drawing so that the text styles are available in future drawings.

Let's practice using the text styles:

1. From the menu bar, select **File | New.** When the AutoCAD LT Today window appears, select **Template** from the Select how to begin list.
2. In the file list, click **I** and scroll to the end of the list.

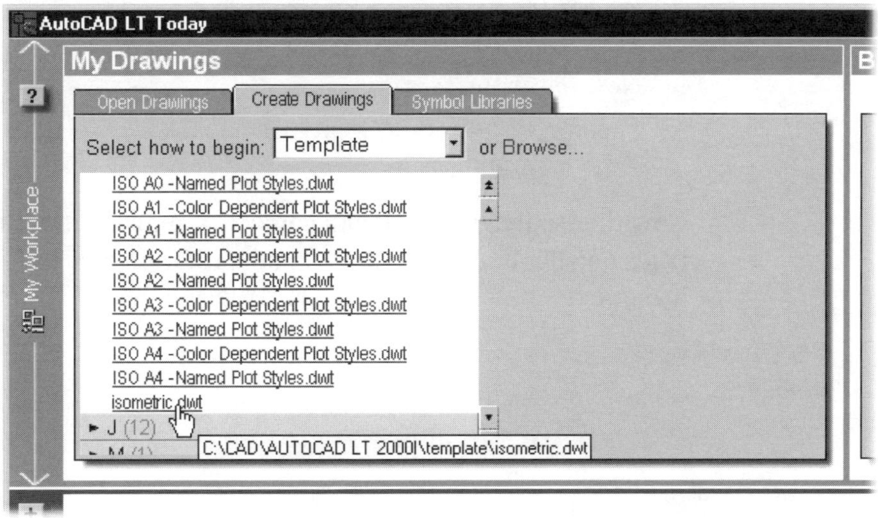

Selecting the Isometric.Dwt template file.

3. Select **Isometric.Dwt.** Notice that the new drawing has the familiar isometric grid pattern.
4. Use the Isoplane command to ensure the left isoplane is set, as follows:

   Command: isoplane
   Current isoplane: Right
   Enter isometric plane setting [Left/Top/Right] <Left>: [Enter]
   Current isoplane: Left

5. To help orient yourself, draw an isometric circle using the Ellipse command's Isocircle option.

6. Let's place the isometric text with the Text command, as follows:

   Command: text
   Current text style: "txt" Text height: 0.2000

7. Change the style to LeftIso with the Style option:

   Specify start point of text or [Justify/Style]: s
   Enter style name or [?] <txt>: leftiso

8. Pick the text starting point and specify a rotation angle of –30 degrees, as follows:

   Current text style: "leftiso" Text height: 0.2000
   Specify start point of text or [Justify/Style]: [pick]
   Specify height <0.2000>: [Enter]
   Specify rotation angle of text <0>: –30

9. Now enter the text:

   Enter text: Left isoplane text.
   Enter text: [Enter]

Now the isometric text looks correct. It may seem bizarre to you that it takes an obliquing angle of –30 and a rotation angle of –30 degrees to make text look correct in the left isoplane, but it works!

Text in the left isoplane.

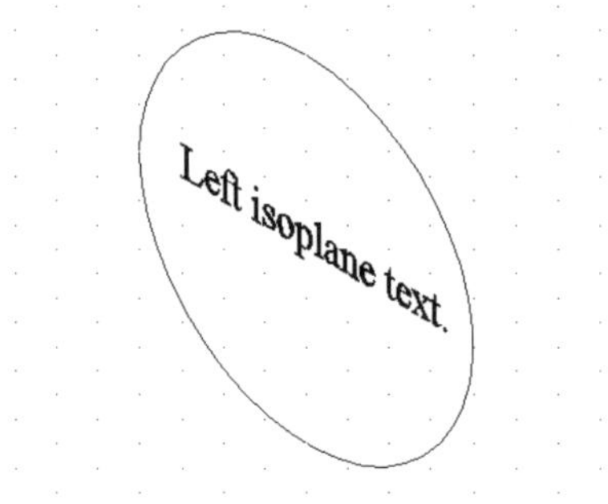

To place text in the top isoplane:

- Press **F5** to switch to the top isoplane.
- Start the Text command, and use the Style option to select the TopIso text style.
- Use the Rotation option to specify a rotation angle of 30 degrees.

To place text in the right isoplane:

- Press **F5** to switch to the right isoplane.
- Start the Text command, and use the Style option to select the RightIso text style.
- Use the Rotation option to specify a rotation angle of 30 degrees.

Text in the three isoplanes.

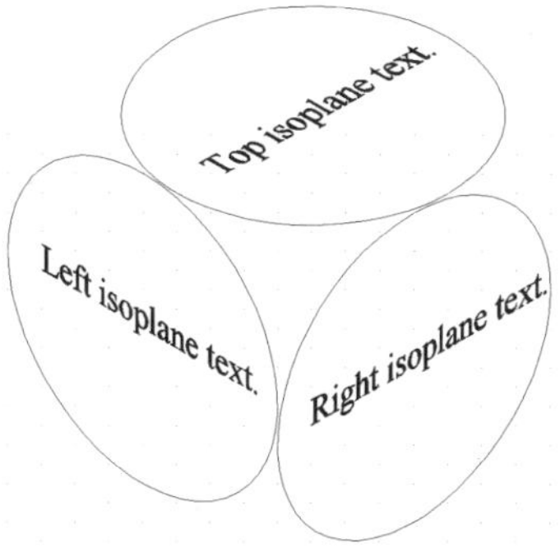

Creating Isometric Dimension Styles

Similarly, using normal dimension text and extension lines just doesn't look "right" in isometric drawings. The solution is to use the isometric text styles along with the DimOblique command to slant the extension lines.

Tips: Use the QUAdrant object snap when dimensioning isocircles and arcs. AutoCAD correctly snaps to the appropriate quadrant point on the isocircle.

The DimRadius and DimDiameter commands cannot find the center of isometric circles and arcs. Use the Leader and QLeader commands, instead.

In summary, you create three dimension styles, one for each isoplane. Each dimstyle uses the corresponding text style. (If you want to be fussy, you could create isometric arrowheads, too!) Then, after the dimensions are placed in each isoplane, use the DimEdit Oblique command to slant the extension lines, as described by the following table:

Isoplane	Text Style	DimEdit: Oblique
Left	IsoLeft	30°
Top	IsoTop	–30°
Right	IsoRight	–30°

To create a dimension style, you use the DimStyle command. Ensure you are in the Isometric.Dwt drawing (from the menu bar, select **Window | Isometric.Dwt**).

1. From the menu bar, select **Format | Dimension Style**. Notice the Dimension Style Manager dialog box.
2. Click **New**. Notice the Create New Dimension Style dialog box.
3. Fill in the following data:
   - New Style Name: **LeftIso**
   - Start with: **Standard**
   - Use for: **All dimensions**
   - Click **Continue**.

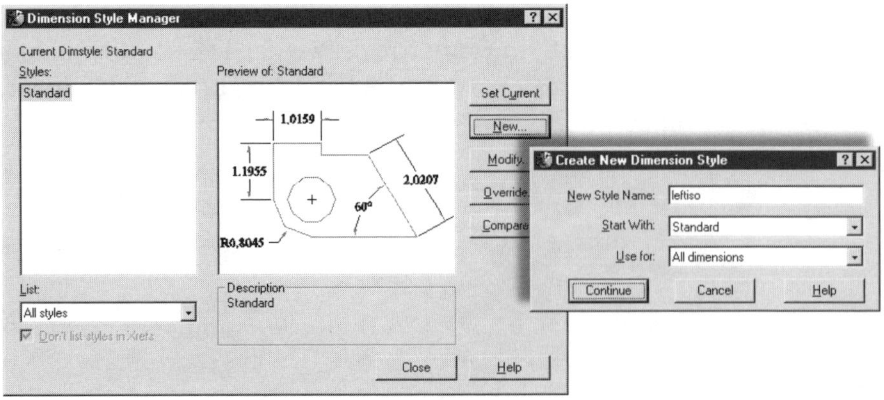

The Create New Dimension Style dialog box.

4. Click the **Text** tab of the Modify Dimension Style dialog box:
   - Text style: **LeftIso**
   - Text alignment: **Aligned with dimension line.**
5. Click the **Primary Units** tab:
   - Precision: **0.0**
   - Click **OK.**

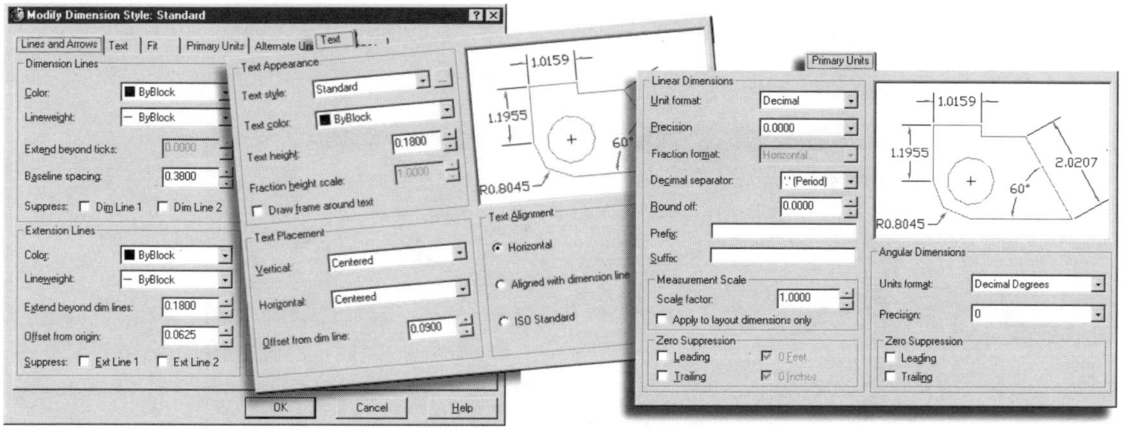

The Text and Primary Units tabs of the Modify Dimension Style dialog box.

6. Repeat for the top isoplane:
   - Click **New.**
   - New Style Name: **TopIso**
   - Start with: **LeftIso**
   - Click **Continue** and select the **Text** tab.
   - Text style: **TopIso**
   - Click **OK.**
7. And once more for the right isoplane:
   - Click **New.**
   - New Style Name: **RightIso**
   - Start with: **LeftIso**

- Click **Continue** and select the **Text** tab.
- Text style: **RightIso**
- Click **OK.**

8. Notice that the Styles list now contains four dimension styles: the original Standard and the three new isometric styles. Set the LeftIso dimstyle as current, as follows:
   - Select **LeftIso.**
   - Click **Set Current.**
   - Click **Close.**

Setting the LeftIso dimstyle as current.

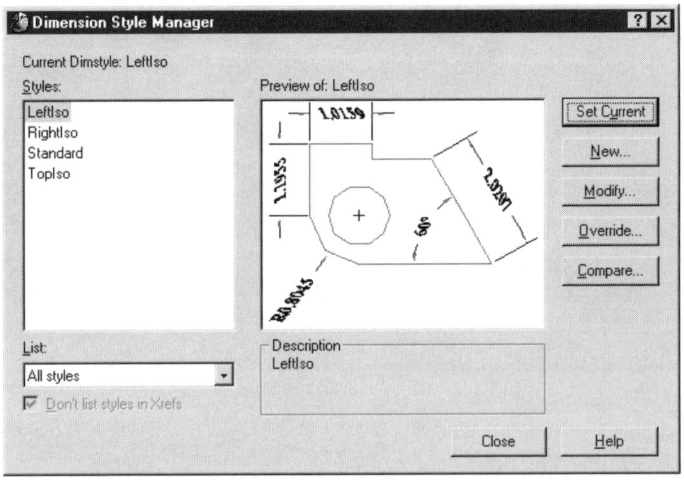

9. Once again, it is a good idea to save the Isometric.Dwt template drawing to preserve your hard work. From the menu, select **File | Save.**

Let's try using the dimension styles to see what they look like!

1. Draw isocircles in each of the isoplanes.
2. Use the Isoplane command to ensure the drawing is displaying the left isoplane.
3. From the menu bar, select **Dimension | Aligned.**
4. Pick two points to dimension the left isocircle:

```
Command: dimaligned
Specify first extension line origin or <select object>: [pick]
Specify second extension line origin: [pick]
```

Specify dimension line location or [Mtext/Text/Angle]: [pick]
Dimension text = 3.0

Placing an aligned dimension in the left isoplane.

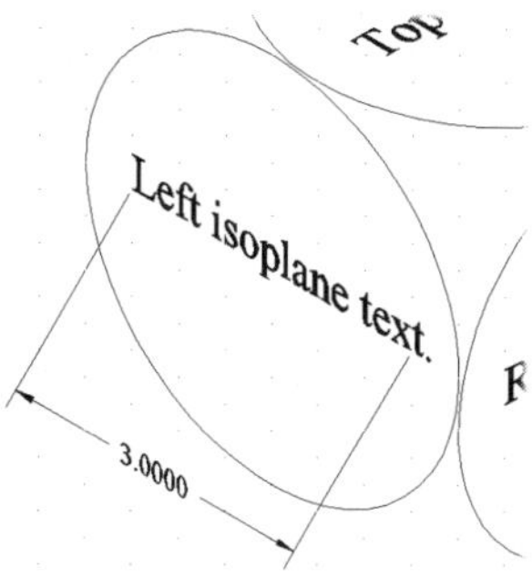

5. Hmm... the dimension looks a bit odd. After the dimensions are placed, you must use the DimEdit command to oblique the extension lines. For the left isoplane, you apply an obliquing angle of +30 degrees, as follows:

Command: dimedit
Enter type of dimension editing [Home/New/Rotate/Oblique]
<Home>: oblique
Select objects: [pick one or more dimensions]
1 found Select objects: [Enter]
Enter obliquing angle (press ENTER for none): 30

Ahh... that's better!

Obliquing the extension lines.

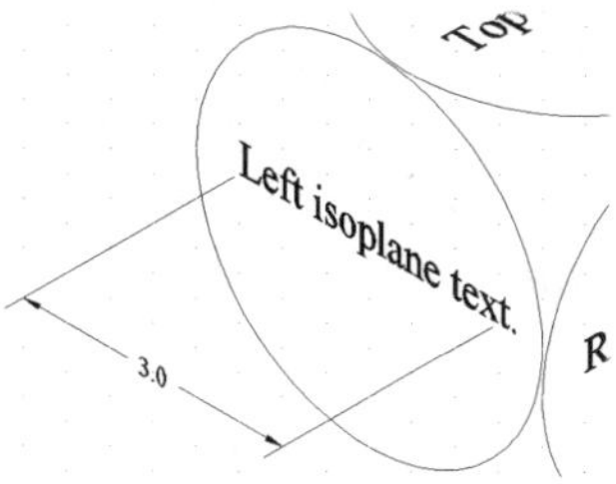

6. To dimension in the other two isoplanes, you follow this pattern:
   - Press **F5** to switch to the next isoplane.
   - Use the DimStyle command to change the dimension style for the isoplane.
   - Use the DimAligned command to place the dimensions.
   - Use the Oblique option of the DimEdit command to change the angle of the extension lines; specify **-30** degrees for both the top and right isoplanes.

   Clearly, you want to do as much dimensioning as possible in one isoplane, before going on to the next.

Dimensions in all three isometric planes.

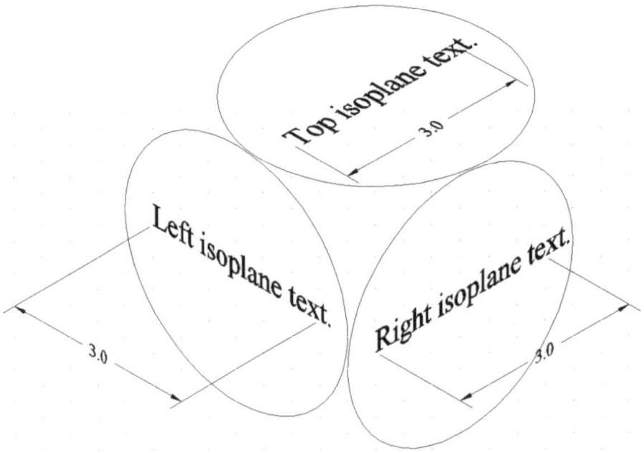

Isometric Drafting Tutorial

The previous sections of this chapter showed you how to set up a drawing for isometric drafting, and how to draw isometric circles and arcs. Let's put this preparatory work to good use with a tutorial that draws and dimensions a simple part in isometric view. The part is a right-angle clip with a hole.

In general, the approach to creating an isometric drawing is: draw as much in one isoplane as possible, then move on to the next isoplane. Because of this, it helps to plan the drafting process in advance. To draw the clip, we will work in stages, as shown by the illustration:

The stages in drawing the clip as an isometric drawing.

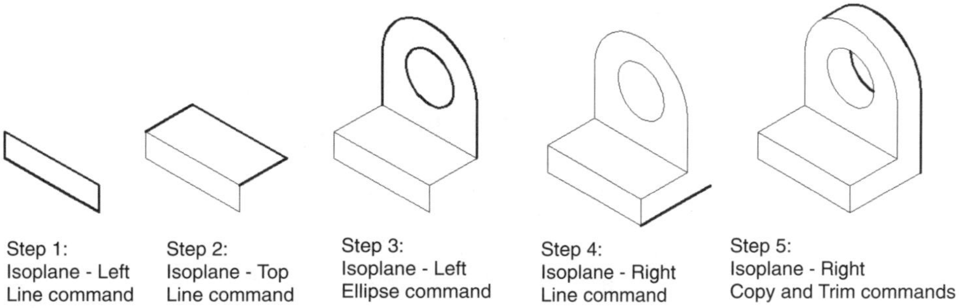

1. Start a new drawing, using Isometric.Dwt as the template drawing. Ensure the following settings:
   - Isoplane: **Left**
   - Ortho: **On**
   - Snap: **On**
   - Grid: **On**
2. Use the Line command to draw Step 1, as shown in the illustration. The rectangle is 2 inches wide and 0.5 inches tall.

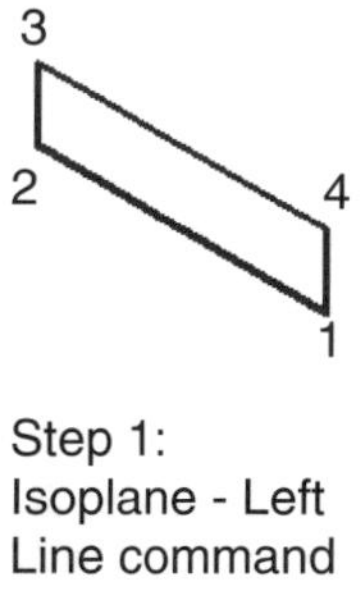

The first stage of drawing the clip.

```
Command: line
Specify first point: [pick 1]
Specify next point or [Undo]: [pick 2, 2" away]
Specify next point or [Undo]: [pick 3, 0.5" away]
Specify next point or [Close/Undo]: [pick 4, 2" away]
Specify next point or [Close/Undo]: Close
```

Notice that you have drawn a rhombus, with two corners at 120 degrees and two at 60 degrees. But since this is in isometric mode, it looks like a 2" x 0.5" rectangle with square corners (90 degrees). This shows you that isometric drafting is basically an illusion!

Note, too, how the ortho and snap modes helped you draw precisely in the isoplane.

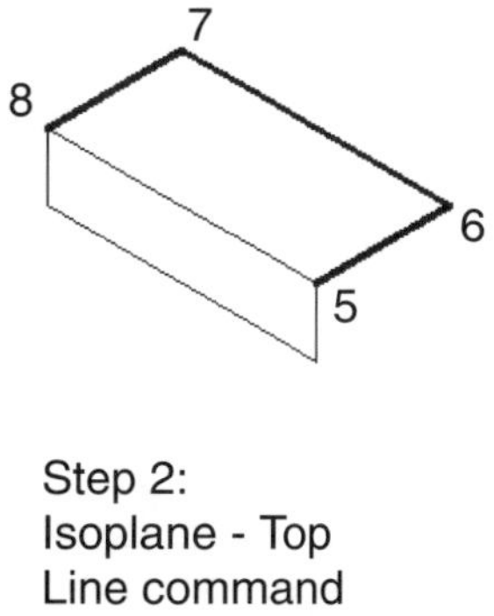

The second stage of drawing the clip.

3. Switch to the top isoplane, and draw the lines for Step 2. The rectangle is 1 inch wide.

```
Command: F5 <Isoplane Top>
```

Command: line
Specify first point: [pick 5]
Specify next point or [Undo]: [pick 6, 1″ away]
Specify next point or [Undo]: [pick 7, 2″ away]
Specify next point or [Close/Undo]: [pick 8, 1″ away]
Specify next point or [Close/Undo]: [Enter]

4. Switch back to the left isoplane (press **F5** twice), and draw the two vertical lines for Step 3. The two vertical straight lines are 1 inch tall:

 Command: F5 <Isoplane Right> F5 <Isoplane Left>

 Command: line
 Specify first point: [pick 9]
 Specify next point or [Undo]: [pick 10, 1″ away]
 Specify next point or [Undo]: [Enter]

 Command: [Spacebar]
 LINE Specify first point: [pick 11]
 Specify next point or [Undo]: [pick 12, 1″ away]
 Specify next point or [Undo]: [Enter]

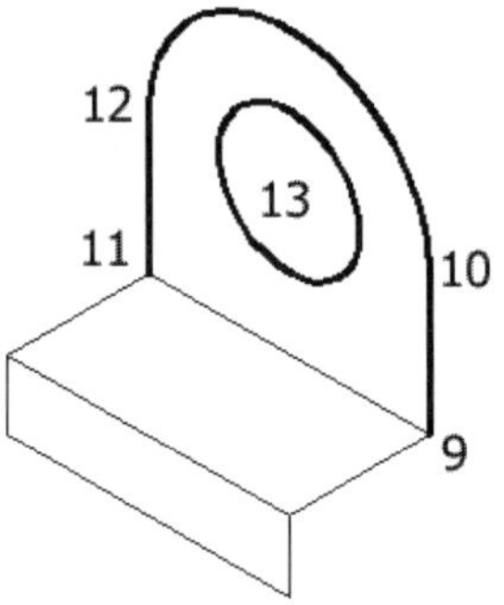

Step 3:
Isoplane - Left
Ellipse command

The third stage draws the isometric circle and arc.

5. Use the Ellipse command to draw the 0.5"-radius "hole," as follows:

 Command: ellipse
 Specify axis endpoint of ellipse or [Arc/Center/Isocircle]: i
 Specify center of isocircle: [pick 13]
 Specify radius of isocircle or [Diameter]: [pick 10]

6. Repeat the Ellipse command to draw the arc, as follows:

 Command: [spacebar]
 ELLIPSE Specify axis endpoint of ellipse or [Arc/Center/Isocircle]: a

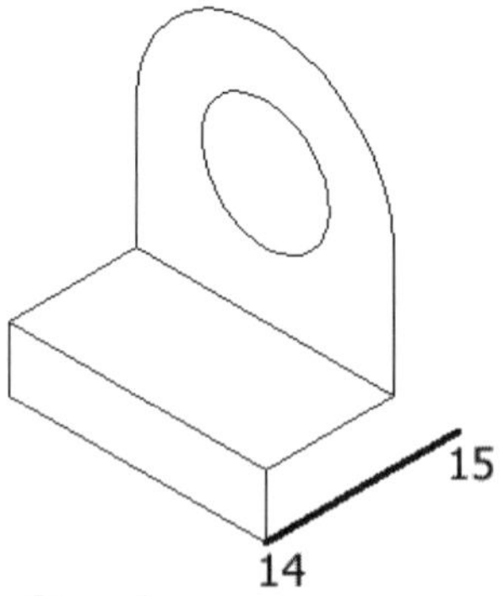

Step 4:
Isoplane - Right
Line command

Drawing the line in the right isoplane.

Specify axis endpoint of elliptical arc or [Center/Isocircle]: i
Specify center of isocircle: [pick 13]
Specify radius of isocircle or [Diameter]: [pick 10]
Specify start angle or [Parameter]: [pick 10 again]
Specify end angle or [Parameter/Included angle]: [pick 12]

7. Draw the line shown in Step 4, as follows:

Command: F5 <Isoplane Top> F5 <Isoplane Right>

Command: line
Specify first point: [pick 14]
Specify next point or [Undo]: [pick 15, 1.5″ away]
Specify next point or [Undo]: [Enter]

8. The easiest way to create the hole effect is with the Copy and Trim commands. The Copy command creates a copy of the isometric circle and arc; the Trim command cuts them off to create the illusion of hidden-line removal.

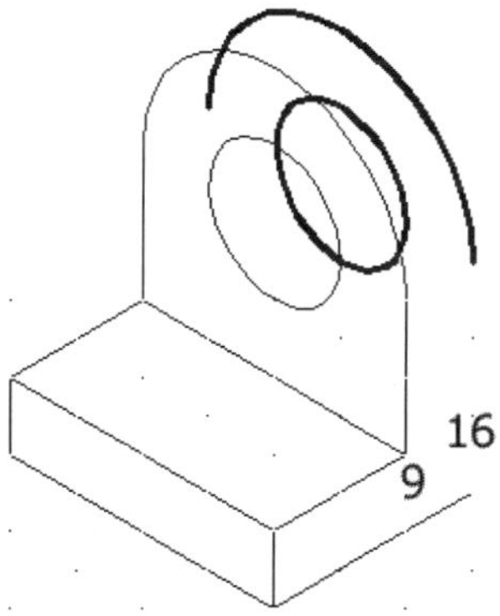

Making a copy of the isocircle and arc.

Command: copy
Select objects: [pick isocircle]
1 found Select objects: [pick isoarc]
1 found, 2 total Select objects: [Enter]
Specify base point or displacement, or [Multiple]: [pick 9]
Specify second point of displacement or <use first point as displacement>: [pick 16, 0.5″ away]

9. Draw the line that connects the copied elements. You will find it useful to employ the QUAdrant object snap, as follows:

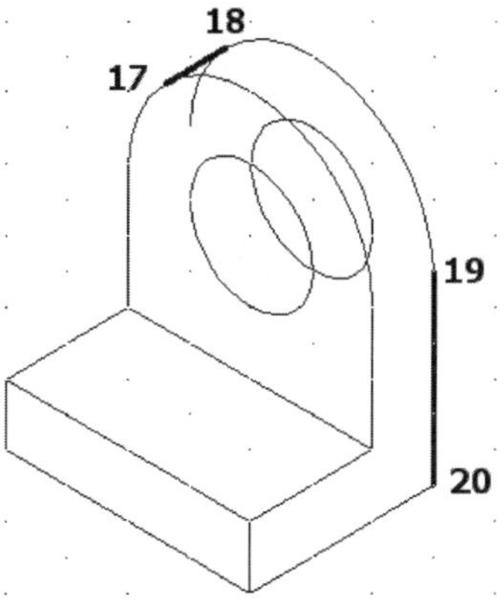

Adding the connecting lines.

Command: line
Specify first point: qua
of [pick 17]
Specify next point or [Undo]: qua
of [pick 18]
Specify next point or [Undo]: [Enter]
Command: [spacebar]
LINE Specify first point: [pick 19]

Specify next point or [Undo]: qua
of [pick 20]
Specify next point or [Undo]: [Enter]

10. Now we'll use the Trim command to clean up the drawing:

Command: trim
Current settings: Projection=UCS
Edge=None
Select cutting edges ...
Select objects: [pick line 17-18]
1 found Select objects: [Enter]
Select object to trim or [Project/Edge/Undo]: [pick end of arc at 21]
Select object to trim or [Project/Edge/Undo]: [Enter]

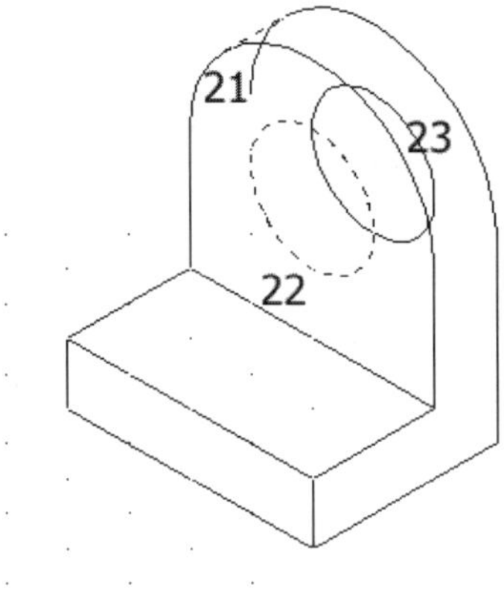

Cleaning up the drawing with the Trim command.

Command: [spacebar]
TRIM Current settings: Projection=UCS
Edge=None
Select cutting edges ...
Select objects: [pick isocircle at 22]
1 found Select objects: [Enter]
Select object to trim or [Project/Edge/Undo]:
[pick isocircle at 23]
Select object to trim or [Project/Edge/Undo]:
[Enter]

11. The isometric drawing is complete! Save your work. In the next section, you apply text and dimensions.

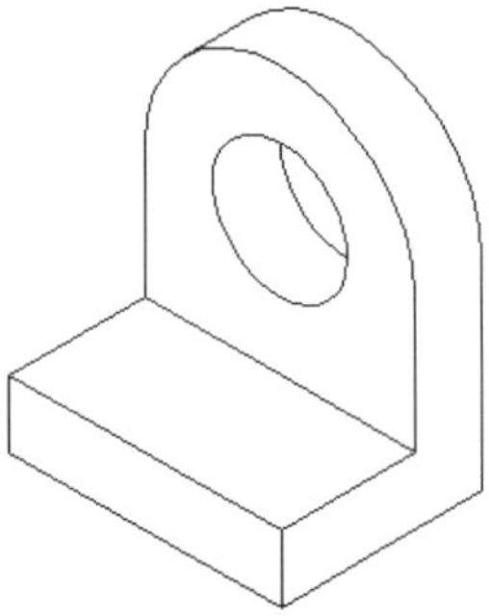

The completed isometric drawing.

Note: An isometric drawing is not a 3D drawing; it is a 3D illusion created in 2D. For this reason, you cannot apply 3D-only commands, such as Hide or Shade.

Applying Isometric Dimensions

Earlier in this chapter, you learned how create text and dimension styles suitable for isometric drawings. Let's apply linear and radial dimension to our drawing of the clip.

1. In preparation, open the Dimension toolbar: right-click any toolbar, and from the shortcut menu select Dimension. Notice that the Dimension toolbar contains a droplist of dimension style names. Selecting a dimstyle name from this list is much faster than opening the Dimension Style Manager dialog box. (There is a similar toolbar for selecting text styles: click the **Properties** tab of the Multiline Text Editor dialog box. Access it with **Draw | Text | Multiline Text** from the menu bar.)

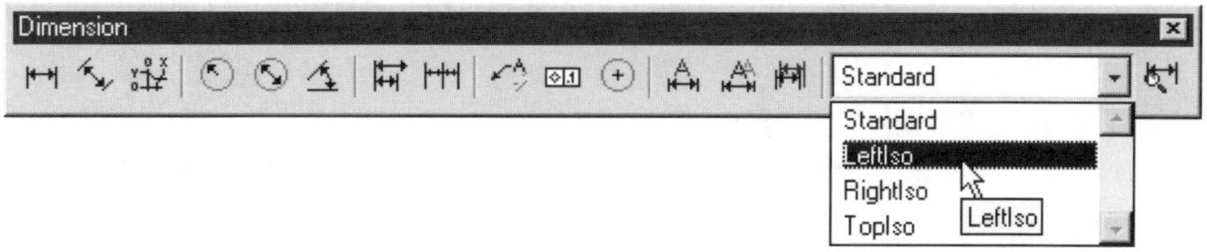

The Dimension toolbar displaying dimension style names.

- From the Dimension toolbar, select the **LeftIso** dimension style.
- With the Isoplane command, set the isoplane to **Left.**
- Look at the status bar, and ensure that **SNAP, GRID,** and **ORTHO** are turned on.

2. Using the DimAligned command, dimension the front rectangular face of the clip:

   Command: dimaligned
   Specify first extension line origin or <select object>: [pick 1]
   Specify second extension line origin: [pick 2]
   Specify dimension line location or [Mtext/Text/Angle]: [pick 3]
   Dimension text = 2.0

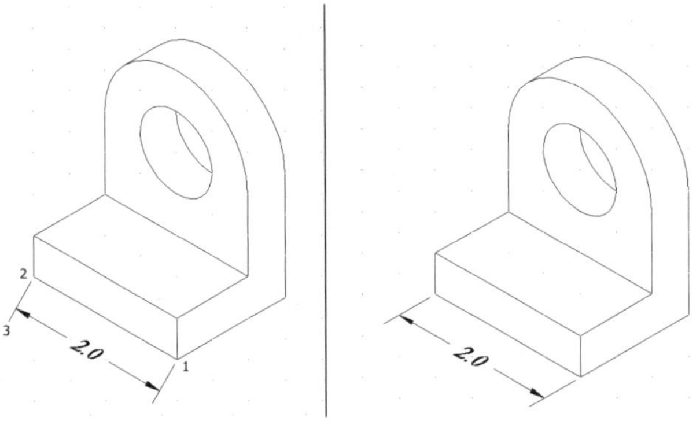

Placing a dimension in the left isoplane: before (left) and after changing the angle of the extension lines.

3. Use the DimEdit command to "straighten out" the extension lines, as follows:

   Command: dimedit
   Enter type of dimension editing [Home/New/Rotate/Oblique]
   <Home>: oblique
   Select objects: last
   1 found Select objects: [Enter]
   Enter obliquing angle (press ENTER for none): 30

4. Place a dimension in each of the other isoplanes. For the top isoplane:

   - Press **F5** to switch to the top isoplane.
   - From the Dimension toolbar, select **TopIso** dimstyle.
   - Use the DimAligned command to place the dimension.

   For the right isoplane:

   - Press **F5** to switch to the right isoplane.
   - From the Dimension toolbar, select **RightIso** dimstyle.
   - Use the DimAligned command to place the dimension.

   Once dimensions are placed in the top and right isoplanes, you use the DimEdit command to oblique all their extension lines by -30 degrees.

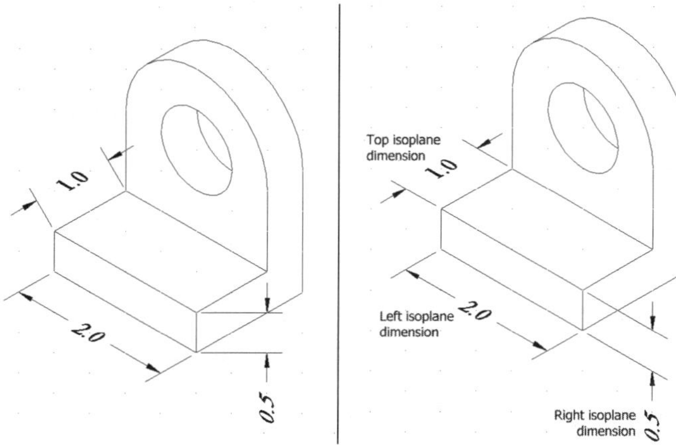

Placing a dimension in the top and right isoplanes: before (left) and after changing the angle of the extension lines.

You cannot use the DimRadius or DimDiameter commands to dimension the hole, because these two commands do not work with ellipses and elliptical arcs. Recall that isometric circles and arcs are drawn as ellipses and elliptical arcs.

For that reason, you need to use the QLeader (or Leader) command to place the "radial" dimensions. The two commands are similar, except that QLeader displays a dialog box for its options, whereas the Leader command displays the options in the command prompt area.

5. Before drawing the leader dimension, switch to the left isoplane.

```
Command: leader
Specify leader start point: [pick center of hole]
Specify next point: [pick 1]
Specify next point or [Annotation/Format/Undo] <Annotation>: [pick 2]
Specify next point or [Annotation/Format/Undo] <Annotation>: a
Enter first line of annotation text or <options>: [Enter]
Enter an annotation option [Tolerance/Copy/Block/None/Mtext] <Mtext>: [Enter]
```

Notice that AutoCAD displays the Multiline Text Editor dialog box.

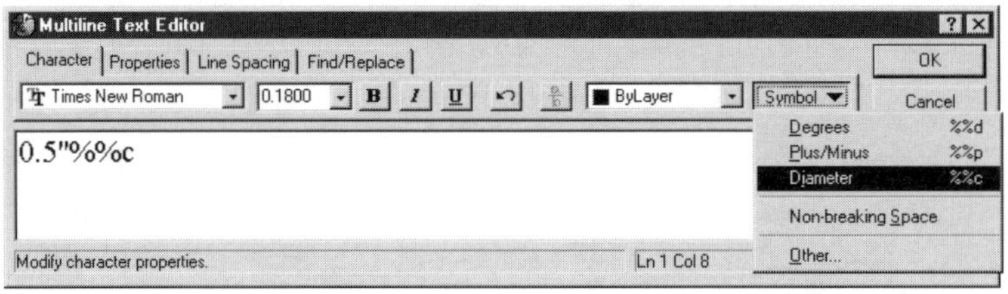

Selecting the diameter symbol in the Multiline Text Editor dialog box.

6. Type the dimension, such as 0.5”. To insert the diameter symbol (Ø), click **Symbol** and select **Diameter**. Notice that AutoCAD adds %%c, which is the code for drawing the diameter symbol.
7. Click the **Properties** tab:
   - Style: **LeftIso**
   - Rotation: **330**
   - Click **OK.**

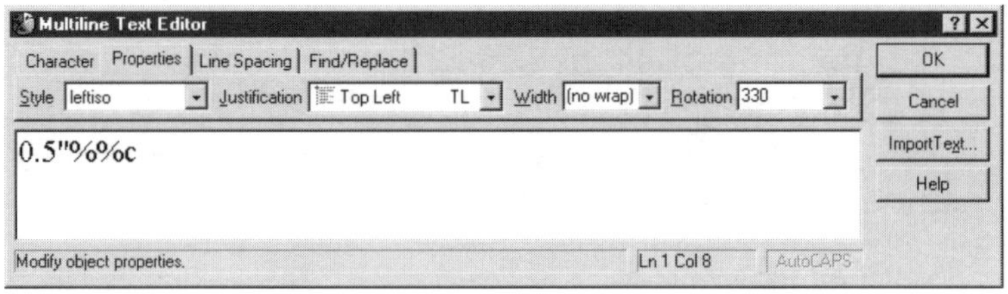

Selecting the text style and rotation angle in the Multiline Text Editor dialog box.

Notice that AutoCAD places the leader text as **0.5”Ø.**

8. Repeat to dimension the arc. Use 0.1”Ø as the dimension.

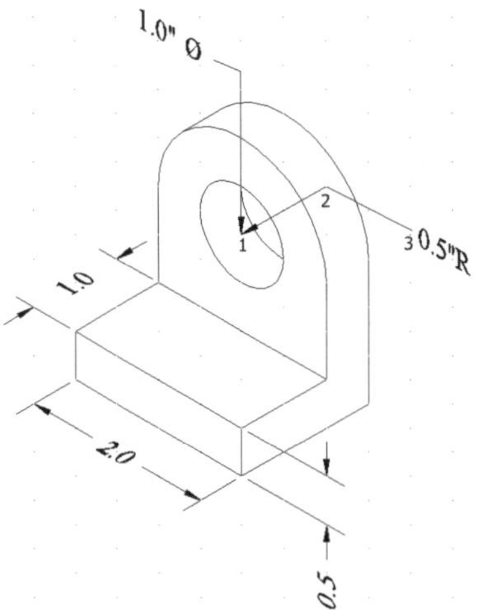

Dimensioning a hole in isometric mode.

9. If you wish, add some more dimensions to the clip. Save your work. This completes the tutorial on creating an isometric drawing.

This concludes *Learn AutoCAD LT 2000i*. I trust you found this tour through AutoCAD LT useful and enjoyable. If you have any questions about the material in this book, feel free to e-mail me at **ralphg@xyzpress.com**.

Review Questions

1. AutoCAD LT can display 3D solid models created by AutoCAD.

   True / False

2. The best meaning for isometric is:

   a. Exercise system.
   b. Equal projection.
   c. Equal height of objects.
   d. Metric system defined by International Standards Organization.

3. Match the isometric shape with its best equivalent shape:

a.	Ellipse	i.	Rectangle
b.	Rhombus	ii.	Circle
c.	Ellipical arc	iii.	Arc

4. Isometric planes are at _____ degrees to each other:

   a. +30
   b. -30
   c. 60
   d. 120

5. To switch between isoplanes, press:

   a. F1
   b. Ctrl+I
   c. Ctrl+E
   d. Alt+I

6. To draw an isometric circle, you use the following command:

   a. Isocircle
   b. Ellipse

c. Isoline
d. Iso

7. To make text look "correct" in an isometric drawing, it must be obliqued by _____ degrees.

a. 30
b. 45
c. 60
d. 120

8. To make a dimension look "correct" in an isometric drawing, it must be obliqued with the:

a. DimIsometric command
b. DimEdit command
c. DimExtension command
d. DimRotate command

9. The DimRadius command together with correct object snap finds the center of an isocircle.

True / False

10. You can successfully use the Hide command on an isometric drawing.

True / False

Appendix A

Setting Up AutoCAD LT 2000i

You must set up the AutoCAD LT 2000i software package before you can use it for computer-aided drafting. This appendix describes the computer hardware needed to run AutoCAD LT and shows you how to install the AutoCAD LT software package. Please note that AutoCAD LT 2000i is not included on the companion CD.

Computer Requirements

Before installing AutoCAD, first check that your computer has the hardware required to run Windows and AutoCAD.

OPERATING SYSTEM—AutoCAD LT 2000i works with Microsoft Windows 95, 98, ME, and Windows NT and 2000. AutoCAD LT does not work with Windows v3.x or MS-DOS.

CPU—To run AutoCAD LT 2000i, your computer must be an IBM-compatible that uses an Intel Pentium-compatible CPU. While Autodesk says that a 133MHz Pentium should be considered the slowest CPU to use with LT, I have found that it works with a 486, in a pinch.

MEMORY—If your computer is able to run Windows, then it probably has enough memory to run AutoCAD LT. AutoCAD LT is, however, a program that uses a lot of memory. Thus, while AutoCAD LT will load on a computer with 8MB RAM memory, it will run slowly. You will have better AutoCAD (and Windows) performance when your computer has at least 64MB RAM and, if you can afford it, even more RAM.

DISK DRIVES—The computer must have a CD-ROM drive for loading the software. To store the AutoCAD LT program, your computer's hard drive needs 73MB of free space on the hard disk for the LT program and support files, plus an additional 45MB if you are going to allow the installation of Microsoft Internet Explorer. While running LT, your computer's hard disk should have 64MB to 100MB of free space for the swap file created by Windows. Summing up, your computer should have roughly 250MB disk space available before beginning the installation of AutoCAD LT.

Tip: When your computer is short on disk space, use the Compact option during installation, which takes up less disk space.

PERIPHERALS—AutoCAD LT works with any peripheral that works with Windows. *Peripherals* include the graphics board, mouse, and printer. While Windows and LT work with graphics boards that have a SVGA resolution (800x600), I recommend a higher resolution—at least 1024x768. The graphics board should display at least 256 colors. Indeed, the new AutoCAD LT Today window is wider than 800 pixels.

AutoCAD LT supports digitizing tablets, provided the tablet is compatible with the WinTab spec. Autodesk includes a tablet overlay with many of LT's commands.

AutoCAD LT supports the Windows system printer, as well as plotters from Hewlett-Packard, CalComp, Oce, Xerox, and those compatible with PostScript.

Tip: Many plotters are compatible with the Hewlett-Packard brand. If AutoCAD LT does not appear to support your plotter, try configuring AutoCAD LT with an HP plotter driver (although this may result in output problems).

Installing AutoCAD LT 2000i

AutoCAD LT comes on a CD-ROM, which has a setup program that copies the contents of the CD-ROM onto the computer's hard drive. The Setup program runs the Microsoft Installer, which installs the AutoCAD LT program, its support files, and sample drawing files. In addition, Microsoft's Internet Explorer is installed, and you have the option of installing the AutoCAD LT Learning Assistance CD. Total installation time is about 10 to 20 minutes.

1. If your computer has more than one hard drive, set up AutoCAD on the drive with the most free disk space. Do not install AutoCAD at all unless one hard drive has at least 250MB free disk space.

   Also, you should close all programs that are running. If you do not, the Windows Installer closes them for you, then reboots the computer. Thus, it is faster to close the programs yourself, and save waiting for the reboot.

2. The installation program should start automatically when you insert the CD-ROM in your computer's CD-ROM drive.

   If not, then click **Start** on the Windows taskbar, then select **Run**. When the Run dialog box appears, type

   Open: d:\setup

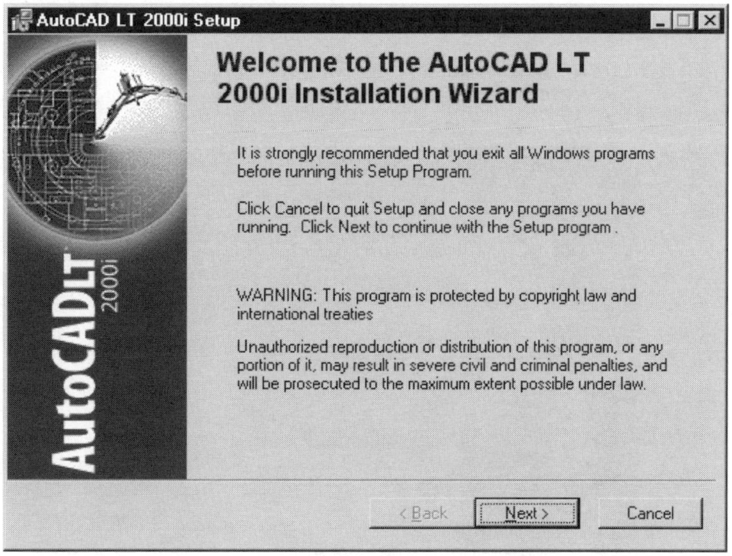

   where "d:" is the disc drive containing the CD-ROM (typically e:).

3. Click **OK**. Notice that the Setup program installs the Windows Installer, before installing AutoCAD LT.

4. You are asked to read the software license. Even if you do not agree with it, there is nothing you can do to change the terms. Click **I accept** to continue with the software installation.

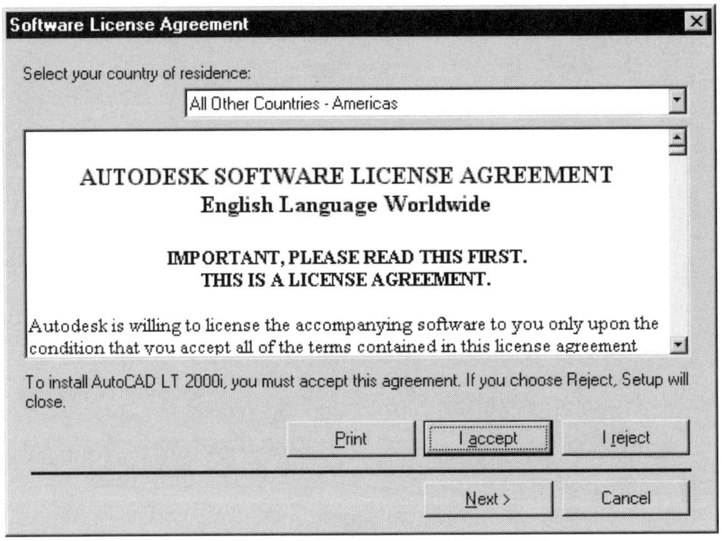

5. When prompted, type the software Serial Number and CD-ROM CD Key. Both numbers are found on the AutoCAD LT package.

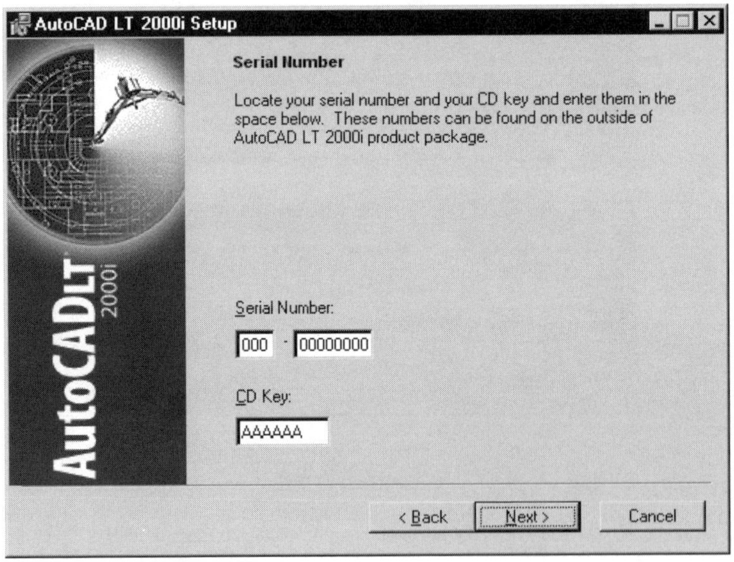

6. The installation program asks you for your Name and Organization. This information is displayed by AutoCAD LT. In this way, Autodesk hopes to keep you from stealing the software.

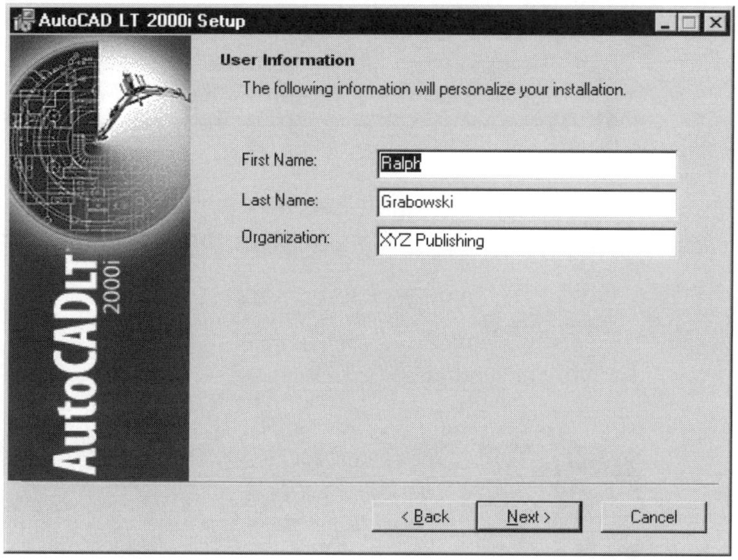

7. If an earlier version of AutoCAD LT already exists on your computer, the installation program gives you the choice to either: (1) overwrite the old copy, so that only AutoCAD LT 2000i exists on your computer; or (2) place LT 2000i in another location so that you have both the old and new versions of LT. Unless your computer is short on disk space, I recommend that you keep both copies on your computer until you become familiar with the LT 2000i and are satisfied it works for you.

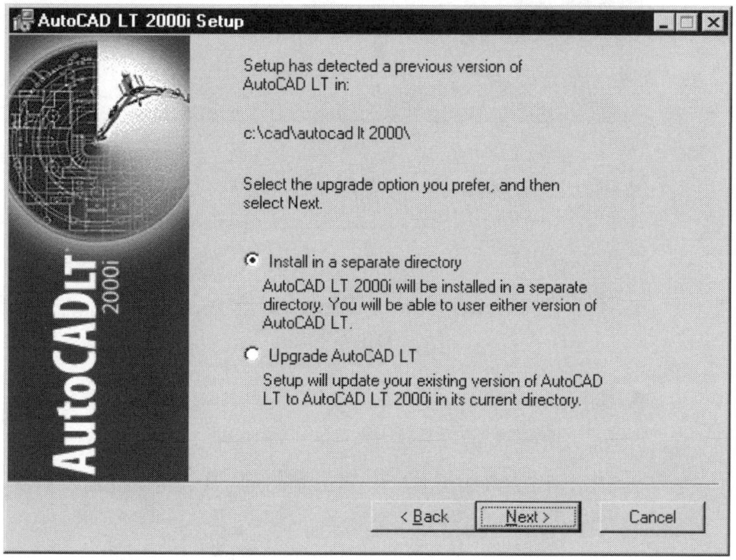

8. Now it's decision time. The installation program performs four kinds of installation:

   - **Typical**: Almost all files are installed; probably the best choice for you.
   - **Compact**: The minimum required files are installed; the best choice for a notebook computer or a computer low on disk space.
   - **Custom**: Gives you some control over what is installed. This is my preference.
   - **Full**: Installs everything.

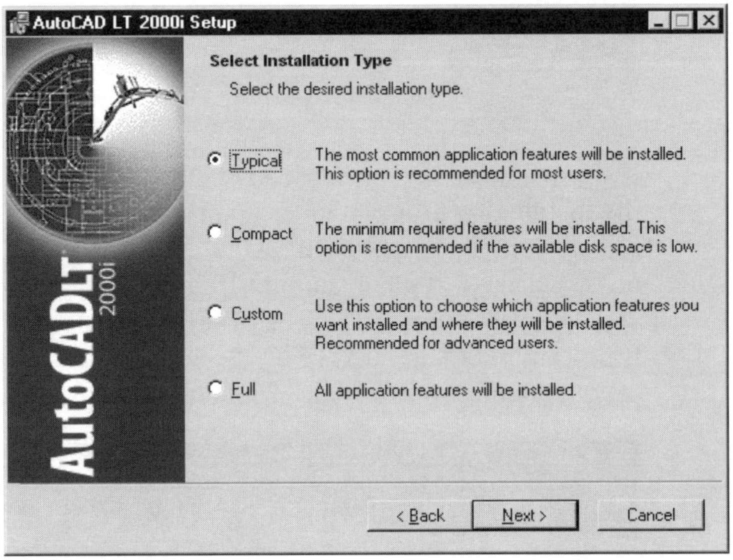

When you select Custom, Setup displays several options, along with the amount of disk space each choice takes up. Here are some considerations:

   - If you are already familiar with AutoCAD, you may want to leave out the Learning Tools.
   - If you already have AutoCAD on your computer or network, you may want to leave out the fonts.
   - If you never plan to use the spell check feature, you can leave out the dictionaries.
   - The samples and templates are useful, and should only be left out if your computer is low on disk space.

To change an option in the "Please select which features you would like to install" list, right-click each disk icon and make a selection from the shortcut menu. I don't care for the new user interface invented by Microsoft, and find it more confusing than earlier interfaces for selecting custom install options.

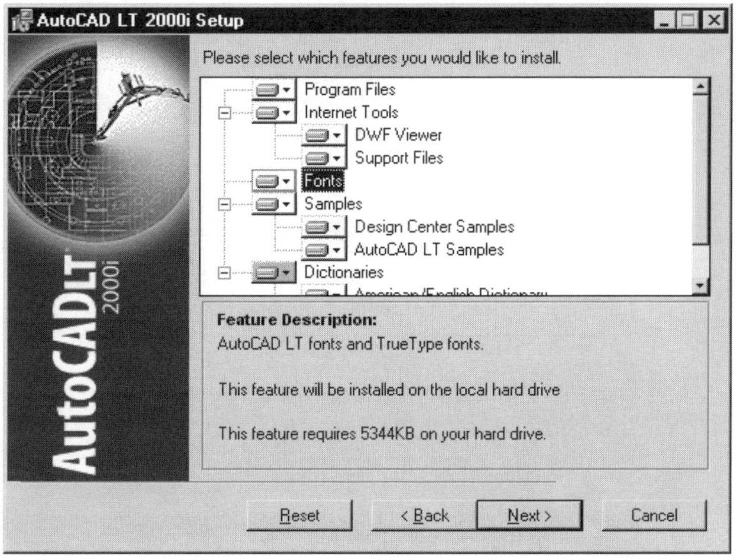

9. The installation program suggests placing the software in C:\Program Files\AutoCAD LT 2000i. That means hard disk drive C:, and folder AutoCAD LT 98 within folder Program Files. If your computer has just one hard drive, then your only choice is C:. If your computer has more than one hard drive, you have a choice; change the drive name to one with the most amount of free disk space.

   Change the destination folder, if necessary, by clicking **Browse.** For example, I keep all my CAD software in a folder called CAD; my copy of LT is located in C:\CAD\AutoCAD LT 2000i.

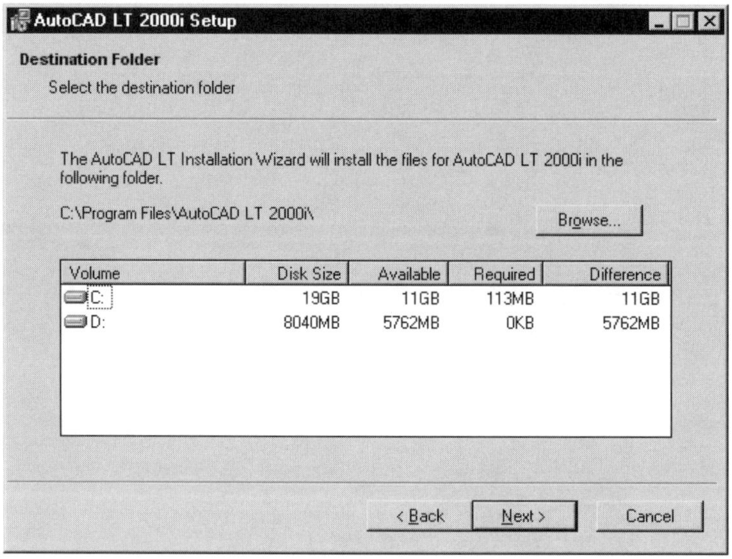

10. The installation program gives you one more chance to change your mind before it begins copying files. (You can always rerun Setup after installation to add and remove files).

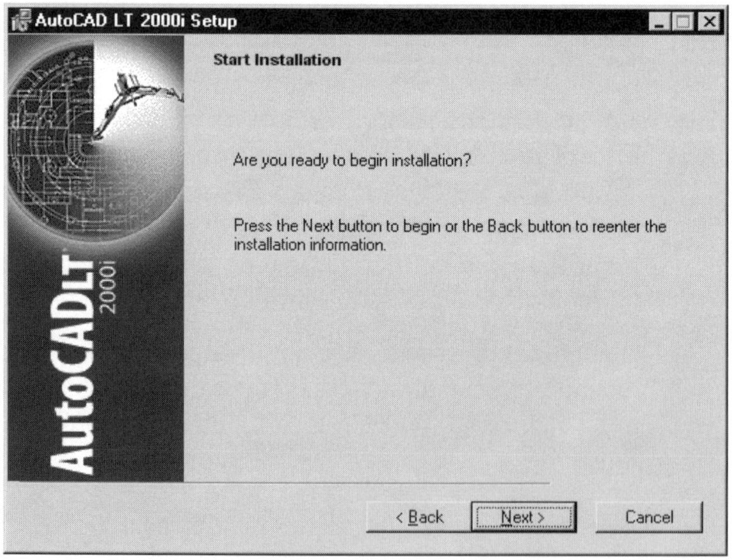

While the files are being copied, the installation program entertains you with a bar graph. Don't rely on the "Time remaining" display; I find it reads 0 or 1 minutes for the entire five minutes it takes LT to install.

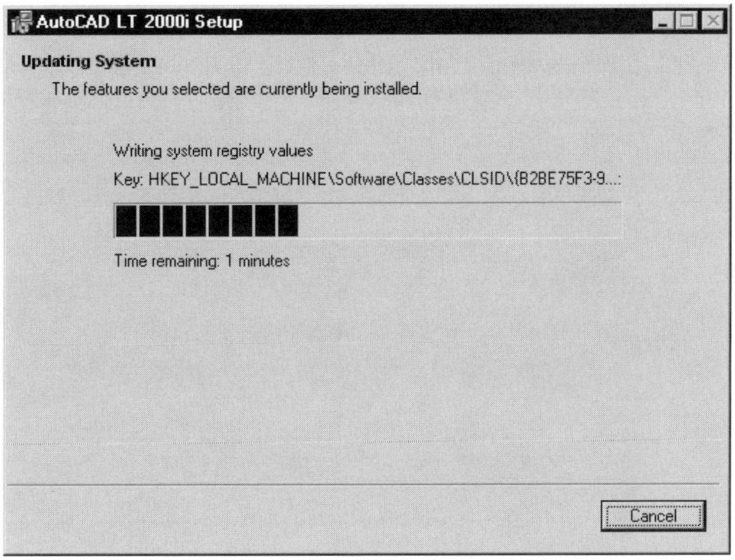

The AutoCAD LT program and its file extensions are registered with the Windows registry. That means you can double-click a DWG file in the Windows Explorer (or File Manager) and Windows will know to launch AutoCAD with that drawing file.

11. Depending on the Microsoft software already installed on your computer, LT might add more, or update some. In my case, the version of Microsoft's NetMeeting was updated, which provides the software needed for LT 2000i's MeetNow command. Other software that might be installed includes Microsoft's Internet Explorer v5.0 if not present, or updates to 5.0 if an older IE is present.

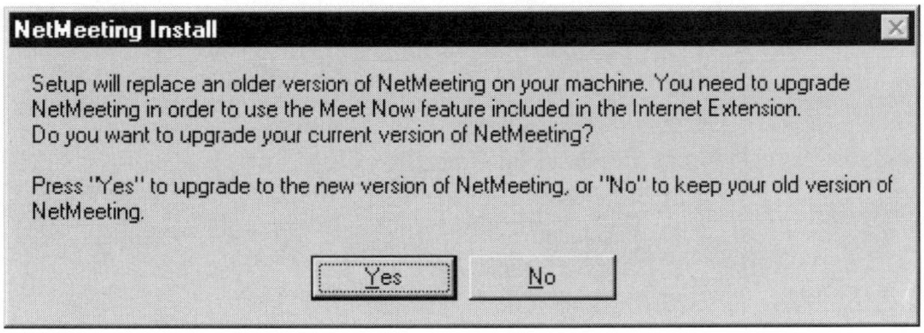

12. After installation is complete, you can read the readme file, which contains important information about what works and what doesn't work in AutoCAD LT—information not found in the printed documentation or online help.

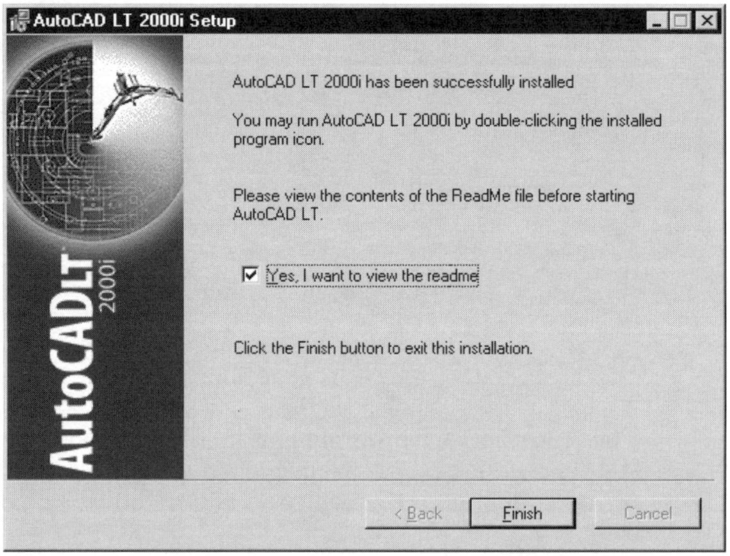

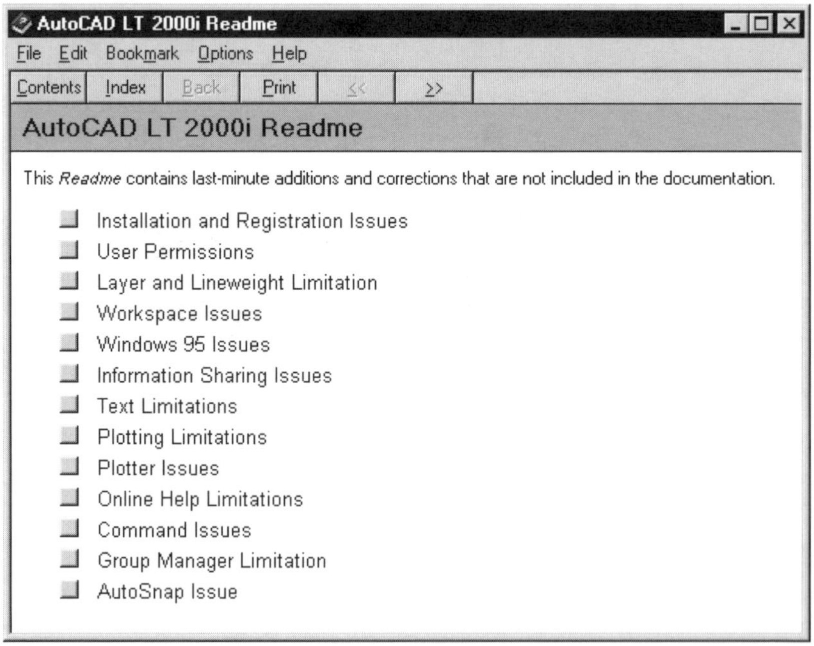

For example, the readme file contains the following useful information regarding the installation of AutoCAD LT:

- If you plan to use both AutoCAD LT 2000i and AutoCAD 2000i on a workstation, it is important that you install AutoCAD 2000i before you install AutoCAD LT 2000i. If you already have AutoCAD LT 2000i on your workstation before you install AutoCAD 2000i, simply reinstall AutoCAD LT 2000i after you perform the AutoCAD 2000i installation.
- If you install AutoCAD LT 2000i on a computer that has AutoCAD LT 2000 installed, and you then uninstall AutoCAD LT 2000, AutoCAD LT 2000i will not launch. You must reinstall AutoCAD LT 2000i to fix this problem.
- If you experience a problem with the AutoCAD LT process not terminating upon exit, it is recommended that you install an updated version of the Microsoft Scripting Engine. The latest scripting engine can be downloaded from the following URL:
- After applying a LiveUpdate service pack for a software licensed version, close AutoCAD LT and choose OK in the Live Update dialog box. A warning dialog box is displayed to explain that the installation cannot continue because Cdilla64.exe is still active. In response, you must close all applications and wait until the process terminates (which can take up to four minutes) before you can resume the installation. You will know that the process has terminated when you choose OK in the warning dialog box and the installation resumes.
- After you uninstall AutoCAD LT 2000i, the WHIP! plug-in may fail when you try to use it with other Autodesk programs. To avoid these problems, reinstall the WHIP! plug-in after you uninstall AutoCAD LT 2000i.

The install program does not exit until you exit the readme file. To do so, select **File | Exit** or press **Alt+F4**. To read the file later, you can find Readme.Hlp in the \AutoCAD LT 2000i\Help folder.

13. When the installation program is finished, remove the CD-ROM from the drive, and store the CD-ROM in a safe place (like your safety deposit box). It may be necessary for Windows to restart your computer.

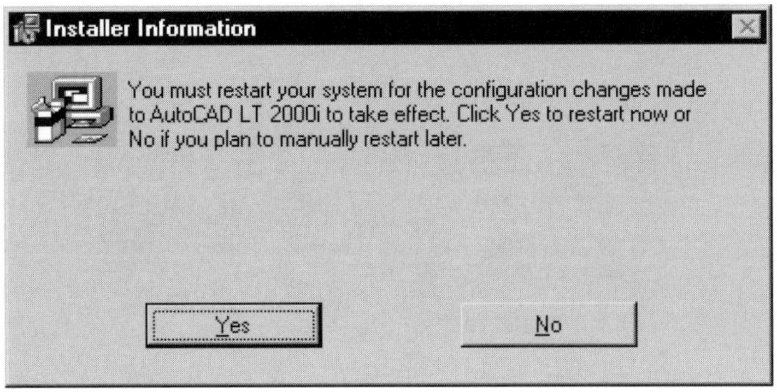

14. Start AutoCAD LT by double-clicking on the icon created by the installation program.

If you are familiar with earlier versions of AutoCAD, relax: you don't need to do anything to configure LT. AutoCAD LT 2000i is now ready to use.

You now start with the lessons in Chapter 1, where you learn about AutoCAD LT's user interface.

Appendix B

Installing and Using the Companion CD-ROM

The inside back cover of this book encloses a companion CD-ROM.

The CD-ROM contains a copy of the drawings you create in this book. As well, the CD-ROM contains these additional files:

- DWF files show what the drawing looks like at the end of each chapter. If you make a mistake during one chapter, you can start the next chapter with a correct drawing.
- The TXT and XLS files created in Chapter 9 for attribute extraction.
- A 30-day trial version of AutoCAD LT 2000i.

This appendix describes the files on the companion CD-ROM, and tells you how to copy the files from the CD-ROM to your computer.

Note: The version of AutoCAD LT 2000i included on the companion CD-ROM is only a trial version. You must purchase the complete product separately from this book.

Drawing Files

The drawings included on the CD-ROM are:

Filename	Contents
Chapter2.Dwg	Drawing at the end of Chapter 2.
Chapter3.Dwg	Drawing at the end of Chapter 3.
Chapter4.Dwg	Drawing at the end of Chapter 4.
Chapter5.Dwg	Drawing at the end of Chapter 5.
Chapter6.Dwg	Drawing at the end of Chapter 6.
Yard.Dwg	Completed yard drawing.

Filename	Contents
Symbols.Dwg	Completed symbols drawing.
Electric.Dwg	Completed electrical schematic drawing.
Chapter10.Dwg	Completed waffle drawing using regions.
Yard.Dwf	Yard drawing exported as a DWF file for Chapter 13.
Isotext.Dwg	Example drawing showing text in all three isometric planes.
Isometric.Dwt	Isometric drawing template file.
Isotutorial.Dwg	Isometric drawing of the clip, without dimensions.
Chapter14.Dwg	Completed isometric clip drawing with dimensions for Chapter 14.

1. Copy the drawings from the companion CD-ROM to your computer's hard disk. Use the Windows Explorer (or File Manager) to copy the file from the CD drive to the hard drive.
2. To load one of the bonus drawings into AutoCAD LT, select **File | Open** from the menu bar. When the dialog box appears, double-click on a filename from the list of drawing names listed. The drawing is displayed on the graphics screen.

TXT Attribute Files

The companion CD-ROM also contains the attribute files created in Chapter 8, "Creating Symbols and Attribute Data." These files are stored in the CD-ROM Symbols folder:

Filename	Contents
Template.Txt	Attribute extraction template file
Electric.Txt	Extracted attributes
Electric.Xls	Excel worksheet with attributes

Copy the attribute files from the companion CD-ROM to your computer's hard disk. If the CD-ROM is in the E: CD drive, and AutoCAD LT resides on the C: hard drive in the folder Autocad lt 2000i, then type the following at the DOS prompt:

```
C:\> copy e:\symbols\*.* c:\autocad lt 2000i\symbols
```

Or,

Use the Windows Explorer (or File Manager) to copy the file from the CD drive to the hard drive.

Appendix C

AutoCAD LT 2000i Commands

In this appendix, you find the complete list of commands found in AutoCAD LT 2000i, along with several methods of accessing the commands. To execute a command, you can:

- Type the complete command name at the Command: prompt.
- Select the command from the menu bar.
- Press a shortcut key, such as Ctrl+C.
- Type a one- or two-letter abbreviation for the command.

This appendix lists all of the ways to execute a command, including commands not documented by Autodesk in the LT 2000i manuals and online help. In addition, this appendix lists commands removed from LT 97, 98, and 2000, and indicates when commands were added to recent versions of LT.

Note:

New—The New column indicates when a command was added to AutoCAD LT:

LT97	Command was added to AutoCAD LT 97.
LT98	Command was added to AutoCAD LT 98.
LT2K	Command was added to AutoCAD LT 2000 (there was no AutoCAD LT 99).
LT2i	Command was added to AutoCAD LT 2000i.

A blank means the command was included with AutoCAD LT Release 1, Release 2, or LT 95 (there was no LT 96).

Command Name—This columns lists the name of the command that you type at the Command: prompt. Some commands can use the following prefixes:

'	Transparent command; it can be invoked in the middle of another command.
-	Forces the command-line version.
+	Prompts you to specify the dialog box's tab number: Tab index <0>.

Menu Selection—Many commands can be selected from the menu bar. The | symbol (vertical bar) separates menu picks. For example, Help | About AutoCAD LT means to: (1) select Help from the menu bar; then (2) select About AutoCAD LT from the menu. You can create your own customized menus with the Customize command.

Shortcut—Some commands use a keystroke shortcut, such as a function key or a Ctrl key. You can create your own shortcuts with the Customize command.

Alias—Most commands have a keyboard shortcut that allows you to type the command without typing the full name. For example, instead of typing "adcenter," type just "adc" to open the AutoCAD DesignCenter window; this saves you a few keystrokes. Autodesk has included some aliases for compatibility. For example, the "ce" and "content" aliases open the AutoCAD DesignCenter window, instead of the old Content Explorer found in LT 98. You can create your own aliases in the Acad.Pgp file.

New	Command Name	Menu Selection	Shortcut	Alias
A				
	'About	Help | About AutoCAD LT	...	...
LT2K	AdcClose	Tools | AutoCAD DesignCenter	Ctrl+2	...
LT2K	AdCenter	Tools | AutoCAD DesignCenter	Ctrl+2	adc
LT2K	AdcNavigate	...	...	...
	Aperture	...	...	...
	Arc	Draw | Arc	...	a
	Area	Tools | Inquiry | Area	...	aa
	Array	Modify | Array	...	ar
LT98	-Array	...	...	-ar
LT2i	Assist	Help | Active Assistance	...	...
LT97	AttachUrl	Removed from LT 2000; replaced by Hyperlink.		
	AttDef	...	....	at, att, dad, ddattdef

New	Command Name	Menu Selection	Shortcut	Alias
A (cont.)				
LT2K	-AttDef	...	...	-at, -att
	AttDisp	View | Display | Attribute Display	...	ad
	AttEdit	Modify | Object | Attribute | Global	...	ae, ate
LT2K	-AttEdit	...	...	-ate, atte
	AttExt	...	...	ax, ddattext
	Audit	File | Drawing Utilities | Audit	...	...
B				
	'Base	Draw | Block | Base	...	ba
	BHatch	Draw | Hatch	...	bh, h
	-Bhatch	...	...	...
	'Blipmode	...	...	bm
	Block	Insert | Block	...	b
LT2K	–Block	...	...	-b
LT2K	BlockIcon	...	...	bupdate
	BmpOut	...	...	...
	Boundary	Draw | Boundary	...	bo
	-Boundary	...	...	-bo
	Break	Modify | Break	...	br
LT97	Browser	Help | Connect To Internet	...	...
LT2i	BuyOnline	Help | Buy Online	...	...
C				
	Chamfer	Modify | Chamfer	...	cha
	Change	...	...	-ch
	ChProp	...	...	...
	Circle	Draw | Circle	...	c
LT2K	Close	File | Close	Ctrl+F4	...
LT2i	CloseAll	Window | Close All		
	Color	...	...	col, ddcolor, colour
	-Color	...	...	...
LT97	Convert	...	...	...
LT2K	ConvertCtb	...	...	...

Appendixes

New	Command Name	Menu Selection	Shortcut	Alias
C (cont.)				
LT2K	ConvertPStyles	. . .	. . .	. . .
	Copy	Modify | Copy	...	cp, co
LT2K	CopyBase	Edit | Copy with Base Point	. . .	. . .
	CopyClip	Edit | Copy	Ctrl + C	...
	CopyHist	In Text Screen: Edit | Copy History	...	...
	CopyLink	Edit | Copy Link	...	cl
LT98	Customize	Tool | Customize | Toolbar	. . .	toolbar
	CutClip	Edit | Cut	Ctrl + X	...
D				
LT2i	DbClkEdit	...	...	...
	DdAttDef	Removed from LT 2000; replaced by AttDef.		
	DdAttE	Removed from LT 2000; replaced by AttEdit.		
	DdAttExt	Removed from LT 2000; replaced by AttExt.		
	DdChProp	Removed from LT 2000; replaced by Properties.		
	DdColor	Removed from LT 2000; replaced by Color.		
	DdEdit	Modify | Object | Text	...	ed
	Ddemodes	Removed from LT 97; replaced by DSettings.		
	DdGrips	Removed from LT 2000; replaced by DSettings.		
	DDim	Removed from LT 2000; replaced by DimStyle.		
	DdInsert	Removed from LT 2000; replaced by Insert.		
	DdLModes	Removed from LT 97; replaced by Layer.		
	DdLType	Removed from LT 97; replaced by Linetype.		
	DdModify	Removed from LT 2000; replaced by Properties.		
	DdOSnap	Removed from LT 97; replaced by DSettings.		
	'DdPType	Format | Point Style	...	...
	DdRename	Removed from LT 2000; replaced by Rename		
	DdRModes	Removed from LT 2000; replaced by DSettings.		
	DdSelect	Removed from LT 2000; replaced by Select.		
	Ddstyle	Removed from LT 97; replaced by Style.		
	DdUcs	Removed from LT 2000; replaced by UCS.		
	DdUcsP	Removed from LT 2000; replaced by UCS.		
	DdUnits	Removed from LT 2000; replaced by Units.		
	DdView	Removed from LT 2000; replaced by View.		
LT2K	DdVPoint	. . .	. . .	vp

New	Command Name	Menu Selection	Shortcut	Alias
D (cont.)				
	'Delay	...	...	...
LT 97	DetachUrl	Removed from LT 2000; replaced by Hyperlink.		
	'Dist	Tools | Inquiry | Distance	...	di
	Divide	Draw | Point | Divide	...	div
	DLine	Draw | Double Line	...	dl
	Donut	Draw | Donut	...	do, doughnut
LT 97	DrawOrder	Tools | Display Order	...	dr
LT2K	DSettings	Tools | Drafting Settings	...	ds, rm, se, polar
	DsViewer	View | Aerial View	...	av
	DText	Removed from LT 2000; replaced by Text.		
	DView	View | 3d Dynamic View	...	dv
LT 97	DwfOut	Removed from LT 2000; replaced by Plot.		
LT 97	DwfOutD	Removed from LT 2000; replaced by Plot.		
LT2K	DwgProps	File | Drawing Properties	...	...
	DxfIn	File | Open	...	dn
	DxfOut	File | Save As	...	dx
Dimension				
	Dim	...	...	...
	Dim1	...	...	d1
	DimAligned	Dimension | Aligned	...	dal
	DimAngular	Dimension | Angular	...	dan
	DimBaseline	Dimension | Baseline	...	dba
	DimCenter	Dimension | Center Mark	...	dce
	DimContinue	Dimension | Continue	...	dco
	DimDiameter	Dimension | Diameter	...	ddi
	DimEdit	Dimension | Oblique	...	ded
	DimLinear	Dimension | Linear	...	dli
	DimOrdinate	Dimension | Ordinate	...	dor
	DimOverride	...	...	dov
	DimRadius	Dimension | Radius	...	dra
	DimStyle	Dimension | Style	...	dst, d, dm, ddim
	DimTEdit	Dimension | Align Text	...	...

New	Command Name	Menu Selection	Shortcut	Alias
E				
	Elev	...	...	..
	Ellipse	Draw | Ellipse	...	el
	End	Removed from LT 97; replaced by Quit.		
LT2i	EndToday	...		...
...				
	Erase	Edit | Clear or Modify | Erase	Del	e
	Explode	Modify | Explode	...	x
	Export	File | Export	...	exp
	Extend	Modify | Extend	...	ex
F				
	'Fill	...	...	...
	Fillet	Modify | Fillet	...	f
LT2K	Filter	...	...	...
LT2K	Find	Edit | Find	...	...
G				
	GetEnv	...	...	...
	'GraphScr	...	F2	...
	'Grid	Double-click GRID on status bar	F7 or Ctrl+G	...
LT98	Group	Tools | Group	...	g
LT98	-Group	...	...	-g
H				
	Hatch	...	...	-h
	HatchEdit	Modify | Object | Hatch	...	he
	-HatchEdit			
	Help	Help | AutoCAD LT Help Topics	F1	?
	Hide	View | Hide	...	hi
LT2K	Hyperlink	Insert | Hyperlink	Ctrl+K	ListURL, InsertURL, DetachURL, AttachUrl
LT2K	Hyperlink Options	...	...	...
I				
	'Id	Tools | Inquiry | ID Point	...	...
LT 97	Image	...	...	im
LT97	-Image	...	...	-im

New	Command Name	Menu Selection	Shortcut	Alias
I (cont.)				
LT 97	ImageFrame	...	...	...
	Import	Insert | File	...	imp
LT 97	InetCfg	Removed from LT 2000; not replaced.		
	Insert	Insert | Block	...	i, ddinsert, inserturl
LT2K	-Insert	...	...	-i
	InsertObj	Insert | OLE Object	...	io
LT 97	InsertUrl	Removed from LT 2000; replaced by Hyperlink.		
	Intersect	Modify | Region | Intersect	...	in
	'Isoplane	...	F5 or Ctrl+E	Is
L				
	'Layer	Format | Layer	...	la, ddlmodes
	-Layer			
-la				
LT2K	Layout	Insert | Layout | New Layout	...	...
LT2K	-Layout	...	...	lo
LT2K	LayoutWizard	Insert | Layout | Layout Wizard	...	...
	Leader	...	...	...
	Lengthen	Modify | Lengthen	...	len
	'Limits	Format | Drawing Limits	...	lm
	Line	Draw | Line	...	l
	'Linetype	Format | Linetype	...	lt, ltype
	-Linetype	...	...	-lt, -ltype
	List	Tools | Inquiry | List	...	li, ls
LT 97	ListUrl	Removed from LT 2000; not replaced.		
	LogFileOff	...	Ctrl+Q	...
	LogFileOn	...	Ctrl+Q	...
	'LtScale	...	...	lts
LT2K	LWeight	Format | Lineweight	...	lineweight, lw
M				
	MakePreview	Removed from LT 97; not replaced.		
LT98	MassProp	Tools | Inquiry | Mass Properties	...	...
LT2K	MatchProp	Modify | Match Properties	...	ma, painter
	Measure	Draw | Point | Measure	...	me
LT2i	MeetNow	Tools | Meet Now	...	...
	MenuLoad	Tools | Customize Menu	...	...

New	Command Name	Menu Selection	Shortcut	Alias
M (cont.)				
	MenuUnload	...	...	...
	Mirror	Modify | Mirror	...	mi
LT2K	Model	. . .	. . .	. . .
	Move	Modify | Move	...	m
LT98	MRedo	. . .	. . .	mr
	MSlide	...	...	...
	MSpace	View | Model Space (Floating) or double-click PAPER on status bar	...	ms
	MText	Draw | Paragraph Text	...	mt, t
	-MText	...	...	-mt, -t
	MtProp	...	...	...
	Multiple	...	...	...
	MView	View | Floating Viewports	...	mv
N				
	New	File | New	Ctrl+N	n
O				
	Offset	Modify | Offset	...	o, of
	OleLinks	Edit | OLE Links	...	...
LT2K	OleScale	. . .	. . .	. . .
	Oops	...	...	oo
	Open	File | Open	Ctrl+O	OpenURL
LT 97	OpenUrl	Removed from LT 2000; replaced by Open.		
LT2K	Options	Tools | Options	. . .	gr, op, pr, ddgrips, preferences
	'Ortho	Tools | PolarSnap Settings or double-click ORTHO on status bar	F8 or Ctrl+L	or
	OSnap	Tools | Object Snap Settings or double-click OSNAP on status bar	F3 or Ctrl+F	os, ddosnap
	-OSnap	...	...	-os
P				
LT2K	PageSetup	File | Page Setup	. . .	. . .
	'Pan	View | Pan | Realtime	...	p, rtpan
	-Pan	View | Pan	...	-p
LT2K	PasteBlock	Edit | Paste as Block	. . .	. . .
	PasteClip	Edit | Paste	Ctrl +V	...

New	Command Name	Menu Selection	Shortcut	Alias
P (cont.)				
LT2K	PasteOrig	Edit | Paste to Original Coordinates	...	...
	PasteSpec	Edit | Paste Special	...	pa
LT2K	PcInWizard	Tools | Wizards | Import AutoCAD Plot Settings	...	...
	PEdit	Modify | Object | Polyline	...	pe
LT2K	PkFstGroup	Tools | Group Manager	...	g, group
LT2K	-PkFstGroup	Tools | Group	...	-g, -group
	Plan	View | 3D Viewpoint | Plan View	...	...
	PLine	Draw | Polyline	...	pl
	Plot	File | Print	Ctrl + P	pp, print
LT2i	PlotStamp	...	...	...
LT2i	-PlotStamp	...	...	...
LT2K	PlotStyle	...	...	...
LT2K	PlotterManager	File | Plotter Manager	...	...
	Point	Draw | Point	...	po, pt
LT97	Polar	Removed from LT 2000; replaced by DSettings.		
	Polygon	Draw | Polygon	...	pol
	Preferences	Removed from LT 2000; replaced by Options.		
LT97	Preview	File | Print Preview	...	pre
LT2K	Properties	Tools | Properties	Ctrl+1	ch, mo, props, ddchmod, ddmodify
LT2K	PropertiesClose	Tools | Properties	Ctrl+1	prclose
LT2K	PSetupIn	...	...	...
	PsOut	File | Export | EPS	...	...
	PSpace	View | Paper Space or double-click MODEL on status bar	...	ps
	PsUpdate	Removed from LT 2000; not replaced.		
LT2i	PublishToWeb	File | Publish to Web	...	ptw
	Purge	File | Drawing Utilities | Purge	...	pu
LT98	-Purge	...	...	-pu
Q				
LT2K	QkUngroup	Tools | Ungroup	...	...
LT2K	QLeader	Dimension | Leader	...	le
	QSave	File | Save	Ctrl+S	...
LT2K	QSelect	Tools | Quick Select	...	...

New	Command Name	Menu Selection	Shortcut	Alias
Q (cont.)				
	QText	...	...	qt
	Quit	File | Exit	Alt+F4	exit
R				
	Ray	Draw | Ray	...	...
	Recover	File | Drawing Utilities | Recover	....	...
	Rectangle	Draw | Rectangle	...	rec, rectang
	Redo	Edit | Redo	Ctrl+Y	...
	'Redraw	View | Redraw	...	r, redrawall
	Regen	View | Regen	...	re
LT2K	RegenAll	View | Regen All	. . .	rea
LT98	Region	Draw | Region	. . .	reg
	ReInit	...	...	ri
	Rename	. . .	...	ren, ddrename
LT2K	-Rename	. . .	. . .	-ren
	'Resume	...	...	...
LT98	RevCloud	Tools | Revision Cloud	. . .	rc
	RevDate	Tools | Time And Date Stamp	...	rd, rev
LT2i	RmlIn	Insert | Markup	...	...
	Rotate	Modify | Rotate	...	ro
	RScript	...	...	...
S				
	Save	. . . .	Ctrl+S	sa
	SaveAs	File | Save As	...	saveurl
LT97	SaveUrl	Removed from LT 2000; replaced by SaveAs.		
	Scale	Modify | Scale	...	sc
	Script	Tools | Run Script	...	scr
	Select	Tools | Selection	...	...
LT97	SelectUrl	Edit | Select URLs	...	...
	SetEnv	. . .	. . .	. . .
	'SetVar	...	...	set
	Shade	View | Shade	...	sh, sha
LT2K	ShadeMode	. . .	. . .	. . .
	Snap	Double-click SNAP on status bar	F9 or Ctrl+B	sn
	Solid	...	...	so

New	Command Name	Menu Selection	Shortcut	Alias
S (cont.)				
	Spell	Tools | Spelling	...	sp
	Spline	Draw | Spline	...	spl
	SplinEdit	Modify | Object | Spline	...	spe
	Stretch	Modify | Stretch	...	s
	Style	Format | Text Style	...	st
	-Style	...	...	...
LT2K	StylesManager	File | Plot Style Manager	...	...
LT98	Subtract	Modify | Region | Subtract	...	su
	SysWindows	Control menu	Alt+Enter	...
T				
	Tablet	Tools | Tablet	F4 or Ctrl+T	ta
	Tbconfig	Removed from LT 97; replaced by Customize.		
	Text	Draw | Text | Single Line Text	...	tx, dt, dtext
LT2i	-Text	...	...	...
	'TextScr	View | Display | Text Window	F2	...
	Thickness	Format | Thickness	...	th
	Tilemode	...	...	ti, tm
	'Time	Tools | Inquiry | Time	...	...
LT2i	Today	Tools | Today	...	...
	Tolerance	Dimension | Tolerance	...	tol
	-Toolbar	...	...	...
	Trim	Modify | Trim	...	tr
U				
	U	Edit | Undo	Ctrl+Z	...
	Ucs	Tools | Ucs	...	uc, ucp, ddučs, ddučsp
	UcsIcon	View | Display | UCS Icon	...	...
LT2K	UcsMan	Tools | Named UCS	...	...
	Undo	...	...	...
	Unlock	Removed from LT 97; not replaced.		
LT98	Union	Modify | Region | Union	...	uni
	'Units	Format | Units	...	un, ddunits
LT2K	-Units	...	...	-un

Appendixes

New	Command Name	Menu Selection	Shortcut	Alias
V				
	'View	View | Named Views	...	v, ddview
LT2K	-View	...	...	-v

Tip: The -View command has four hidden options—swiso, seiso, neiso, and nwiso—which display the southwest isometric view, etc.

New	Command Name	Menu Selection	Shortcut	Alias
LT2K	ViewRes	...	...	...
	Viewtoolbar	Removed from LT 97; replaced by Customize.		
	VpLayer	...	...	vl
	VPoint	View | 3D Viewpoint	...	vp, -vp
	VPorts	View | Tiled Viewports	...	viewports
	VSlide	...	...	vs
W				
	WBlock	...	...	w
LT2K	-Wblock	...	...	-w
LT2K	WhoHas	...	...	...
	WmfIn	Insert | File | WMF	...	wi
	WmfOpts	Insert | File | WMF | Options	...	...
	WmfOut	File | Export | WMF	...	wo
X				
LT 97	XAttach	Insert | External Reference | Attach	...	xa
	XBind	Modify | Object | External Reference Bind	xb	
	-XBind	...	...	-xb
	XLine	Draw | Construction Line	...	xl
	XRef	Insert | External Reference	...	xr
	-Xref	...	...	-xr
Z				
	'Zoom	View | Zoom	...	z, rtzoom
3				
	3dPoly	Draw | 3D Polyline	...	3p
Miscellaneous				
	Cancel	...	Esc	...
	Coordinate toggle	Double-click coordinates on status bar	F6 or Ctrl+D	...
	Menu bar	...	Alt or F10	...

Special Commands

The following commands are designed for use by menu and toolbar macros, but can also be typed at the keyboard:

New	Command Name	Menu Selection	Meaning
A			
LT2K	Ai_Browse	Help | Augi	Starts Web browser with a URL.
LT2K	Ai_Deselect	Edit mode context menu: Deselect All	Removes all objects from the selection set.
LT2K	Ai_Dim_TextAbove	Dimension context menu:	Moves dimension text above dimension line.
		Dim Text position | Above dim line	
LT2K	Ai_Dim_TextCenter	Dimension context menu:	Moves dimension text to center position.
		Dim Text position | Centered	
LT2K	Ai_Dim_TextHome	Dimension context menu:	Moves dimension text to dimension line.
		Dim Text position | Home text	
LT2K	AiDimPrec	Dimension context menu:	Sets the dimension value to specified decimal places.
		Precision | 0, 0.0, 0.00, 0.000, etc	Options: 0 1 2 3 4 5 6
LT2K	AiDimStyle	Dimension context menu:	Saves the properties of the selected dimension in up to six dimension styles.
		Dim Style | Save as New Style, etc	Options: Other Save 1 2 3 4 5 6
LT2K	AiDimTextMove	Dimension context menu:	Move text with dimension line (0), with leader (1), or alone (2).
		Dim Text position | Move text alone, etc	Options: 0 1 2
LT98	Ai_DrawOrder	Tools | Display Order	Changes the display order of overlapping objects.
			Options: Front Back Above Underneath
LT97	Ai_Fms	Utility tablet menu	Switches to layout mode.
LT2K	Ai_Learning_Assistance	Help | Learning Assistance	Starts the Learning Assistance.
LT98	Ai_Molc	Object Properties toolbar	Makes the layer of a selected object the current layer.
LT2K	Ai_Pan	View | Pan | Left, Right, etc.	Moves the view.
			Options: left right up down
LT97	Ai_PSpace	Removed from LT2K	Switches to paper space without prompting you to turn off Tilemode.
	Ai_SelAll	Edit | Select All or Ctrl+A	Selects all objects in the drawing.
LT2K	Ai_StartApp	Tools | Wizards | Add Plotter, etc.	Starts an external application.

New	Command Name	Menu Selection	Meaning
A			
			Options: addplwiz.exe styshwiz.exe
LT2K	Ai_Support_Assistance	Help | Support Assistance	Displays window with support assistance and tips.
LT98	Ai_Tab3	Tablet menu area 3.	Loads alternate tablet area 3 for metric units.
H			
LT2i	HyperlinkBack	Web toolbar	Navigates to the previous hyperlinked document.
LT2i	HyperlinkFwd	Web toolbar	Navigates to the next hyperlinked document.
LT2i	HyperlinkStop	Web toolbar	Cancels the current navigation selection.
R			
LT2K	R14PenWizard	Tools | Wizards | Add Color-Dependent Plot Style Table	Creates a color-dependent plot style table.

Appendix D

Compatibility with AutoCAD 2000i

AutoCAD LT 2000i is more compatible with AutoCAD 2000i than any other CAD package on the market. There remain, however, some incompatibilities. This incompatibility is due to: (1) missing features; and (2) different features.

Both editions of AutoCAD contain features unique to each other. In fact, there are approximately 180 commands that AutoCAD 2000i and LT 2000i don't have in common with each other.

Features Unique to AutoCAD 2000i

The following AutoCAD 2000i features are not found in AutoCAD LT 2000i:

- Most 3D drawing commands, such as EdgeSurf and Pface
- All 3D surface primitives, such as Ai_Box
- All 3D editing commands such as Align and Mirror3D
- ACIS 3D solids modeling
- ASE (AutoCAD SQL Extension) database functions
- LS (landscape)-related commands
- MLine-related commands
- Render-related commands
- AutoLISP, Visual LISP, VBA, ObjectDBX, and ObjectARX programming
- Batch plotting

The Image and ImageFrame commands are included in LT 2000i, but images are not displayed (only the frame of the image is displayed); no other image-related commands are included.

Commands Missing from LT 2000i

The following AutoCAD 2000i commands are not found in LT 2000i:

Command	Meaning
AppLoad	Loads AutoLISP, Visual LISP, VBA, and ObjectARX routines
Cal	Vector calculator
Compile	Compiles SHP files
DragMode	Toggles display of the drag image
Filter	Filters a selection set
Load	Loads SHX shape file
MInsert	Array insert
MtEdit	MText editor
MvSetup	Multiple viewpoint setup
Redefine	Returns a command name's definition
RedrawAll	Redraws all viewports
RegenAuto	Control regeneration
Shell	Temporarily exits to the operating system
Sketch	Sketches in real time
Status	Displays status about the drawing and AutoCAD
Trace	Draws wide lines
TreeStat	Controls the memory structure of the drawing
Undefine	Temporarily removes the definition of a command name
ViewRes	Controls the resolution of curve objects
XClip	Clips the displayable area of externally referenced drawing

Features Unique to LT 2000i

In addition to the special commands found in Appendix C, the following LT 97, LT 98, LT 2000, and LT 2000i commands are not found in AutoCAD 2000i:

Command Name	Meaning
DLine	Draws parallel (double) lines.
GetEnv	Gets the value of an environment variable from the Windows Registry.
PsFstGroup	Displays the Group Manager window.
QkUngroup	Removes the group definition.
RevCloud	Draws a revision cloud.
RevDate	Inserts (or revises) user name, company name, drawing name, date, and time on layer TITLE_BLOCK.
SetEnv	Sets the value of an environment variable in the Windows Registry.

Differences in User Interface

The structure of the menus and toolbars differs somewhat between LT 2000i and AutoCAD 2000i. The most dramatic difference is in the Insert menu. The illustration shows the Insert menu for AutoCAD 2000i (left) and LT 2000i (right):

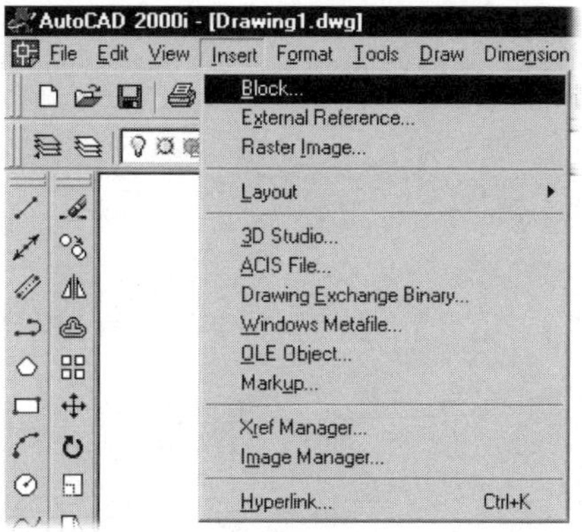

AutoCAD

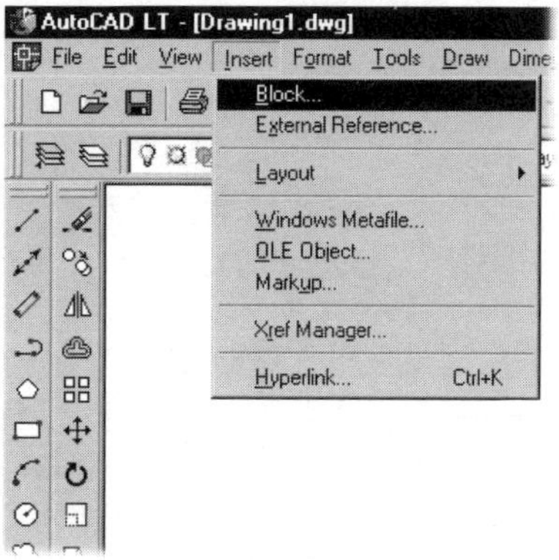

AutoCAD LT

Other user interface differences include:

- In LT 2000i, the status bar displays 2D x,y-coordinates, rather than the 3D x,y,z-coordinates displayed by AutoCAD 2000i.
- The shortcut Ctrl+A toggles groups in AutoCAD 2000i, but selects all objects in LT 2000i.
- AutoCAD 2000i allows multiple UCSs per viewport; LT 2000i is restricted to a single UCS per viewport.
- AutoCAD 2000i has seven options for the ShadeMode command; only two are available in AutoCAD LT: 2D Wireframe and Hidden. Once ShadeMode turns off rendering in LT, it cannot be turned back on, except via the Undo command.
- AutoCAD LT's Group command displays the non-modal Group Manager window; in AutoCAD 2000i, the Group command displays the modal Object Grouping dialog box (see illustration).

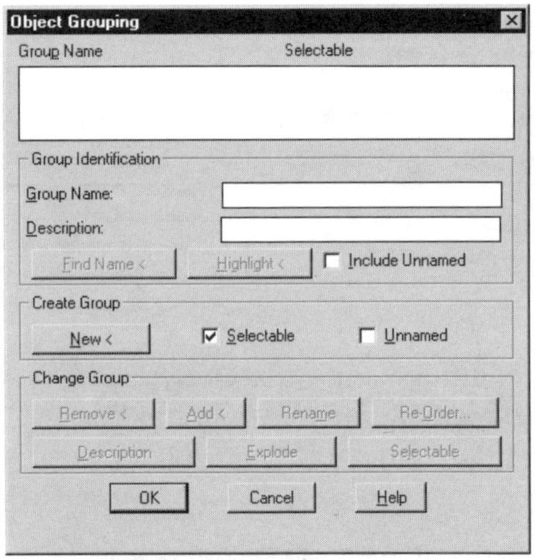

AutoCAD

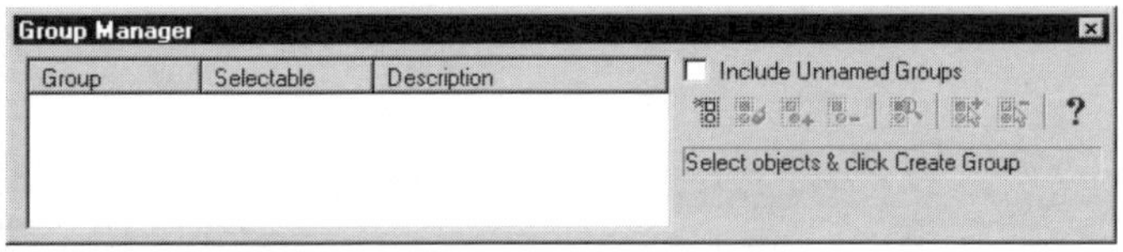

AutoCAD LT

DWG File Compatibility

For the most part, LT 2000i and AutoCAD 2000i can read each other's DWG files, with some exceptions.

- LT 2000i displays, but cannot edit, xref clipping paths, images, 3D solid models, and landscape objects.

- LT 2000i does not display multiple user coordinate systems in a single drawing file; and 2D and 3D solid object shading.

Compatibility with Earlier LTs

Objects specific to AutoCAD LT 2000 and 2000i are converted to a format compatible with earlier releases of AutoCAD LT:

LT 98 and earlier

- Dimension text is converted to text objects.
- Additions to the dimension style table and associated system variables are ignored.
- New system variables are erased.

LT 97 and earlier

- Groups listed by the List command.
- Regions identified as regions, but cannot be manipulated.
- Revision cloud converted to a series of polylines.

LT 95 and earlier

- Ellipse converted to a series of polyline arcs or a circle.
- Hatch pattern associativity lost.

LT Release 2 and earlier

- Xline converted to a line, cut off at the drawing extents.
- Ray converted to a line, cut off at the drawing extents.
- Spline converted to a series of polylines.
- TrueType fonts converted to txt.shx font.
- Mtext (paragraph text) converted to multiple text objects; individual formatting may be lost.

LT Release 1

- Leader converted to lines or polylines; arrows are converted to filled or unfilled circles, polygons, or polylines.
- Global linetype scale ignored.
- Shape definitions in complex linetypes ignored.

- OLE objects ignored.
- Overlays ignored.
- Preview image ignored.
- Tolerance converted to polylines and text.

Appendix E

Configuring LT 2000i for a Plotter

AutoCAD LT 2000 significantly changed the way in which plotters are configured. This appendix describes the kinds of plotting devices LT 2000i works with, and how to create a custom configuration for your plotter.

System Printers

After AutoCAD LT 2000i is set up on your computer, it can print its drawings to any *system printer*. A system printer is any printer connected to your computer, whether directly or via a network. Using the system printer is the preferred method.

A *direct* connection means the printer is connected directly to your computer via a serial (rare nowadays), parallel (most common), USB (increasingly common), or infrared port.

A *network* connection means the printer is connected indirectly to your computer, and you access it over your network. The printer is connected either to another computer (and the computer is connected to the network), or is on the network (the printer contains a network card).

When you use the Plot command, AutoCAD LT's Plot dialog box lists all the system printers AutoCAD knows about. The list is usually identical to the list of printers you see from any other Windows application. In fact, AutoCAD LT simply uses the Windows list.

The printers most commonly used in an office environment include:

- **Laser printers** typically produce high-speed, black and white prints.
- **Inkjet printers** usually produce color prints.
- **Fax** "printers" send the drawing as a fax.

Other Output Devices

Anything that LT outputs a drawing to is called an *output device*. This includes pen plotters, inkjet plotters, large format laser printers, milling machines, and 3D stereolithography devices.

To communicate with printers, Windows applications use software called a *driver*. The driver translates the AutoCAD LT drawing into the format understood by the printer. For example, some HP printers use a language called PCL (short for "printer control language"), while Epson printers use a language called ESC.

Windows includes drivers for hundreds of printers and plotters. In addition, AutoCAD LT includes its own drivers for the following brands of printers and plotters:

- Adobe PostScript Level 1, 1 Plus, and 2
- CalComp pen, LED, inkjet, and electrostatics plotters using 907/PCI/CGL,
- Hewlett Packard pen plotters (HPGL, short for Hewlett Packard graphics language), inkjet plotters (HPGL/2), and laser printers (HPGL/2).
- Oce plotters (HPGL/2)
- Xerox plotters (HPGL/2)
- DXB Format

To set up AutoCAD LT with one of these output devices, use the PlotterManager command, as described below.

One other output device is a *file*; AutoCAD LT can print the drawing to a file on disk, which can be used by other software. For example, it is common to plot a drawing to disk in HPGL (short for Hewlett Packard Graphics Language), then import the file into desktop publishing software.

Using the Raster File Formats driver, AutoCAD LT can output drawings to file in the following raster formats:

- CALS used for scanned drawings.
- JPEG (Joint Photographic Experts Group) used for images on the Internet.
- BMP (bitmap) used by Windows.

- PNG (Portable Network Graphics) used as a royalty-free alternative to GIF.
- TIFF (Tagged Image File Format) used in desktop publishing.
- TGA (Targa) used in high-end imaging.
- PCX used by paint programs.

Use the AutoCAD DXB File driver to export the drawing in DXB (drawing interchange binary) format. The Autodesk ePlot (DWF) drive exports the drawing in DWF (drawing Web format) and JPEG format (an extremely compressed raster format).

Using the PlotterManager Command

The PlotterManager command (File | Plotter Manager) runs the Add-A-Plotter wizard, which steps you through the process of selecting parameters for a newly added plotter. The parameters are saved in a file with the extension PC3 (short for "plotter configuration v3").

Once a plotter configuration is added to AutoCAD, you use the Plotter Configuration Editor to modify parameters, if necessary.

AutoCAD also allows you to set up multiple configurations for a single printer. This allows you, for example, to plot a drawing several different ways—color plot, gray scale plot, fax—without having to select and reset parameters in the Plot dialog box; you simply select the saved configuration.

Tutorial: Creating a Custom Plotter Configuration

This tutorial shows you how to create a custom plotter configuration for AutoCAD LT 2000i. This tutorial accomplishes two tasks (1) add an HPGL plotter to AutoCAD's list of plot devices; and (2) configure the HPGL plotter to plot to a file.

1. Start AutoCAD LT 2000i. From the menu bar, select **File | Plotter Manager**. Notice that AutoCAD displays the Plotters window. Double-click **Add-A-Plotter Wizard**.

The Plotters window.

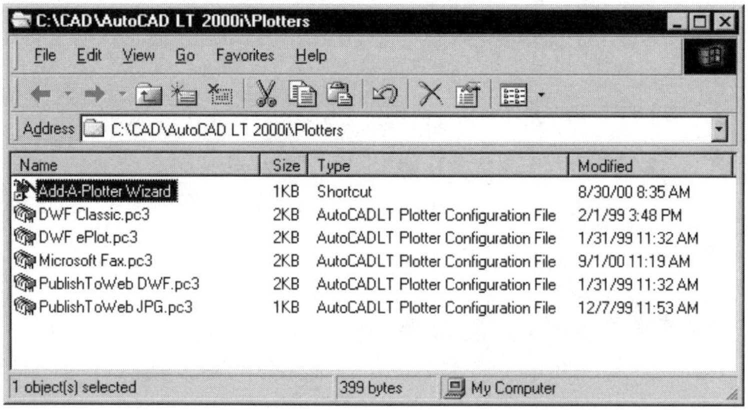

Note: You can get to the Add-A-Plotter wizard from several different places. An alternative from within AutoCAD LT is to select Tools | Wizards | Add Plotter from the menu bar.

Another alternative is to select Tools | Options. In the Options dialog box, select the Plotting tab, then click the Add or Configure Plotters button.

External to LT, you can access the wizard from the Windows task bar: select Start | Settings | Control Panel. In the Control Panel, double-click Autodesk Plotter Manager.

If you have more than one version of AutoCAD on your computer, the **AutoCAD Hardcopy System** dialog box asks you to select one version. Select **AutoCAD LT 2000i**, and click **Continue**.

Select a version of AutoCAD to configure.

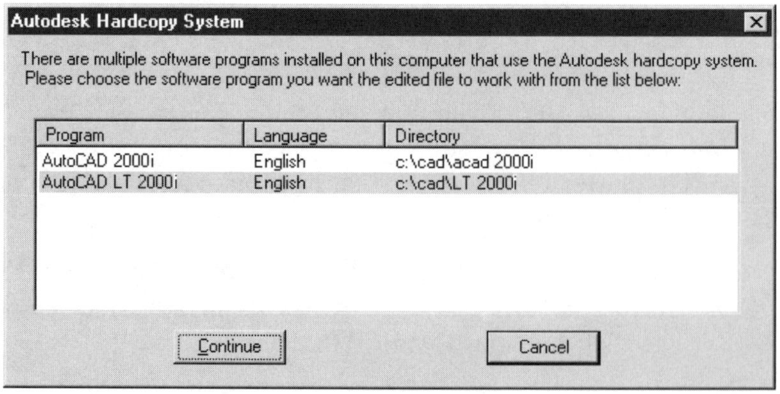

2. AutoCAD displays the Add Plotter - Introduction Page. Click **Next**.

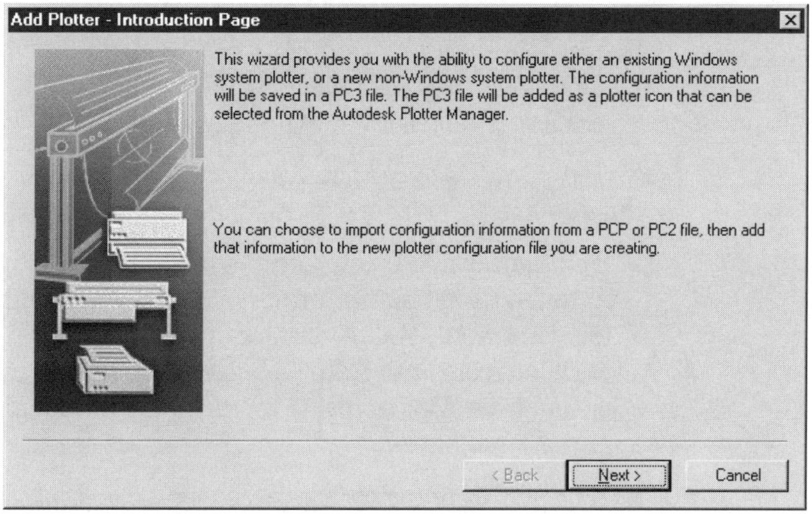

The introduction to the Add Plotter wizard.

3. AutoCAD displays the Begin page. You can select from three kinds of printers (plotters):

   - **My Computer** applies to any printer you want to configure specifically for AutoCAD LT.
   - **Network Plotter Server** applies to printers that use a network plotter server, such as Novel or Windows NT.
   - **System Printer** allows you to create multiple configurations for any of the system printers.

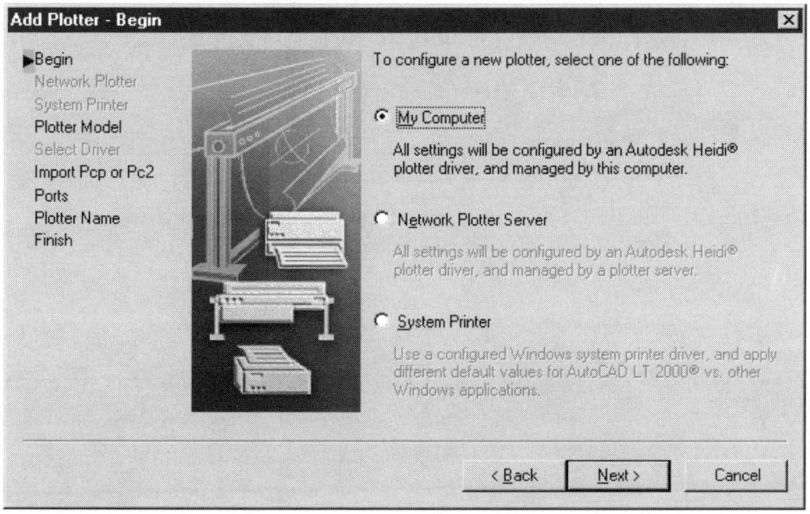

Determining where the plotter settings will be stored.

For this tutorial, select **My Computer**.

4. In the Plotter Model page, you select the manufacturer and the model that matches your plotter. For example, if your computer has an HP LaserJet 5, you would select **Hewlett-Packard** from the Manufacturers list. Then, scroll down the Model list until you see **LaserJet 5**.

   If your plotter's brand name and/or model number don't match any on the two lists, you have several choices:

   - Your plotter might come with a diskette or CD-ROM containing AutoCAD 2000 plotter drivers. (The documentation may refer to the drivers as "Heidi" drivers or "HDI" drivers; you cannot use "ADI" drivers with AutoCAD 2000i, which are meant for older versions of AutoCAD). Insert the disk in your computer, and click the **Have Disk** button.
   - Your plotter probably emulates one of the brand names listed here. Most plotters, for example, are compatible with HPGL or HPGL/2. Any plotter with PostScript can use the Adobe PostScript driver. Read your plotter's documentation to find out which models it emulates.
   - Be careful, though; emulation is getting worse each year as manufacturers update their drivers, and abandon older formats. Some old devices don't properly support the HPGL specification, which causes output problems. There is a simplified HPGL option in LT 2000i for supporting really old devices where emulation is a problem and there is no known Windows System driver.

For this tutorial, select **Hewlett-Packard** from the Manufacturers list.

Select **7550A** from the Models list.

Click **Next**.

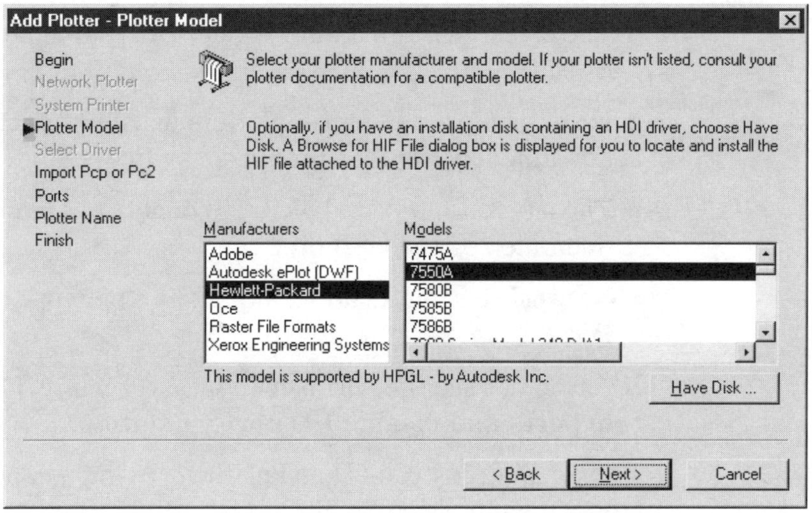

Selecting the plotter brand and model.

5. Unless you have PCP or PC2 plotter configuration files from earlier versions of AutoCAD, you can skip the Import Pcp or Pc2 page by clicking **Next.**

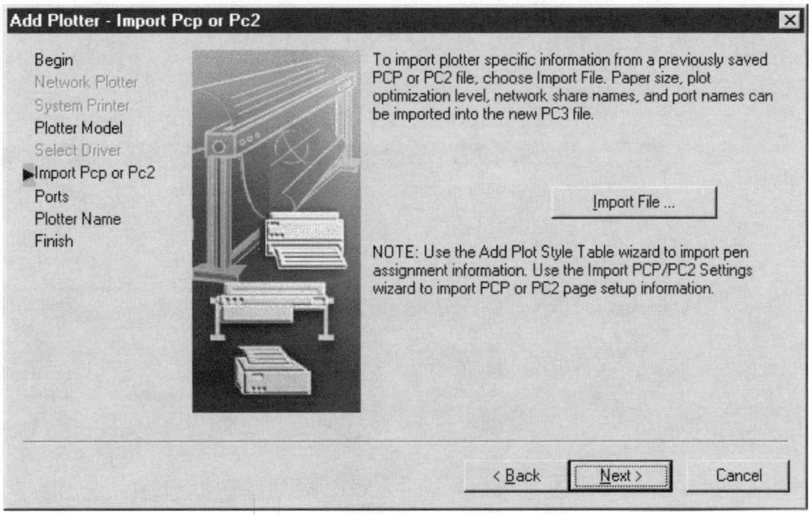

Importing PCP and PC2 files.

6. The Ports page asks you where the plotter is connected:

   Plot to a port: When your plotter is connected to your computer, select the specific port. AutoCAD helps you out by short-listing the probable ports, as well as listing the names of printers already connected to ports:

- The oldest plotters always connect to a serial port (identified by the COM prefix).
- Laser printers and newer plotters usually connect to a parallel port (identified by the LPT prefix).
- The latest printers and plotters connect to the universal serial bus (identified by the USB prefix).
- Networked printers are identified by a network name and "Network Port."

If you do not see the port name that you require, click **Show all system ports and disable I/O port validation.**

Plot to File: Instead of sending the drawing to a plotter, AutoCAD stores the plot data in a file on disk.

AutoSpool: You can have AutoCAD send the plot data to a temporary file on disk. Another program, called a spooler, reads the file and sends it to the plotter. This option helps speed up plotting to older plotters that have a small amount of memory.

For this tutorial, select **Plot to File.**

Click **Next.**

Selecting the port.

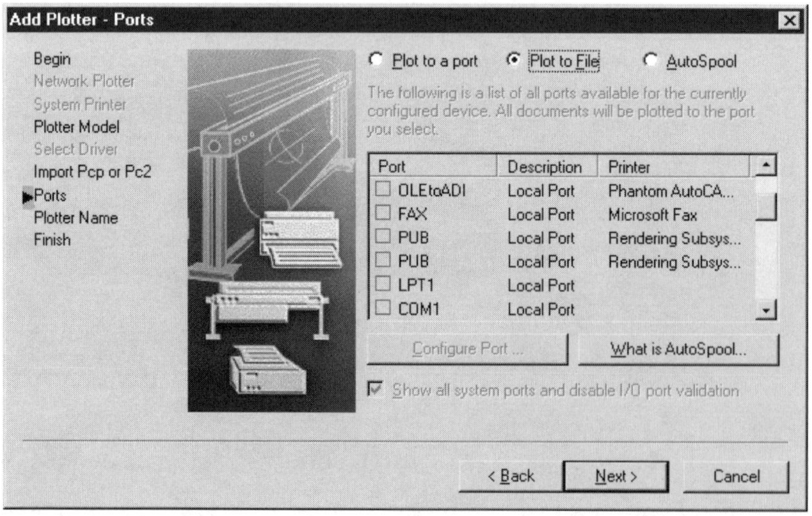

7. Here you give the plotter configuration a name. The first time you create the configuration, the name is not that important; AutoCAD offers a default that might be good enough for you. (AutoCAD calls this the "Plotter Name"; I think "Plotter Configuration Name" would be more accurate.)

   When you create more than one configuration for a single plotter, then you will want to make the name more descriptive, such as "Color plots" and "Faxed plots."

For this tutorial, enter **HPGL Plot to File** next to Plotter Name, since that's what this plotter configuration does.

Click **Next**.

Giving the plotter configuration a name.

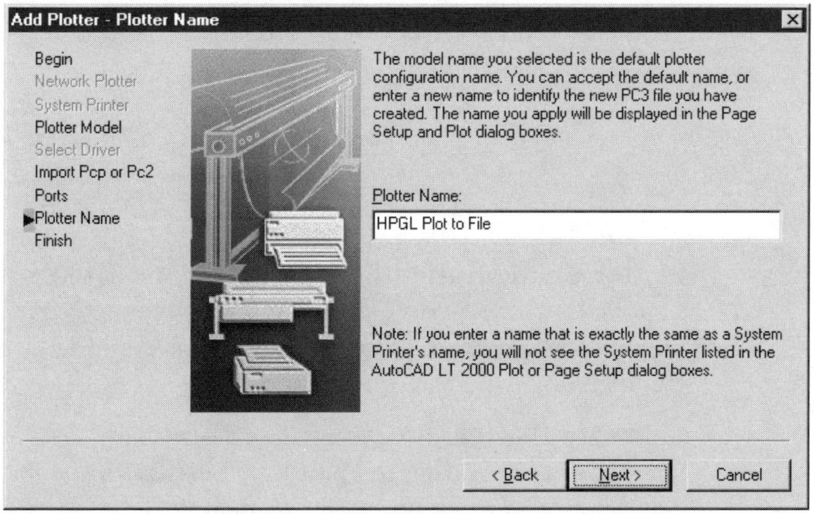

8. The name of the Finish page might lead you to think that you are finished with configuring the plotter. You are, if you click **Finish**.

   You aren't finished if you click either of the other two buttons on this page:

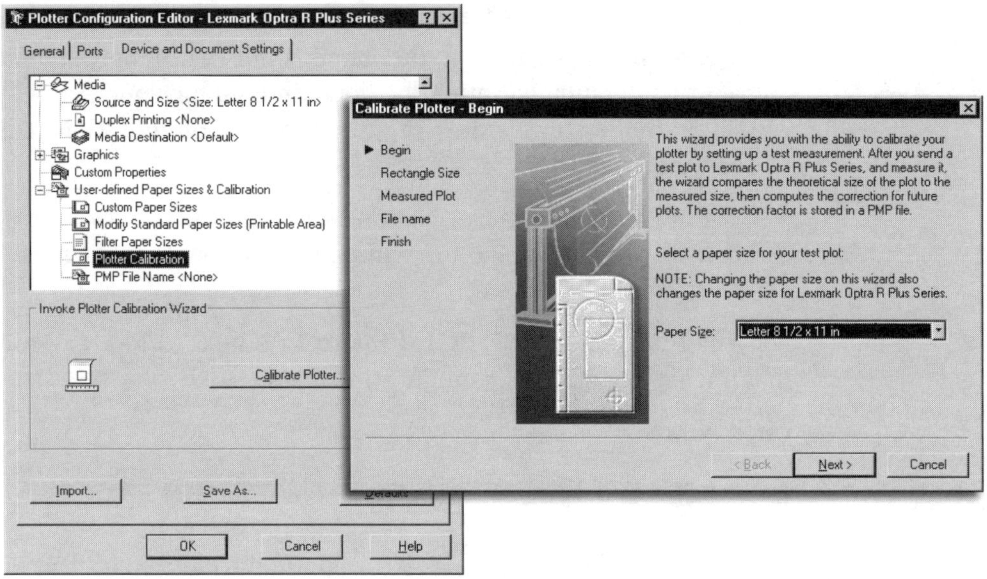

Plotter Configuration Editor dialog box (at left), and Calibrate Plotter wizard (at right).

Plotter Configuration Editor displays one of the most confusing dialog boxes I have ever seen. Its purpose is to let you specify additional plotter-specific parameters, such as paper size, margins, and pen matching.

Calibrate Plotter takes you to another wizard that leads you through the steps of calibrating the plotter. This is required for ultra-precise plots; the process ensures that an inch in AutoCAD plots as 1.0000000 inches on the plotter.

For this tutorial, click **Finish**.

Finishing the plotter configuration.

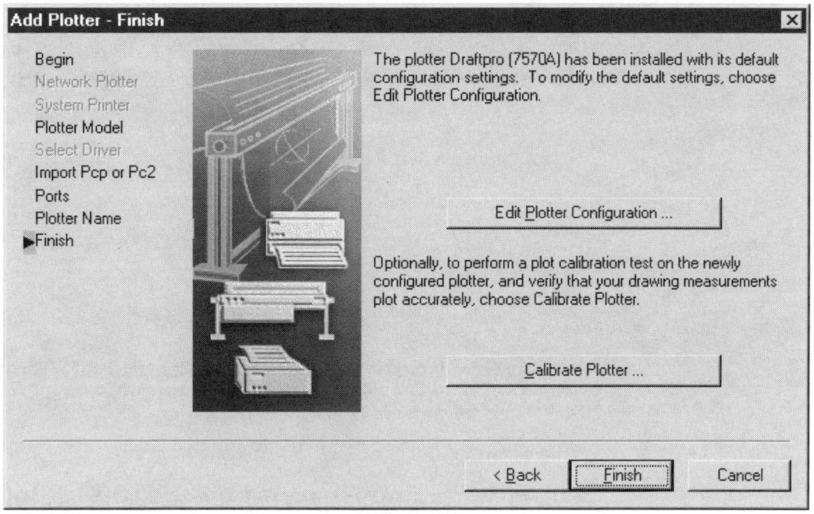

9. You have completed the plotter configuration. Let's try it out now. Back in AutoCAD LT 2000i, open a drawing. From the menu bar, select **File | Plot.** Notice that the Plot to file box is unchecked.

   When the Plot dialog box appears, select the **Plot Device** tab.

   In the Plotter configuration area, click the **Name** list box.

   Scroll down to the end, and select **HPGL Plot to File.pc3**. Notice that the Plot to file box is now checked.

   If you wish, click **OK** to plot the current drawing to a file on disk.

Selecting a plotter configuration.

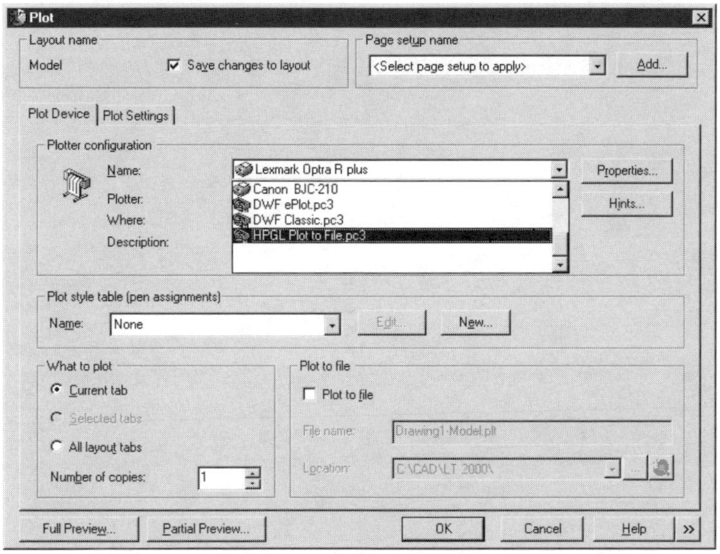

Tip: Here is a list of Web links for AutoCAD 2000, LT 2000, and LT 2000i. If you are experiencing plotting problems, take one step at a time.

- *10 Steps to Plotting in AutoCAD 2000.* This resource has gotten many users over the learning curve.

  http://www.autodesk.com/support/filelib/acad2000/plot10st.htm

- *AutoCAD 2000i Plotting Knowledge Base.*

  http://www3.autodesk.com/adsk/support/troubleshoot/0,,123882--273818-273412_21845,00.html

- *Online AutoCAD 2000 Plotting Tutorials.*

  http://www.autodesk.com/autocad2000-tutorials

- Autodesk Print-plot discussion group.

  news://discussion.autodesk.com/autodesk.autocad.print-plot

- *General AutoCAD 2000 Knowledge Base.*

  http://www.autodesk.com/autocad2000-solutions

- AutoCAD LT 2000 Support.

  http://www.autodesk.com/autocadlt-support

Appendix F

Answers

This appendix contains answers to the end of chapter review questions.

Chapter 1

1. True
2. (b.) The command opens a dialog box.
3. (a.) An icon.
4. True
5. False
6. (a.) ByLayer — (iii.) Objects take on the layer's color.
   (b.) Lineweight — (i.) Defines the width of lines.
   (c.) ByBlock — (iv.) Objects take on the block's color.
   (d.) Plot style — (ii.) Defines how lines are plotted.
   (e.) Layout — (v.) Defines how a drawing is plotted.
7. (d.) Is at the center of the crosshair cursor, and shows the point you are picking.
8. (a.) U — (v.) Undo
   (b.) UCS — (iii.) User-defined coordinate system
   (c.) F1 — (i.) Help
   (d.) Ctrl — (ii.) Control key
   (e.) Esc — (iv.) Escape key
9. True
10. False

Chapter 2

1. (a.) Units — (v.) Variety of measurement styles.
   (b.) Snap — (iii.) Constrains mouse movement to discrete distances.

(c.) Grid — (ii.) Visual guide consisting of an array of dots.

(d.) Limits — (i.) Specifies the nominal limits of a drawing, and constrains the grid marks.

(e.) Layers — (iv.) Organizes the drawing by separating elements.

2. (a.) ' — (iv.) feet
   (b.) " — (i.) inches
   (c.) in — (v.) inches
   (d.) DWG — (ii.) Identifies the file as an AutoCAD drawing.
   (e.) BAK — (iii.) Backup copy of an AutoCAD drawing file.

3. True

4. False

5. (c.) East

6. False

7. (d.) 20'

8. (c.) Both of the above.

9. (b.) AutoCAD displays the grid at the same spacing as the snap distance.

10. True

Chapter 3

1. True

2. (a.) Origin — (v.) Located at 0', 0' usually the lower-left corner of the drawing.
   (b.) Ortho — (iv.) Constrains cursor movement to the vertical and horizontal.
   (c.) Pick cursor — (ii.) Cursor that indicates AutoCAD's search area for selecting objects.
   (d.) Polar coordinates — (iii.) Specifies a distance and an angle.
   (e.) Relative coordinates — (i.) Measurement is made relative to the last point.

3. False

4.

(a.) Absolute coordinates	(iii.)	Measurement made relative to the origin.
(b.) Aperture	(iv.)	Square cursor that indicates AutoCAD's search area for object snaps.
(c.) Direct Distance Entry	(v.)	Point in a direction, then type a distance.
(d.) Extents	(i.)	Invisible rectangle that encompasses the entire drawing.
(e.) Object Snap	(ii.)	Snaps to a geometric feature, such as the center of a circle.

5. (a.) x=20, y=30

6. (c.) Distance =20, angle = 30

7. True

8. True

9.

(a.) cen	(ii.)	Center of an arc.
(b.) end	(i.)	End of a polyline.
(c.) int	(iv.)	Intersection of a line and a circle.
(d.) qua	(v.)	90-degree point on an arc.
(e.) per	(iii.)	Perpendicular to a line.

10. False

Chapter 4

1. (c.) Three points on the circle's circumference.

2. (b.) Redraw the polyline segments as a cubic Bezier curve based on the polyline's frame.

3.

(a.) Array	(iv.)	Creates a repeating rectangular or circular pattern of objects.
(b.) PEdit	(v.)	Edits the look of a polyline.
(c.) Offset	(i.)	Copies an object parallel to an existing object.
(d.) BHatch	(ii.)	Adds an associative hatch pattern to a closed area.
(e.) DsViewer	(iii.)	Displays an overview of the entire drawing.

4. (a.) Block — (iv.) Draw
   (b.) Insert — (v.) Insert
   (c.) Pan — (iii.) View
   (d.) Array — (i.) Modify
   (e.) Save — (ii.) File
5. False
6. (a.) Pedit — (iii.) Polyline Edit
   (b.) Pline — (iv.) Polyline
   (c.) Bhatch — (v.) Boundary Hatch
   (d.) Ltype — (i.) Linetype
   (e.) TTR — (ii.) Tangent, tangent, radius
7. True
8. (d.) Pickbox
9. (a.) Hot grip — (iv.) Red square
   (b.) Cold grip — (v.) Blue square
   (c.) Verb — (ii.) Command
   (d.) Non-modal — (i.) Noun-verb editing
   (e.) Noun — (iii.) Object
10. True

Chapter 5

1. False
2. (b.) 0.8x
3. (a.) Dist — (ii.) Measures the 3D distance and angle between two points.
   (b.) List — (iii.) Lists object properties in the Text window.
   (c.) Lengthen — (iv.) Makes open objects longer and shorter.
   (d.) Stretch — (v.) Moves and stretches objects.
   (e.) Tracking — (i.) Locates an x or y coordinate relative to the previous point.
4. (c.) Load the linetype pattern from Aclt.Lin file into the drawing.
5. (c.) Takes on the linetype assigned to its layer.
6. False

7. False
8. (d.) The value cannot be changed.
9. True
10. (c.) Both of the above.

Chapter 6

1. (a.) Text — (ii.) Places one line of text at a time.
   (b.) MText — (v.) Places a paragraph of text.
   (c.) QText — (iv.) Changes text to rectangles.
   (d.) Style — (iii.) Specifies the look of the text.
   (e.) DdEdit — (i.) Edits the text.
2. (b.) Color
3. True
4. (b.) Txt
5. (a.) F — (v.) Fit
   (b.) R — (iv.) Right
   (c.) TL — (ii.) Top left
   (d.) MC — (iii.) Middle center
   (e.) BR — (i.) Bottom right
6. (a.) Enter
7. (d.) Rectangles are displayed in place of the text.
8. (c.) Both of the above.
9. (a.) The dimension distance, as calculated by AutoCAD.
10. False

Chapter 7

1. True
2. (b.) Double-click the title bar.
3. (c.) Both of the above.
4. (d.) Display a list of previously opened files.
5. (a.) Displays the drawings (and their content) currently open in AutoCAD LT.

6. (b.) Display all the drives, folders, and files on your computer, local network, and the Internet.
7. (d.) Display proxy data when specific ObjectARX applications are running.
8. True
9. False
10. False

Chapter 8

1. True
2. Matching answers:

(a.) Attribute	(iv.) Text data attached to a block.
(b.) Insertion point	(v.) Specifies where the block is inserted in the drawing.
(c.) Tag	(ii.) Identifies the attribute.
(d.) Prompt	(i.) Attribute prompt displayed by AutoCAD during block insertion.
(e.) Value	(iii.) Default value of the attribute.

3. (a.) The block drawn to fit a 1 x 1 area.
4. (b.) Attributes can be attached only to blocks.
5. False
6. (c.) 256
7. (b.) Block
8. (a.) Disappear from the screen.
9. True
10. True

Chapter 9

1. True
2. (b.) AttExt
3. Matching answers:

(a.) TXT	(iii.) Extension for an ASCII text file
(b.) CDF	(v.) Comma-delimited format
(c.) SDF	(iv.) Space-delimited file
(d.) DXF	(i.) Drawing interchange file

	(e.) XLS	(ii.) Excel spreadsheet file

4. False

5. (a.) BL	(v.) Block
(b.) C	(iv.) Characters
(c.) N	(ii.) Numbers
(d.) N012002	(iii.) 12-digit number and two decimal places
(e.) C012000	(i.) 12-character word

6. (a.) All attributes not set to invisible.
7. (c.) Commas
8. True
9. True
10. False

Chapter 10

1. (c.) A 2D closed area with no intersecting curves.
2. (b.) Draw the region's outline with drawing commands, then convert with the Region command.
3. (a.) Explode command.
4. (d.) Erase objects after they are converted to a region.
5. True

6. (a.) Intersect	(v.) Remove all but overlapping areas of two regions.
(b.) Subtract	(iv.) Remove one region from another region.
(c.) Union	(i.) Join two regions into a single region.
(d.) AND	(ii.) Same as Intersect
(e.) OR	(iii.) Same as Union

7. True
8. (a.) Array
9. (d.) Area command.
10. (d.) Center of mass of the region

Chapter 11

1. (b.) A specialized Web site that provides information specific to your industry.
2. (d.) A software plug-in that allows AutoCAD to edit proxy objects.
3. False
4. (c.) Accounting
5. True
6. False
7. (a.) Phrase that reminds you of a forgotten password.
8. True
9. True
10. (c.) All of the above.

Chapter 12

1. (b.) Hyperlink
2. (a.) -Hyperlink
3. True
4. (c.) Open
5. True
6. False
7. (d.) FTP
8. (c.) Both of the above
9. (b.) Ctrl+K
10. (a.) SelectURL

Chapter 13

1. True
2. (a.) DWF — (iii.) Drawing Web format
   (b.) JPEG — (i.) Joint Photographic Experts Group
   (c.) HTML — (iv.) Hypertext Markup Language

(d.) WHIP (ii.) Windows High Performance
(e.) DWG (v.) Drawing

3. (b.) Becomes pixelated when zoomed in.
4. (a.) DWF and JPEG
5. (d.) Can be edited by AutoCAD LT.
6. True
7. False
8. True
9. False
10. (d.) Embed

Chapter 14

1. True
2. (b.) Equal projection.
3. (a.) Ellipse (ii.) Circle
   (b.) Rhombus (i.) Rectangle
   (c) Ellipical arc (iii.) Arc
4. (d.) 120
5. (c.) Ctrl+E
6. (b.) Ellipse
7. (a.) 30
8. (b.) DimEdit command
9. False
10. False

Index

About the CD

The companion CD-ROM contains the example files used in this book, along with a 30-day trial version of AutoCAD LT 2000i from Autodesk and Real Architect LITE from Straight Forward Software.

When you insert the CD into your CD drive, the AutoCAD LT 2000i trial version will autorun. Follow the on-screen installation instructions.

Real Architect LITE is a complete architectural drafting application designed to run under AutoCAD LT 2000. It provides all the tools necessary for producing full sets of working drawings, including elevations and sections. Among its features are self-cleaning walls, parametric windows and doors, advanced layer control, block libraries, tags and labeling, cabinets, and roof plans. It supports both imperial and metric units.

Use Windows Explorer to access the RA2000 folder. Installation instructions are in the readme.txt file on the root, and a tutorial is in the JumpStart folder. To read and print this document, double-click on the Jump Start.pdf file. (You must have the Adobe Acrobat Reader installed to view and print this document. To install it, use Windows Explorer to access the Acrobat Reader folder on the companion CD and double-click Setup.exe. Follow the on-screen instructions.) For more information about Real Architect LITE, visit www.realarchitect.com.

Use Windows Explorer to access the examples folder. See Appendix B for information about these files.

Please note that the full version of AutoCAD LT 2000i is not included on the CD.

Warning:
Opening the CD package makes this book non-returnable.

CD/Source Code Usage License Agreement